This item must be returned or renewed by the last date shown
below. The loan period may be shortened if it is reserved by
another reader.   A fine will be due if it is not returned on time.

DATE OF RETURN

Springer Topics in Signal Processing

Volume 3

Series Editors

J. Benesty, Montreal, Québec, Canada
W. Kellermann, Erlangen, Germany

Springer Topics in Signal Processing

Edited by J. Benesty and W. Kellermann

Vol. 1: Benesty, J.; Chen, J.; Huang, Y.
Microphone Array Signal Processing
250 p. 2008 [978-3-540-78611-5]

Vol. 2: Benesty, J.; Chen, J.; Huang, Y.; Cohen, I.
Noise Reduction in Speech Processing
240 p. 2009 [978-3-642-00295-3]

Vol. 3: Cohen, I.; Benesty, J.; Gannot, S. (Eds.)
Speech Processing in Modern Communication
360 p. 2010 [978-3-642-11129-7]

Israel Cohen · Jacob Benesty
Sharon Gannot (Eds.)

Speech Processing in Modern Communication

Challenges and Perspectives

 Springer

Prof. Israel Cohen
Technion - Israel Institute of
Technology
Dept. Electrical Engineering
32000 Haifa
Technion City
Israel
E-mail: icohen@ee.technion.ac.il

Dr. Sharon Gannot
Bar-Ilan University
School of Engineering
52900 Ramat-Gan
Bdg. 1103
Israel
E-mail: gannot@eng.biu.ac.il

Prof. Dr. Jacob Benesty
Université de Quebec
Inst. National de la Recherche
Scientifique (INRS)
800 de la Gauchetiere Ouest
Montreal QC H5A 1K6
Canada
E-mail: benesty@emt.inrs.ca

ISBN 978-3-642-11129-7   e-ISBN 978-3-642-11130-3

DOI 10.1007/978-3-642-11130-3

Springer Topics in Signal Processing   ISSN 1866-2609

          e-ISSN 1866-2617

Library of Congress Control Number: 2009940137

Cover Design: WMXDesign GmbH, Heidelberg

Printed in acid-free paper

9 8 7 6 5 4 3 2 1

springer.com

Preface

More and more devices for human-to-human and human-to-machine communications, where sound pickup and rendering is necessary, require some sophisticated algorithms. This is due to the fact that the acoustic environment in which we live in and communicate is extremely challenging. The difficult problems encountered in this environment are very well known and they are mainly acoustic echo cancellation, interference and noise suppression, and dereverberation. More than ever, these fundamental problems need to be tackled rigorously. This is the objective of this edited book, which contains twelve chapters that are briefly summarized below.

Chapter 1 addresses the problem of linear system identification in the short-time Fourier transform (STFT) domain. Identification of linear systems is of major importance in diverse applications of signal processing, including acoustic echo cancellation, relative transfer function (RTF) identification, dereverberation, blind source separation, and beamforming in reverberant environments. In this chapter, the authors introduce three models for linear system identification and investigate the influence of model order on the estimation accuracy. The three models are based on either crossband filters between subbands, multiplicative transfer functions, or cross-multiplicative transfer functions. It is shown both analytically and experimentally that the estimation accuracy does not necessarily improve by increasing the model order.

The problem of RTF identification between sensors is addressed in Chapter 2. This transfer function represents the coupling between two sensors with respect to a desired or interfering source. The authors describe an alternative representation of time domain convolution with convolutive transfer functions in the STFT domain, and show improved results compared to existing RTF identification methods.

In low-cost hands-free telecommunication systems the loudspeaker signal may contain a certain level of nonlinear distortions, which necessitate nonlinear modeling of the acoustic echo path. Chapter 3 describes a novel approach for nonlinear system identification in the STFT domain. It introduces Volterra filters in the STFT domain and considers the identification of quadratically nonlinear systems. It shows that a significant reduction in computational cost as well as substantial improvement in estimation accuracy can be achieved over a time-domain Volterra model, particularly when long-memory systems are considered.

Chapter 4 presents a family of non-parametric variable step-size (VSS) algorithms, which are particularly suitable for realistic acoustic echo cancellation (AEC) scenarios. The VSS algorithms are developed based on another objective of AEC application, i.e., to recover the near-end signal from the error signal of the adaptive filter. As a consequence, these algorithms are equipped with good robustness features against near-end signal variations, like double-talk.

Speech enhancement in transient noise environments is addressed in Chapter 5. An estimate of the desired signal is obtained under signal presence uncertainty using a simultaneous detection and estimation approach. This method facilitates suppression of transient noise with a controlled level of speech distortion. Cost parameters control the tradeoff between speech distortion, caused by missed detection of speech components, and residual musical noise resulting from false-detection.

Chapter 6 describes a model-based approach for combined dereverberation and denoising of speech signals. This approach is developed by using a multichannel autoregressive model of room acoustics and a time-varying power spectrum model of clean speech signals.

Chapter 7 investigates separation of speech and music signals from single-sensor audio mixtures. It describes codebook approaches and a Bayesian probabilistic framework for source modeling and source estimation. The source models include Gaussian scaled mixture models, codebooks of auto regressive models, and Bayesian non negative matrix factorization (BNMF).

Microphone arrays are becoming increasingly more common in the acquisition and denoising of acoustic signals. Additional microphones allow us to apply spatiotemporal filtering methods, which are significantly more powerful than conventional temporal filtering techniques. Chapter 8 is concerned with beamformer designs tailored to the specific nature of microphone array environments, i.e., broadband signals and reverberant channels. A distinction is made between wideband and narrowband metrics, and the relationships between broadband performance measures and the corresponding component narrowband measures are analyzed.

Chapter 9 presents some new insights into the minimum variance distortionless response (MVDR) beamformer. It analyzes the tradeoff between dereverberation and noise reduction achieved by using the MVDR beamformer, and discusses relations between the MVDR and other optimal beamformers.

Chapter 10 addresses the problem of extracting several desired speech signals from multi-microphone measurements, which are contaminated by non-stationary and stationary interfering signals. A linearly constrained minimum variance (LCMV) beamformer is designed with two sets of linear constraints:

one for maintaining the desired signals and one for mitigating both the stationary and non-stationary interferences.

Spherical microphone arrays have been recently studied for spatial sound recording, speech communication, and sound field analysis for room acoustics and noise control. Complementary studies presented progress in beamforming methods. Chapter 11 reviews beamforming methods recently developed for spherical arrays, from the widely used delay-and-sum and Dolph-Chebyshev, to the more advanced optimal methods, typically performed in the spherical harmonics domain.

Finally, Chapter 12 presents a family of broadband source localization algorithms based on parameterized spatiotemporal correlation, including the popular and robust steered response power (SRP) algorithm. It develops source localization methods based on minimum information entropy and temporally constrained minimum variance.

This book has been edited for engineers, researchers, and graduate students who work on speech processing for communication applications. We hope that the readers will find many new and interesting concepts that are presented in this text useful and inspiring.

We deeply appreciate the efforts, willingness, and enthusiasm of all the contributing authors. Without their commitment, this book would not have been possible.

We would like to take this opportunity to thank again Christoph Baumann, Carmen Wolf, and Petra Jantzen from Springer (Germany) for their wonderful help in the preparation and publication of this manuscript. Working with them is always a pleasure and a wonderful experience.

Finally, we would like to dedicate this edited book to our parents.

Haifa/ Montreal/ Ramat-Gan   *Israel Cohen*
Nov. 2009   *Jacob Benesty*
       Sharon Gannot

Contents

1 Linear System Identification in the Short-Time Fourier Transform Domain 1
Yekutiel Avargel and Israel Cohen
1.1 Introduction 2
1.2 Problem Formulation 4
1.3 System Identification Using Crossband Filters............. 6
    1.3.1 Crossband Filters Representation 6
    1.3.2 Batch Estimation of Crossband Filters............ 8
    1.3.3 Selecting the Optimal Number of Crossband Filters 11
1.4 System Identification Using the MTF Approximation 14
    1.4.1 The MTF Approximation 14
    1.4.2 Optimal Window Length........................ 15
1.5 The Cross-MTF Approximation 18
    1.5.1 Adaptive Estimation of Cross-Terms 19
    1.5.2 Adaptive Control Algorithm..................... 20
1.6 Experimental Results 22
    1.6.1 Crossband Filters Estimation................... 23
    1.6.2 Comparison of the Crossband Filters and MTF
           Approaches 23
    1.6.3 CMTF Adaptation for Acoustic Echo Cancellation . 25
1.7 Conclusions 28
Appendix 28
References 29

2 Identification of the Relative Transfer Function between Sensors in the Short-Time Fourier Transform Domain 33
Ronen Talmon, Israel Cohen, and Sharon Gannot
2.1 Introduction 33
2.2 Identification of the RTF Using Multiplicative Transfer
    Function Approximation 34

2.2.1   Problem Formulation and the Multiplicative
        Transfer Function Approximation 35
2.2.2   RTF Identification Using Non-Stationarity 36
2.2.3   RTF Identification Using Speech Signals 37
2.3   Identification of the RTF Using Convolutive Transfer
      Function Approximation 38
2.3.1   The Convolutive Transfer Function Approximation . 39
2.3.2   RTF Identification Using the Convolutive Transfer
        Function Approximation 40
2.4   Relative Transfer Function Identification in Speech
      Enhancement Applications 41
2.4.1   Blocking Matrix 42
2.4.2   The Transfer Function Generalized Sidelobe Canceler 44
2.5   Conclusions 45
References 46

3   Representation and Identification of Nonlinear Systems in the Short-Time Fourier Transform Domain 49

Yekutiel Avargel and Israel Cohen

3.1   Introduction 49
3.2   Volterra System Identification 51
3.3   Representation of Volterra Filters in the STFT Domain 55
3.3.1   Second-Order Volterra Filters 55
3.3.2   High-Order Volterra Filters 59
3.4   A New STFT Model For Nonlinear Systems 61
3.4.1   Quadratically Nonlinear Model 61
3.4.2   High-Order Nonlinear Models 65
3.5   Quadratically Nonlinear System Identification 65
3.5.1   Batch Estimation Scheme 67
3.5.2   Adaptive Estimation Scheme 72
3.6   Experimental Results 76
3.6.1   Performance Evaluation for White Gaussian Inputs 77
3.6.2   Nonlinear Undermodeling in Adaptive System
        Identification 79
3.6.3   Nonlinear Acoustic Echo Cancellation Application . 81
3.7   Conclusions 83
Appendix 83
References 84

4   Variable Step-Size Adaptive Filters for Echo Cancellation   89

Constantin Paleologu, Jacob Benesty, and Silviu Ciochină

4.1   Introduction 90
4.2   Non-Parametric VSS-NLMS Algorithm 92
4.3   VSS-NLMS Algorithms for Echo Cancellation 96
4.4   VSS-APA for Echo Cancellation 103
4.5   VFF-RLS for System Identification 106

4.6   Simulations 112
       4.6.1   VSS-NLMS Algorithms for AEC 112
       4.6.2   VSS-APA for AEC 117
       4.6.3   VFF-RLS for System Identification 121
4.7   Conclusions 123
References 124

**5   Simultaneous Detection and Estimation Approach for
       Speech Enhancement and Interference Suppression** 127
Ari Abramson and Israel Cohen
5.1   Introduction 127
5.2   Classical Speech Enhancement in Nonstationary Noise
       Environments 129
5.3   Simultaneous Detection and Estimation for Speech
       Enhancement 131
       5.3.1   Quadratic Distortion Measure 134
       5.3.2   Quadratic Spectral Amplitude Distortion Measure . 137
5.4   Spectral Estimation Under a Transient Noise Indication 140
5.5   A Priori SNR Estimation 142
5.6   Experimental Results 144
       5.6.1   Simultaneous Detection and Estimation 146
       5.6.2   Spectral Estimation Under a Transient Noise
                Indication 147
5.7   Conclusions 148
References 149

**6   Speech Dereverberation and Denoising Based on Time
       Varying Speech Model and Autoregressive Reverberation
       Model** 151
Takuya Yoshioka, Tomohiro Nakatani, Keisuke Kinoshita, and
Masato Miyoshi
6.1   Introduction 151
       6.1.1   Goal 152
       6.1.2   Technological Background 153
       6.1.3   Minimum Mean-Squared Error Signal Estimation
                and Model-Based Approach 154
6.2   Dereverberation Method 156
       6.2.1   Heuristic Derivation of Weighted Prediction Error
                Method 156
       6.2.2   Reverberation Model 160
       6.2.3   Clean Speech Model 165
       6.2.4   Clean Speech Signal Estimator and Parameter
                Optimization 165
6.3   Combined Dereverberation and Denoising Method 167
       6.3.1   Room Acoustics Model 168
       6.3.2   Clean Speech Model 171

6.3.3   Clean Speech Signal Estimator 172
6.3.4   Parameter Optimization 173
6.4   Experiments 179
6.5   Conclusions 180
References 181

**7   Codebook Approaches for Single Sensor Speech/Music
     Separation** 183
Raphaël Blouet and Israel Cohen
7.1   Introduction 183
7.2   Single Sensor Source Separation 185
      7.2.1   Problem Formulation 185
      7.2.2   GSMM-Based Source Separation 186
      7.2.3   AR-Based Source Separation 187
      7.2.4   Bayesian Non-Negative Matrix Factorization 188
      7.2.5   Learning the Codebook 190
7.3   Multi-Window Source Separation 190
      7.3.1   General Description of the Algorithm 190
      7.3.2   Choice of a Confidence Measure 191
      7.3.3   Practical Choice of the Thresholds 192
7.4   Estimation of the Expansion Coefficients 193
      7.4.1   Median Filter 193
      7.4.2   Smoothing Prior 194
      7.4.3   GMM Modeling of the Amplitude Coefficients 195
7.5   Experimental Study 195
      7.5.1   Evaluation Criteria............................. 195
      7.5.2   Experimental Setup and Results 195
7.6   Conclusions 196
References 197

8   Microphone Arrays: Fundamental Concepts 199
Jacek P. Dmochowski and Jacob Benesty
8.1   Introduction 199
8.2   Signal Model 200
8.3   Array Model... 202
8.4   Signal-to-Noise Ratio 203
8.5   Array Gain 204
8.6   Noise Rejection and Desired Signal Cancellation......... 206
8.7   Beampattern 207
      8.7.1   Anechoic Plane Wave Model....................... 208
8.8   Directivity 210
      8.8.1   Superdirective Beamforming...................... 210
8.9   White Noise Gain 211
8.10 Spatial Aliasing 212
      8.10.1   Monochromatic Signal 215
      8.10.2   Broadband Signal 217

    8.11   Mean-Squared Error 218
        8.11.1 Wiener Filter 220
        8.11.2 Minimum Variance Distortionless Response 221
    8.12   Conclusions 222
    References 222

9   The MVDR Beamformer for Speech Enhancement 225

Emanuël A. P. Habets, Jacob Benesty, Sharon Gannot, and Israel Cohen

    9.1   Introduction 226
    9.2   Problem Formulation 227
    9.3   From Speech Distortion Weighted Multichannel Wiener
           Filter to Minimum Variance Distortionless Response Filter . 230
        9.3.1 Speech Distortion Weighted Multichannel Wiener
              Filter 230
        9.3.2 Minimum Variance Distortionless Response Filter .. 232
        9.3.3 Decomposition of the Speech Distortion Weighted
              Multichannel Wiener Filter....................... 234
        9.3.4 Equivalence of MVDR and Maximum SNR
              Beamformer 235
    9.4   Performance Measures 235
    9.5   Performance Analysis 237
        9.5.1 On the Comparison of Different MVDR Beamformers 237
        9.5.2 Local Analyzes 239
        9.5.3 Global Analyzes 241
        9.5.4 Non-Coherent Noise Field 242
        9.5.5 Coherent plus Non-Coherent Noise Field 243
    9.6   Performance Evaluation 244
        9.6.1 Influence of the Number of Microphones 245
        9.6.2 Influence of the Reverberation Time.............. 245
        9.6.3 Influence of the Noise Field 247
        9.6.4 Example Using Speech Signals 249
    9.7   Conclusions 251
    Appendix 251
    References 253

10 Extraction of Desired Speech Signals in Multiple-Speaker Reverberant Noisy Environments 255

Shmulik Markovich, Sharon Gannot, and Israel Cohen

    10.1   Introduction 256
    10.2   Problem Formulation 259
    10.3   Proposed Method 261
        10.3.1 The LCMV and MVDR Beamformers 261
        10.3.2 The Constraints Set 262
        10.3.3 Equivalent Constraints Set 263
        10.3.4 Modified Constraints Set........................ 264

10.4 Estimation of the Constraints Matrix 266
    10.4.1 Interferences Subspace Estimation 267
    10.4.2 Desired Sources RTF Estimation 269
10.5 Algorithm Summary 270
10.6 Experimental Study 271
    10.6.1 The Test Scenario 272
    10.6.2 Simulated Environment 274
    10.6.3 Real Environment 276
10.7 Conclusions 277
References 278

11 Spherical Microphone Array Beamforming 281
Boaz Rafaely, Yotam Peled, Morag Agmon, Dima Khaykin, and
Etan Fisher
11.1 Introduction 281
11.2 Spherical Array Processing 282
11.3 Regular Beam Pattern 284
11.4 Delay-and-Sum Beam Pattern 286
11.5 Dolph-Chebyshev Beam Pattern 288
11.6 Optimal Beamforming 290
11.7 Beam Pattern with Desired Multiple Nulls 293
11.8 2D Beam Pattern and its Steering 295
11.9 Near-Field Beamforming 297
11.10 Direction-of-Arrival Estimation 300
11.11 Conclusions 303
References 303

**12 Steered Beamforming Approaches for Acoustic Source
Localization** 307
Jacek P. Dmochowski and Jacob Benesty
12.1 Introduction 307
12.2 Signal Model 309
12.3 Spatial and Spatiotemporal Filtering 310
12.4 Parameterized Spatial Correlation Matrix (PSCM) 311
12.5 Source Localization Using Parameterized Spatial Correlation 313
    12.5.1 Steered Response Power 313
    12.5.2 Minimum Variance Distortionless Response 315
    12.5.3 Maximum Eigenvalue 316
    12.5.4 Broadband MUSIC 318
    12.5.5 Minimum Entropy 320
12.6 Sparse Representation of the PSCM 326
12.7 Linearly Constrained Minimum Variance 329
    12.7.1 Autoregressive Modeling 331
12.8 Challenges 333
12.9 Conclusions 334
References 335

Index 339

List of Contributors

Ari Abramson
Technion–Israel Institute of Technology, Israel
e-mail: aari@tx.technion.ac.il

Morag Agmon
Ben-Gurion University of the Negev, Israel
e-mail: moraga@ee.bgu.ac.il

Yekutiel Avargel
Technion–Israel Institute of Technology, Israel
e-mail: kutiav@tx.technion.ac.il

Jacob Benesty
INRS-EMT, QC, Canada
e-mail: benesty@emt.inrs.ca

Raphaël Blouet
Audionamix, France
e-mail: raphael.blouet@audionamix.com

Silviu Ciochină
University Politehnica of Bucharest, Romania
e-mail: silviu@comm.pub.ro

Israel Cohen
Technion–Israel Institute of Technology, Israel
e-mail: icohen@ee.technion.ac.il

Jacek P. Dmochowski
City College of New York, NY, USA
e-mail: jdmochowski@ccny.cuny.edu

Etan Fisher
Ben-Gurion University of the Negev, Israel
e-mail: `fisher@ee.bgu.ac.il`

Sharon Gannot
Bar-Ilan University, Israel
e-mail: `gannot@eng.biu.ac.il`

Emanuël A. P. Habets
Imperial College, UK
e-mail: `e.habets@imperial.ac.uk`

Dima Khaykin
Ben-Gurion University of the Negev, Israel
e-mail: `khaykin@ee.bgu.ac.il`

Keisuke Kinoshita
NTT Communication Science Laboratories, Japan
e-mail: `kinoshita@cslab.kecl.ntt.co.jp`

Shmulik Markovich
Bar-Ilan University, Israel
e-mail: `shmulik.markovich@gmail.com`

Masato Miyoshi
NTT Communication Science Laboratories, Japan
e-mail: `miyo@cslab.kecl.ntt.co.jp`

Tomohiro Nakatani
NTT Communication Science Laboratories, Japan
e-mail: `nak@cslab.kecl.ntt.co.jp`

Constantin Paleologu
University Politehnica of Bucharest, Romania
e-mail: `pale@comm.pub.ro`

Yotam Peled
Ben-Gurion University of the Negev, Israel
e-mail: `yotamp@ee.bgu.ac.il`

Boaz Rafaely
Ben-Gurion University of the Negev, Israel
e-mail: `br@ee.bgu.ac.il`

Ronen Talmon
Technion–Israel Institute of Technology, Israel
e-mail: `ronenta2@techunix.technion.ac.il`

Takuya Yoshioka
NTT Communication Science Laboratories, Japan
e-mail: `takuya@cslab.kecl.ntt.co.jp`

Chapter 1
Linear System Identification in the Short-Time Fourier Transform Domain

Yekutiel Avargel and Israel Cohen

Abstract [1]Identification of linear systems in the short-time Fourier transform (STFT) domain has been studied extensively, and many efficient algorithms have been proposed for that purpose. In this chapter, we introduce three models for linear system identification in the STFT domain, and investigate the influence of model order on the estimation accuracy. The first model, which forms a perfect STFT representation of linear systems, includes crossband filters between the subbands. We investigate the influence the these filters on a system identifier, and show that as the length or power of the input signal increases, a larger number of crossband filters should be estimated to achieve the minimal mean-squared error (mse). The second model discussed in this chapter is the multiplicative transfer function (MTF) approximation, which relies on the assumption of a long STFT analysis window. We analytically show that the mse performance does not necessarily improve by increasing the window length. We further prove the existence of an optimal window length that achieves the minimal mse. Finally, we introduce the cross-MTF model and propose an adaptive-control algorithm that achieves a substantial improvement in steady-state performance over the MTF approach, without compromising for slower convergence. Experimental results validate the theoretical derivations and demonstrate the effectiveness of the proposed approaches.

Yekutiel Avargel

Technion–Israel Institute of Technology, Israel, e-mail: `kutiav@tx.technion.ac.il`

Israel Cohen

Technion–Israel Institute of Technology, Israel, e-mail: `icohen@ee.technion.ac.il`

[1] This work was supported by the Israel Science Foundation under Grant 1085/05.

1.1 Introduction

Identification of linear systems has been studied extensively and is of major importance in diverse fields of signal processing, including acoustic echo cancellation [1, 2, 3], relative transfer function (RTF) identification [4], dereverberation [5, 6], blind source separation [7, 8] and beamforming in reverberant environments [9, 10]. In acoustic echo cancellation applications, for instance, a loudspeaker-enclosure-microphone (LEM) system needs to be identified in order to reduce the coupling between loudspeakers and microphones. Traditionally, the identification process has been carried out in the time domain using batch or adaptive methods [11]. However, when long-memory systems are considered, these methods may suffer from slow convergence and extremely high computational cost. Moreover, when the input signal is correlated, which is often the case in acoustic echo cancellation applications, the estimation process may lead to high estimation-error variance and to slow convergence of the adaptive algorithm [12].

To overcome these problems, block processing techniques have been introduced [13, 12]. These techniques partition the input data into blocks and perform the adaptation in the frequency domain to achieve computational efficiency. However, block processing introduces a delay in the signal paths and reduces the time-resolution required for control purposes. Alternatively, subband (multirate) techniques [14] have been proposed for improved system identification (e.g., [15, 16, 17, 18, 19, 20, 21]). Accordingly, the desired signals are filtered into subbands, then decimated and processed in distinct subbands. Some time-frequency representations, such as the short-time Fourier transform (STFT), are employed for the implementation of subband filtering (see Fig. 1.1) [22, 23, 24].

The main motivation for subband methods is the reduction in computational cost compared to time-domain methods, due to processing in distinct subbands. Together with a reduction in the spectral dynamic range of the input signal, the reduced complexity may also lead to a faster convergence of adaptive algorithms. Nonetheless, because of the decimation, subband techniques produce aliasing effects, which necessitates crossband filters between the subbands [19, 25]. It has been found [19] that the convergence rate of subband adaptive algorithms that involve crossband filters with critical sampling is worse than that of fullband adaptive filters. Several techniques to avoid crossband filters have been proposed, such as inserting spectral gaps between the subbands [15], employing auxiliary subbands [18], using polyphase decomposition of the filter [20] and oversampling of the filter-bank outputs [16, 17]. Spectral gaps impair the subjective quality and are especially annoying when the number of subbands is large, while the other approaches are costly in terms of computational complexity.

A widely-used approach to avoid the crossband filters is to approximate the transfer function as multiplicative in the STFT domain [26]. This approximation relies on the assumption that the support of the STFT analysis window is sufficiently large compared with the duration of the system impulse response. As the length of the analysis window increases, the multiplicative transfer function (MTF) approximation becomes more accurate. Due to its computational efficiency, the MTF approximation is useful in many real-world applications (e.g., [27, 24, 4, 9]). However, since such applications employ finite length windows, the MTF approximation is never accurate. Recently, the cross-MTF (CMTF) approximation, which extends the MTF approximation by including cross-terms between subbands, was introduced [28]. This approach substantially improves the steady-state mean-squared error (mse) achieved by the MTF approximation, but suffers from slower convergence.

In this chapter, we discuss three different models for linear system identification in the STFT domain. We start by considering the influence of crossband filters on the performance of a system identifier implemented in the STFT domain [25]. We introduce an offline estimation scheme, and derive analytical relations between the input signal-to-noise ratio (SNR), the length of the input signal, and the number of crossband filters which are useful for system identification. We show that increasing the number of crossband filters not necessarily implies a lower mse in subbands. As the power of input signal increases or as the time variations in the system become slower, a larger number of crossband filters should be estimated to achieve the minimal mse (mmse). We proceed with investigating the influence of the STFT analysis window length on the performance of a system identifier that utilizes the MTF approximation [26]. The MTF in each frequency bin is estimated offline using a least squares (LS) criterion. We show that the performance does not necessarily improve by increasing the window length. The optimal window length, which achieves the mmse, depends on the SNR and the input signal length. Finally, we introduce the CMTF model which includes cross-terms between distinct subbands for improved system identification [28]. We propose an algorithm that adaptively controls the number of cross-terms to achieve the mmse at each iteration [29]. The resulting algorithm achieves a substantial improvement in steady-state performance over the conventional MTF approach, without compromising for slower convergence. Experimental results with white Gaussian signals and real speech signals validate the theoretical results derived in this chapter and demonstrate the effectiveness of the proposed approaches for subband system identification.

The chapter is organized as follows. In Section 1.2, we address the problem of system identification in the time and STFT domains. In Section 1.3, we introduce the crossband filters representation for linear systems, and investigate the influence of crossband filters on the performance of a system identifier implemented in the STFT domain. In Section 1.4, we present the MTF approximation and investigate the influence of the analysis window length on the mse performance. In Section 1.5, we introduce the CMTF model and propose an adaptive-control algorithm for estimating the model parameters. Finally, in Section 1.6, we present experimental results which demonstrate the proposed approaches to subband system identification.

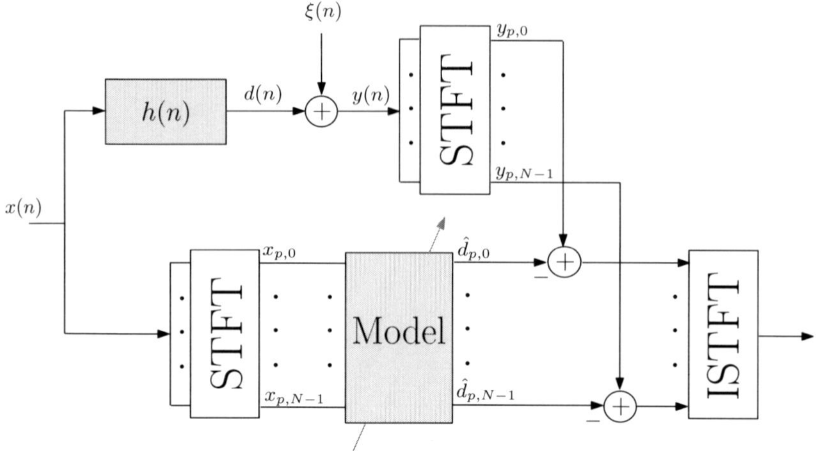

Fig. 1.1 Linear system identification scheme in the STFT domain. The unknown time-domain (linear) system $h(n)$ is estimated using a given model in the STFT domain.

1.2 Problem Formulation

Let an input $x(n)$ and output $y(n)$ of an unknown linear time-invariant (LTI) system be related by (see Fig. 1.1)

$$y(n) = \sum_{m=0}^{N_h-1} h(m)x(n-m) + \xi(n) = d(n) + \xi(n), \qquad (1.1)$$

where $h(n)$ represents the (causal) impulse response of the system, N_h is its length, $\xi(n)$ is a corrupting additive noise signal, and $d(n)$ is the clean output signal. The "noise" signal $\xi(n)$ may sometimes include a useful signal, e.g., the local speaker signal in acoustic echo cancellation. The problem of system identification in the time-domain can be formulated as follows: Given an input signal $x(n)$ and noisy observation $y(n)$, construct a model for describing the input-output relationship, and select its parameters so that the model output $\hat{d}(n)$ best estimates (or predicts) the measured output signal [11]. In many real-world applications, though, the number of model parameters (often referred to as the *model order*) is extremely large, which may lead to high computational complexity and slow convergence of time-domain estimation algorithms.

Consequently, system identification algorithms often operate in the time-frequency domain, achieving both computational efficiency and improved convergence rate due to processing in distinct subbands. In this chapter, we concentrate on STFT-based estimation methods. In order to estimate the system in the STFT domain, an appropriate model that relates the input and output signals in that domain should be defined. To do so, let us first briefly review the representation of digital signals in the STFT domain.

The STFT representation of a signal $x(n)$ is given by [30]

$$x_{p,k} = \sum_m x(m)\tilde{\psi}^*_{p,k}(m)\,, \tag{1.2}$$

where

$$\tilde{\psi}_{p,k}(n) \triangleq \tilde{\psi}(n-pL)e^{j\frac{2\pi}{N}k(n-pL)}\,, \tag{1.3}$$

$\tilde{\psi}(n)$ denotes an analysis window (or analysis filter) of length N, p is the frame index, k represents the frequency bin index, L is the translation factor (in filter bank interpretation, L denotes the decimation factor), and $*$ denotes complex conjugation. The inverse STFT, i.e., reconstruction of $x(n)$ from its STFT representation $x_{p,k}$, is given by

$$x(n) = \sum_p \sum_{k=0}^{N-1} x_{p,k}\psi_{p,k}(n)\,, \tag{1.4}$$

where

$$\psi_{p,k}(n) \triangleq \psi(n-pL)e^{j\frac{2\pi}{N}k(n-pL)} \tag{1.5}$$

and $\psi(n)$ denotes a synthesis window (or synthesis filter) of length N. Substituting (1.2) into (1.4), we obtain the so-called completeness condition:

$$\sum_p \psi(n-pL)\tilde{\psi}(n-pL) = \frac{1}{N} \qquad \text{for all } n\,. \tag{1.6}$$

Given analysis and synthesis windows that satisfy (1.6), a signal $x(n) \in \ell_2(\mathbb{Z})$ is guaranteed to be perfectly reconstructed from its STFT coefficients $x_{p,k}$. However, for $L \leq N$ and for a given synthesis window $\psi(n)$, there might be an infinite number of solutions to (1.6); therefore, the choice of the analysis window is generally not unique [31, 32].

Using the linearity of the STFT, $y(n)$ in (1.1) can be written in the time-frequency domain as

$$y_{p,k} = d_{p,k} + \xi_{p,k}\,, \tag{1.7}$$

where $d_{p,k}$ and $\xi_{p,k}$ are the STFT representations of $d(n)$ and $\xi(n)$, respectively. Figure 1.1 illustrates an STFT-based system identification scheme, where the output of the desired STFT model is denoted by $\hat{d}_{p,k}$. Note that since the system to be identified is linear, the output of the STFT model should depend linearly on its parameters, i.e.,

$$\hat{d}_{p,k}(\theta_k) = \mathbf{x}_k^T(p)\theta_k\,, \tag{1.8}$$

where θ_k is the model parameter vector in the kth frequency bin, and $\mathbf{x}_k(p)$ is the corresponding input data vector. Once a model structure has been chosen (i.e., the structure of θ_k has been determined), an estimator for the model parameters $\hat{\theta}_k$ should be derived. To do so, conventional linear estimation methods in batch or adaptive forms can be employed. For instance, given P observations in each frequency bin, the LS estimator of θ_k is given by

$$\hat{\theta}_{k,\mathrm{LS}} = \left(\mathbf{X}_k^H \mathbf{X}_k\right)^{-1} \mathbf{X}_k^H \mathbf{y}_k, \tag{1.9}$$

where $\mathbf{X}_k^T = \left[\mathbf{x}_k(0) \ \mathbf{x}_k(1) \ \cdots \ \mathbf{x}_k(P-1) \right]$ and $\mathbf{y}_k$ is the observable data vector in the kth frequency bin. The LS estimate of the clean output signal in the STFT domain is then obtained by substituting $\hat{\theta}_{k,\mathrm{LS}}$ for θ_k in (1.8).

In the following, we introduce three different models for describing the linear system $h(n)$ in the STFT domain, where each model determines different structures for the model parameter vector θ_k and the input data vector $\mathbf{x}_k(p)$. The parameters of each model are estimated using either batch or adaptive methods.

1.3 System Identification Using Crossband Filters

It is well known that in order to perfectly represent a linear system in the STFT domain, crossband filters between subbands are generally required [25, 19]. In this section, we derive explicit relations between the crossband filters in the STFT domain and the system's impulse response in the time domain, and investigate the influence of these filters on a system identifier implemented in the STFT domain.

1.3.1 Crossband Filters Representation

Using (1.1) and (1.2), the STFT of the clean output signal $d(n)$ can be written as

$$d_{p,k} = \sum_{m}\sum_{\ell} h(\ell)x(m-\ell)\tilde{\psi}_{p,k}^*(m) . \tag{1.10}$$

Substituting (1.4) into (1.10), and using the definitions in (1.3) and (1.5), we obtain after some manipulations (see Appendix)

$$d_{p,k} = \sum_{k'=0}^{N-1}\sum_{p'} x_{p',k'} h_{p-p',k,k'} = \sum_{k'=0}^{N-1}\sum_{p'} x_{p-p',k'} h_{p',k,k'} , \tag{1.11}$$

where $h_{p-p',k,k'}$ may be interpreted as a response to an impulse $\delta_{p-p',k-k'}$ in the time-frequency domain (the impulse response is translation-invariant in the time axis and is translation varying in the frequency axis). The impulse response $h_{p,k,k'}$ in the time-frequency domain is related to the impulse response $h(n)$ in the time domain by

$$h_{p,k,k'} = \{h(n) * \phi_{k,k'}(n)\}\big|_{n=pL} \triangleq \bar{h}_{n,k,k'}\big|_{n=pL},\qquad(1.12)$$

where $*$ denotes convolution with respect to the time index n and

$$\phi_{k,k'}(n) \triangleq e^{j\frac{2\pi}{N}k'n}\sum_{m}\tilde{\psi}(m)\psi(n+m)e^{-j\frac{2\pi}{N}m(k-k')}$$

$$= e^{j\frac{2\pi}{N}k'n}\psi_{n,k-k'},\qquad(1.13)$$

where $\psi_{n,k}$ is the STFT representation of the synthesis window $\psi(n)$ calculated with a decimation factor $L = 1$. Equation (1.11) indicates that for a given frequency-bin index k, the temporal signal $d_{p,k}$ can be obtained by convolving the signal $x_{p,k'}$ in each frequency bin k' ($k' = 0, 1, \ldots, N-1$) with the corresponding filter $h_{p,k,k'}$ and then summing over all the outputs. We refer to $h_{p,k,k'}$ for $k = k'$ as a band-to-band filter and for $k \neq k'$ as a crossband filter. Crossband filters are used for canceling the aliasing effects caused by the subsampling. Note that equation (1.12) implies that for fixed k and k', the filter $h_{p,k,k'}$ is noncausal in general, with $\lceil N/L \rceil - 1$ noncausal coefficients. In echo cancellation applications, in order to consider those coefficients, an extra delay of $(\lceil N/L \rceil - 1)L$ samples is generally introduced into the microphone signal [16]. It can also be seen from (1.12) that the length of each crossband filter is given by

$$M = \left\lceil \frac{N_h + N - 1}{L} \right\rceil + \left\lceil \frac{N}{L} \right\rceil - 1.\qquad(1.14)$$

To illustrate the significance of the crossband filters, we apply the discrete-time Fourier transform (DTFT) to the undecimated crossband filter $\bar{h}_{n,k,k'}$ [defined in (1.12)] with respect to the time index n and obtain

$$\bar{H}_{k,k'}(\theta) = \sum_{n}\bar{h}_{n,k,k'}e^{-jn\theta} = H(\theta)\tilde{\Psi}(\theta - \frac{2\pi}{N}k)\Psi(\theta - \frac{2\pi}{N}k'),\qquad(1.15)$$

where $H(\theta)$, $\tilde{\Psi}(\theta)$ and $\Psi(\theta)$ are the DTFT of $h(n)$, $\tilde{\psi}(n)$ and $\psi(n)$, respectively. Had both $\tilde{\Psi}(\theta)$ and $\Psi(\theta)$ been ideal low-pass filters with bandwidth $f_s/2N$ (where f_s is the sampling frequency), a perfect STFT representation of the system $h(n)$ could be achieved by using just the band-to-band filter $h_{p,k,k}$, since in this case the product of $\tilde{\Psi}(\theta - (2\pi/N)k)$ and $\Psi(\theta - (2\pi/N)k')$ is identically zero for $k \neq k'$. However, the bandwidths of $\tilde{\Psi}(\theta)$ and $\Psi(\theta)$ are generally greater than $f_s/2N$ and therefore, $\bar{H}_{k,k'}(\theta)$ and $\bar{h}_{n,k,k'}$ are not zero for $k \neq k'$. One can observe from (1.15) that the energy of a crossband filter from frequency bin k' to frequency bin k decreases as $|k - k'|$ increases, since the overlap between $\tilde{\Psi}(\theta - (2\pi/N)k)$ and $\Psi(\theta - (2\pi/N)k')$ becomes smaller.

As a result, relatively few crossband filters need to be considered in order to capture most of the energy of the STFT representation of $h(n)$.

To demonstrate these points, we use a synthetic room impulse response based on a statistical reverberation model, which assumes that a room impulse response can be described as a realization of a nonstationary stochastic process $h(n) = u(n)\beta(n)e^{-\alpha n}$, where $u(n)$ is a step function [i.e., $u(n) = 1$ for $n \geq 0$, and $u(n) = 0$ otherwise], $\beta(n)$ is a zero-mean white Gaussian noise and α is related to the reverberation time T_{60} (the time for the reverberant sound energy to drop by 60 dB from its original value). In our example, α corresponds to $T_{60} = 300$ ms (where $f_s = 16$ kHz) and $\beta(n)$ has a unit variance. For the STFT, we employ a Hamming synthesis window of length $N = 256$, and a corresponding minimum energy analysis window that satisfies (1.6) for $L = 128$ (50% overlap) [31]. The undecimated crossband filters $\bar{h}_{n,k,k'}$ from (1.12) are then computed for the synthetic impulse response and for an anechoic chamber [i.e., $h(n) = \delta(n)$].

Figures 1.2(a) and (b) show mesh plots of $|\bar{h}_{n,1,k'}|$ and contours at -40 dB (values outside this contour are lower than -40 dB), as obtained for the anechoic chamber and the room reverberation model, respectively. Figure 1.2(c) shows an ensemble averaging of $|\bar{h}_{n,1,k'}|^2$ over realizations of the stochastic process $h(n) = u(n)\beta(n)e^{-\alpha n}$ which is given by

$$E\left\{|\bar{h}_{n,1,k'}|^2\right\} = u(n)e^{-2\alpha n} * |\phi_{1,k'}(n)|^2 . \qquad (1.16)$$

Recall that the crossband filter $h_{p,k,k'}$ is obtained from $\bar{h}_{n,k,k'}$ by decimating the time index n by a factor of L [see (1.12)]. We observe from Fig. 1.2 that most of the energy of $\bar{h}_{n,k,k'}$ (for both impulse responses) is concentrated in the eight crossband filters, i.e., $k' \in \{(k+i) \bmod N |\ i = -4, \ldots, 4\}$; therefore, both impulse responses may be represented in the time-frequency domain by using only eight crossband filters around each frequency bin. As expected from (1.15), the number of crossband filters required for the representation of an impulse response is mainly determined by the analysis and synthesis windows, while the length of the crossband filters (with respect to the time index n) is related to the length of the impulse response.

1.3.2 Batch Estimation of Crossband Filters

In the following, we address the problem of estimating the crossband filters in a *batch* form using an LS optimization criterion for each frequency bin. An *adaptive* estimation of these filters and a detailed mean-square analysis of the adaptation process are given in [33], and the reader is referred to there for further details.

Let $\hat{d}_{p,k}$ be the output of the linear STFT model in (1.11), using only $2K$ crossband filters around the frequency bin k, i.e.,

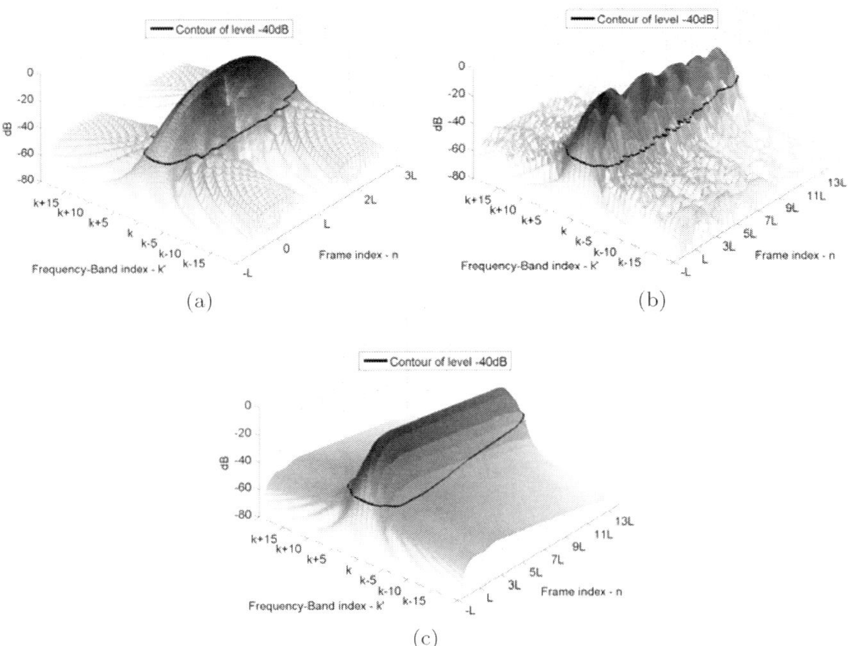

Fig. 1.2 A mesh plot of the crossband filters $|\bar{h}_{n,1,k'}|$ for different impulse responses. (a) An anechoic chamber impulse response: $h(n) = \delta(n)$. (b) A synthetic room impulse response: $h(n) = u(n)\beta(n)e^{-\alpha n}$, where $u(n)$ is a step function, $\beta(n)$ is zero-mean unit-variance white Gaussian noise and α corresponds to $T_{60} = 300$ ms (sampling rate is 16 kHz). (c) An ensemble averaging $E|\bar{h}_{n,1,k'}|^2$ of the impulse response given in (b).

$$\hat{d}_{p,k} = \sum_{k'=k-K}^{k+K} \sum_{p'=0}^{M-1} h_{p',k,k' \bmod N} \, x_{p-p',k' \bmod N} \,, \tag{1.17}$$

where we exploited the periodicity of the frequency bin (see an example illustrated in Fig. 1.3). The value of K controls the undermodeling error caused by restricting the number of crossband filters, such that not all the filters are estimated in each frequency bin[2]. Let N_x denote the time-domain observable data length and let $P \approx N_x/L$ be the number of samples given in a time-trajectory of the STFT representation (i.e., length of $x_{p,k}$ for a given k).

Due to the linear relation in (1.17) between the model output and its parameters, $\hat{d}_{p,k}$ can easily be written in the form of (1.8). Specifically, the model parameter vector θ_k consists of $2K + 1$ filters and is given by

---

[2] Undermodeling errors in system identification problems arise whenever the proposed model does not admit an exact description of the true system [11].

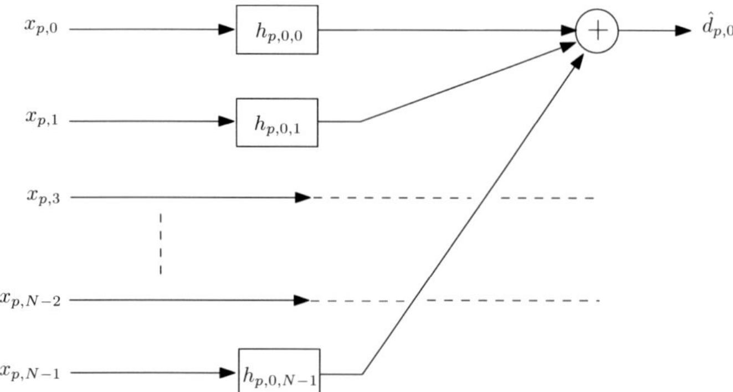

Fig. 1.3 Crossband filters illustration for frequency-bin $k = 0$ and $K = 1$.

$$\theta_k = \left[\mathbf{h}^T_{k,(k-K)\,\mathrm{mod}\,N} \; \mathbf{h}^T_{k,(k-K+1)\,\mathrm{mod}\,N} \; \cdots \; \mathbf{h}^T_{k,(k+K)\,\mathrm{mod}\,N} \right]^T , \qquad (1.18)$$

where $\mathbf{h}_{k,k'} = \left[h_{0,k,k'} \; h_{1,k,k'} \; \cdots \; h_{M-1,k,k'} \right]^T$ is the crossband filter from frequency bin k' to frequency bin k, and the corresponding input data vector $\mathbf{x}_k(p)$ can be expressed as

$$\mathbf{x}_k(p) = \left[\bar{\mathbf{x}}^T_{k,(k-K)\,\mathrm{mod}\,N}(p) \; \bar{\mathbf{x}}^T_{k,(k-K+1)\,\mathrm{mod}\,N}(p) \; \cdots \; \bar{\mathbf{x}}^T_{k,(k+K)\,\mathrm{mod}\,N}(p) \right]^T ,$$
$$(1.19)$$

where $\bar{\mathbf{x}}_k(p) = \left[x_{p,k} \; x_{p-1,k} \; \cdots \; x_{p-M+1,k} \right]^T$. In a vector form, the model output (1.17) can be written as

$$\hat{\mathbf{d}}_k(\theta_k) = \mathbf{X}_k \theta_k , \qquad (1.20)$$

where

$$\mathbf{X}^T_k = \left[\mathbf{x}_k(0) \; \mathbf{x}_k(1) \; \cdots \; \mathbf{x}_k(P-1) \right] ,$$
$$\hat{\mathbf{d}}_k(\theta_k) = \left[\hat{d}_{0,k} \; \hat{d}_{1,k} \; \cdots \; \hat{d}_{P-1,k} \right]^T .$$

Finally, denoting the observable data vector by $\mathbf{y}_k = \left[y_{0,k} \; y_{1,k} \; \cdots \; y_{P-1,k} \right]^T$ and using the above notations, the LS estimate of the crossband filters is given by

$$\hat{\theta}_k = \arg\min_{\theta_k} \| \mathbf{y}_k - \mathbf{X}_k \theta_k \|^2$$
$$= \left(\mathbf{X}^H_k \mathbf{X}_k \right)^{-1} \mathbf{X}^H_k \mathbf{y}_k , \qquad (1.21)$$

where we assume that $\mathbf{X}_k^H \mathbf{X}_k$ is not singular[3]. Note that both $\hat{\theta}_k$ and $\hat{\mathbf{d}}_k(\theta_k)$ depend on the parameter K, but for notational simplicity K has been omitted. Substituting (1.21) into (1.20), we obtain the LS estimate of the system output signal in the STFT domain at the kth frequency bin using $2K + 1$ crossband filters.

Next, we evaluate the computational complexity of the proposed estimation approach. Computing the parameter-vector estimate $\hat{\theta}_k$ requires a solution of the LS normal equations $\left(\mathbf{X}_k^H \mathbf{X}_k\right)\hat{\theta}_k = \mathbf{X}_k^H \mathbf{y}_k$ for each frequency bin. This results in $P\left[(2K+1)M\right]^2 + \left[(2K+1)M\right]^3/3$ arithmetic operations[4] when using the Cholesky decomposition [35]. Computation of the desired signal estimate (1.20) requires additional $2PM(2K+1)$ arithmetic operations. Assuming P is sufficiently large and neglecting the computations required for the forward and inverse STFTs, the complexity associated with the proposed approach is

$$O_{\mathrm{cbf}} \sim O\left\{ NP\left[(2K+1)M\right]^2 \right\}, \tag{1.22}$$

where the subscript cbf stands for crossband filters. Expectedly, we observe that the computational complexity increases as K increases.

1.3.3 Selecting the Optimal Number of Crossband Filters

The number of crossband filters that are estimated in the identification process has a crucial influence on the system estimate accuracy [25]. In the following, we present explicit relations between the mse and the number of crossband filters, and discuss the problem of determining the optimal number of filters that achieves the mmse in each frequency bin. The (normalized) mse in the kth frequency bin is defined by[5]

$$\epsilon_k(K) = \frac{E\left\{ \left\| \mathbf{d}_k - \hat{\mathbf{d}}_k\left(\hat{\theta}_k\right) \right\|^2 \right\}}{E\left\{ \|\mathbf{d}_k\|^2 \right\}}. \tag{1.23}$$

---

[3] In the ill-conditioned case, when $\mathbf{X}_k^H \mathbf{X}_k$ is singular, matrix regularization is required [34].

[4] An arithmetic operation is considered to be any complex multiplication, complex addition, complex subtraction, or complex division.

[5] To avoid the well-known *overfitting* problem [11], the mse defined in (1.23) measures the fit of the optimal estimate $\hat{\mathbf{d}}_k(\hat{\theta}_k)$ to the clean output signal $\mathbf{d}_k$, rather than to the measured (noisy) signal $\mathbf{y}_k$. Consequently, the growing model variability caused by increasing the number of model parameters is compensated, and a more reliable measure for the model estimation quality is achieved.

To derive an explicit expression for the mse, we assume that $x_{p,k}$ and $\xi_{p,k}$ are uncorrelated zero-mean white Gaussian signals with variances σ_x^2 and σ_ξ^2, respectively, and that $x_{p,k}$ is variance-ergodic [36]. The Gaussian assumption of the corresponding STFT signals is often justified by a version of the central limit theorem for correlated signals [37, Theorem 4.4.2], and it underlies the design of many speech-enhancement systems [38, 39]. Denoting the SNR by $\eta = \sigma_x^2/\sigma_\xi^2$ and using the above assumptions, the mse in (1.23) can be expressed as [25]

$$\epsilon_k(K) = \frac{\alpha_k(K)}{\eta} + \beta_k(K), \tag{1.24}$$

where

$$\alpha_k(K) \triangleq \frac{M}{P\left\|\bar{\mathbf{h}}_k\right\|^2}\left(2K+1\right), \tag{1.25}$$

$$\beta_k(K) \triangleq 1 - \frac{M\left(2K+1\right)}{P} - \frac{1}{\left\|\bar{\mathbf{h}}_k\right\|^2}\sum_{m=0}^{2K}\left\|\bar{\mathbf{h}}_{k,(k-K+m)\bmod N}\right\|^2, \tag{1.26}$$

with $\bar{\mathbf{h}}_{k,k'}$ denoting the *true* crossband filter from frequency bin k' to frequency bin k, and $\bar{\mathbf{h}}_k = \left[\bar{\mathbf{h}}_{k,0}^T\ \bar{\mathbf{h}}_{k,1}^T\ \cdots\ \bar{\mathbf{h}}_{k,N-1}^T\right]^T$.

From (1.24), the mse $\epsilon_k(K)$ for fixed k and K values, is a monotonically decreasing function of η, which expectedly indicates that higher SNR values enable a better estimation of the relevant crossband filters. Moreover, it is easy to verify from (1.25) and (1.26) that $\alpha_k(K+1) > \alpha_k(K)$ and $\beta_k(K+1) \leq \beta_k(K)$. Consequently $\epsilon_k(K)$ and $\epsilon_k(K+1)$ are two monotonically decreasing functions of η that satisfy

$$\epsilon_k(K+1) > \epsilon_k(K), \text{ for } \eta \to 0 \ (\text{low SNR}),$$
$$\epsilon_k(K+1) \leq \epsilon_k(K), \text{ for } \eta \to \infty \ (\text{high SNR}). \tag{1.27}$$

Accordingly, these functions must intersect at a certain SNR value, denoted by $\eta_k\left(K+1 \to K\right)$, that is, $\epsilon_k(K+1) \leq \epsilon_k(K)$ for $\eta \geq \eta_k\left(K+1 \to K\right)$, and $\epsilon_k(K+1) > \epsilon_k(K)$ otherwise (see typical mse curves in Fig. 1.4). For SNR values higher than $\eta_k\left(K+1 \to K\right)$, a lower mse value can be achieved by estimating $2(K+1)+1$ crossband filters rather than only $2K+1$ filters. The SNR-intersection point $\eta_k\left(K+1 \to K\right)$ can be obtained from (1.24) by requiring that $\epsilon_k(K+1) = \epsilon_k(K)$, obtaining

$$\eta_k\left(K+1 \to K\right) = \tag{1.28}$$

$$\frac{2M}{2M\left\|\bar{\mathbf{h}}_k\right\|^2 + P\left(\left\|\bar{\mathbf{h}}_{k,(k-K-1)\bmod N}\right\|^2 + \left\|\bar{\mathbf{h}}_{k,(k+K+1)\bmod N}\right\|^2\right)}.$$

From (1.29), we have $\eta_k\left(K \to K-1\right) \leq \eta_k\left(K+1 \to K\right)$, which indicates that the number of crossband filters, which should be used for the system

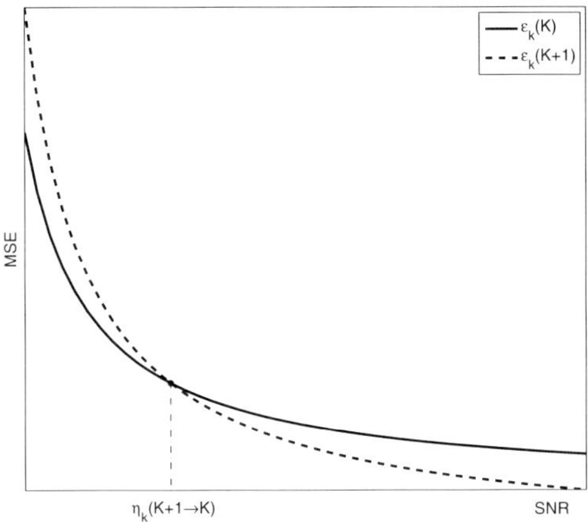

Fig. 1.4 Illustration of typical mse curves as a function of the input SNR showing the relation between $\epsilon_k(K)$ (solid) and $\epsilon_k(K+1)$ (dashed).

identifier, is a monotonically increasing function of the SNR. Estimating just the band-to-band filter and ignoring all the crossband filters yields the mmse only when the SNR is lower than $\eta_k(1 \to 0)$.

Another interesting point that can be concluded from (1.29) is that $\eta_k(K+1 \to K)$ is inversely proportional to P, the length of $x_{p,k}$ in frequency bin k. Therefore, for a fixed SNR value, the number of crossband filters, which should be estimated in order to achieve the mmse, increases as we increase P. For instance, suppose that P is chosen such that the input SNR satisfies $\eta_k(K \to K-1) \leq \eta \leq \eta_k(K+1 \to K)$, so that $2K+1$ crossband filters should be estimated. Now, suppose that we increase the value of P, so that the same SNR now satisfies $\eta_k(K+1 \to K) \leq \eta \leq \eta_k(K+2 \to K+1)$. In this case, although the SNR remains the same, we would now prefer to estimate $2(K+1)+1$ crossband filters rather than $2K+1$. It is worth noting that P is related to the update rate of crossband filters. We assume that during P frames the system impulse response does not change, and its estimate is updated every P frames. Therefore, a small P should be chosen whenever the system impulse response is time varying and fast tracking is desirable. However, in case the time variations in the system are slow, we can increase P, and correspondingly increase the number of crossband filters.

It is worthwhile noting that the results in this section are closely related to the problem of model order selection, where in our case the model order is determined by the number of estimated crossband filters. Selecting the optimal model order for a given data set is a fundamental problem in many system identification applications [11, 40, 41, 42, 43, 44, 45], and many criteria have been proposed for this purpose. The Akaike information criterion (AIC)

[44] and the minimum description length (MDL) [45] are among the most popular choices. As the model order increases, the empirical fit to the data improves [i.e., $\|\mathbf{y}_k - \hat{\mathbf{d}}_k(\hat{\theta}_k)\|^2$ can be smaller], but the variance of parametric estimates increases too [i.e., variance of $\hat{\mathbf{d}}_k(\hat{\theta}_k)$], thus possibly worsening the accuracy of the model on new measurements [11, 40, 41], and increasing the mse, $\epsilon_k(K)$. Hence, the optimal model order is affected by the level of noise in the data and the length of observable data that can be employed for the system identification. As the SNR increases or as more data is employable, the optimal model order increases, and correspondingly additional crossband filters can be estimated to achieve lower mse.

1.4 System Identification Using the MTF Approximation

The MTF approximation is a widely-used approach for modeling linear systems in the STFT domain. It avoids the crossband filters by approximating the transfer function as multiplicative in each frequency bin. In this section, we introduce the MTF approximation and investigate the influence of the analysis window length on the performance of a system identifier that utilizes this approximation.

1.4.1 The MTF Approximation

Let us rewrite the STFT of the clean output signal $d(n)$ from (1.10) as

$$d_{pk} = \sum_m x(m) \sum_\ell h(\ell)\, \tilde{\psi}^*_{pk}(m + \ell)\,. \tag{1.29}$$

Let us assume that the analysis window $\tilde{\psi}(n)$ is long and smooth relative to the system's impulse-response $h(n)$ so that $\tilde{\psi}(n)$ is approximately constant over the duration of $h(n)$. Then $\tilde{\psi}(n - m)\, h(m) \approx \tilde{\psi}(n)\, h(m)$, and by substituting (1.3) into (1.29), $d_{p,k}$ can be approximated as

$$d_{pk} \approx \sum_\ell h(\ell)e^{-j\frac{2\pi}{N}k\ell} \sum_m x(m)\tilde{\psi}(m - pL)e^{-j\frac{2\pi}{N}k(m-pL)}\,. \tag{1.30}$$

Recognizing the last summation in (1.30) as the STFT of $x(n)$, we may write

$$d_{pk} \approx h_k\, x_{pk}\,, \tag{1.31}$$

where $h_k \triangleq \sum_m h(m) \exp\left(-j2\pi mk/N\right)$. The approximation in (1.31) is the well-known MTF approximation for modeling an LTI system in the STFT domain. This approximation is also employed in some block frequency-domain methods, which attempt to estimate the unknown system in the frequency domain using block updating techniques (e.g., [46, 47, 48, 13]). Note that the MTF approximation (1.31) approximates the time-domain linear convolution in (1.1) by a circular convolution of the input-signal's pth frame and the system impulse response, using a frequency-bin product of the corresponding discrete Fourier transforms (DFTs). In the limit, for an infinitely long analysis window, the linear convolution would be exactly multiplicative in the STFT domain. However, since practical implementations employ finite length analysis windows, the MTF approximation is never accurate.

The output of a model that utilizes the MTF approximation can be written in the form of (1.8), with $\theta_k = \theta_k = h_k$ and $\mathbf{x}_k(p) = x_{p,k}$. In this case, the model parameter vector θ_k consists of a single coefficient [in contrast with the parameter vector (1.18) of the crossband-filters approach]. In a vector form, the output of the MTF model can be written as

$$\hat{\mathbf{d}}_k\left(\theta_k\right) = \mathbf{x}_k\,\theta_k\,, \tag{1.32}$$

where $\mathbf{x}_k = \left[\begin{array}{cccc} x_{0,k} & x_{1,k} & \cdots & x_{P-1,k} \end{array}\right]^T$, and $\hat{\mathbf{d}}_k$ is defined similarly. Using these notations, the LS estimate of θ_k is given by

$$\hat{\theta}_k = \arg\min_{\theta_k} \left\|\mathbf{y}_k - \mathbf{x}_k\,\theta_k\right\|^2$$
$$= \frac{\mathbf{x}_k^H\,\mathbf{y}_k}{\mathbf{x}_k^H\,\mathbf{x}_k}\,, \tag{1.33}$$

where $\mathbf{y}_k$ is the observable data vector. Following a similar complexity analysis to that given for the crossband filters approach (see Section 1.3.2), the complexity associated with the MTF approach is

$$O_{\mathrm{mtf}} \sim O\left(NP\right). \tag{1.34}$$

A comparison to (1.22) indicates that the complexity of the crossband filters approach is higher than that of the MTF approach by a factor of $\left[(2K+1)\,M\right]^2$, which proves the computational efficiency of the latter. However, in terms of estimation accuracy, the crossband filters approach is more advantageous than the MTF approach, as will be demonstrated in Section 1.6.

1.4.2 Optimal Window Length

Selecting the length of the analysis window is a critical problem that may significantly affect the performance of an MTF-based system identifier [26]. Clearly, as N, the length of the analysis window, increases, the MTF approximation becomes more accurate. On the other hand, the length of the input signal that can be employed for the system identification must be finite to enable tracking during time variations in the system. Therefore, increasing the analysis window length while retaining the relative overlap between consecutive windows (the overlap between consecutive analysis windows determines the redundancy of the STFT representation), a fewer number of observations in each frequency bin become available (smaller P), which increases the variance of $\hat{\theta}_k$. Consequently, the mse in each subband may not necessarily decrease as we increase the length of the analysis window. An appropriate window length should then be found, which is sufficiently large to make the MTF approximation valid, and sufficiently small to make the system identification performance most satisfactory.

Let us define the mse in the STFT domain as

$$\epsilon = \frac{\sum_{k=0}^{N-1} E\left\{\left\|\mathbf{d}_k - \hat{\mathbf{d}}_k\left(\hat{\theta}_k\right)\right\|^2\right\}}{\sum_{k=0}^{N-1} E\left\{\|\mathbf{d}_k\|^2\right\}}, \tag{1.35}$$

where $\hat{\mathbf{d}}_k\left(\hat{\theta}_k\right)$ is obtained by substituting (1.33) into (1.32). An explicit expression for the mse in (1.35) is derived in [26] assuming that $x(n)$ and $\xi(n)$ are uncorrelated zero-mean white Gaussian signals. It is shown that the mse can be decomposed into two error terms as

$$\epsilon = \epsilon_N + \epsilon_P. \tag{1.36}$$

The first error term ϵ_N is attributable to using a finite-support analysis window. As we increase the support of the analysis window, this term reduces to zero [i.e., $\epsilon_N(N \to \infty) = 0$], since the MTF approximation becomes more accurate. On the other hand, the error ϵ_P is a consequence of restricting the length of the input signal. It decreases as we increase either P or the SNR, and reduces to zero when $P \to \infty$.

Figure 1.5 shows the theoretical mse curves ϵ, ϵ_N and ϵ_P as a function of the ratio between the analysis window length, N, and the impulse response length, N_h, for a signal length of 3 seconds and a 0 dB SNR (the SNR is defined as $E\{|x(n)|^2\}/E\{|\xi(n)|^2\}$). We model the impulse response as a stochastic process with an exponential decay envelope, i.e., $h(n) = u(n)\beta(n)e^{-0.009n}$, where $u(n)$ is the unit step function and $\beta(n)$ is a unit-variance zero-mean white Gaussian noise. The impulse response length is set to 16 ms, a Hamming synthesis window with 50% overlap ($L = 0.5N$) is employed, and the sampling rate is 16 kHz. Expectedly, we observe from Fig. 1.5 that ϵ_N is a monotonically decreasing function of N, while ϵ_P is a monotonically increasing function (since P decreases as N increases). Consequently, the total mse, ϵ, may reach its minimum value for a certain optimal window length N^*, i.e.,

$$N^* = \arg\min_N \epsilon. \tag{1.37}$$

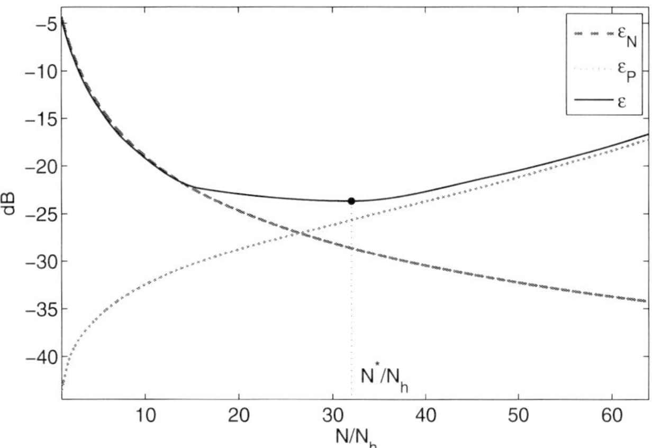

Fig. 1.5 Theoretical mse curves as a function of the ratio between the analysis window length (N) and the impulse response length (N_h), obtained for a 0 dB SNR.

In the example of Fig. 1.5, N^* is approximately $32\,N_h$.

The optimal window length represents the trade-off between the number of observations in time-trajectories of the STFT representation and accuracy of the MTF approximation. Equation (1.36) implies that the optimal window length depends on the relative weight of each error, ϵ_N or ϵ_P, in the overall mse ϵ. Since ϵ_P decreases as we increase either the SNR or the length of the time-trajectories P [26], we expect that the optimal window length N^* would increase as P or the SNR increases. For given analysis window and overlap between consecutive windows (given N and N/L), P is proportional to the input-signal length N_x (since $P \approx N_x/L$). Hence, the optimal window length generally increases as N_x increases. Recall that the impulse response is assumed time invariant during N_x samples, in case the time variations in the system are slow, we can increase N_x, and correspondingly increase the analysis window length in order to achieve a lower mse.

To demonstrate these points, we utilize the MTF approximation for estimating the impulse response from the previous experiment (see Fig. 1.5), using several SNR and signal-length values. Figure 1.6 shows the resulting mse curves (1.35), both in theory and in simulation, as a function of the ratio between the analysis window length and the impulse response length. Figure 1.6(a) shows the mse curves for SNR values of -10, 0 and 10 dB, obtained with a signal length of 3 seconds, and Fig. 1.6(b) shows the mse curves for signal lengths of 3 and 15 sec, obtained with a -10 dB SNR. The experimental results are obtained by averaging over 100 independent runs. Clearly, as the SNR or the signal length increases, a lower mse can be achieved by using a longer analysis window. Accordingly, as the power of the input signal increases or as the time variations in the system become slower (which enables one to use a longer input signal), a longer analysis window should be used

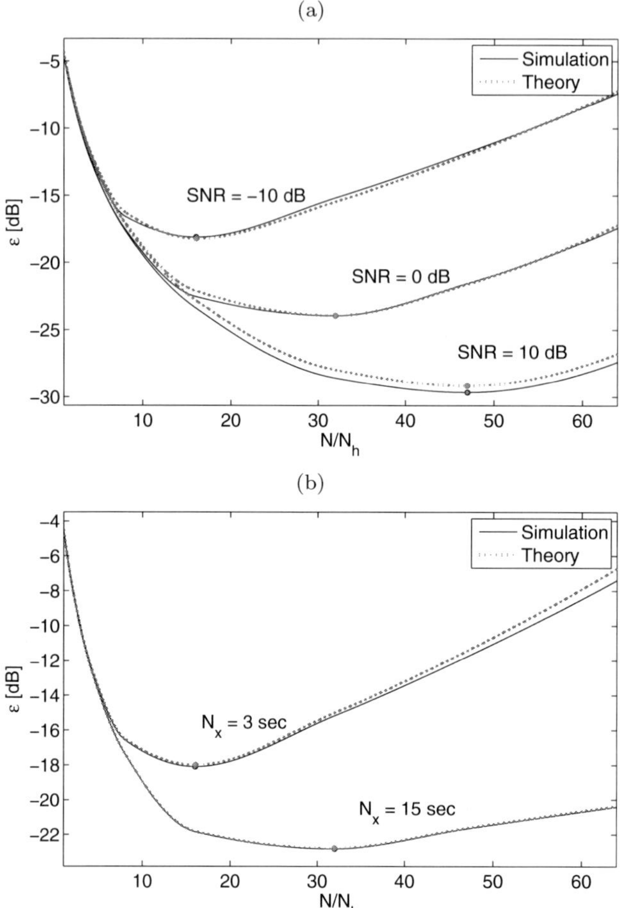

Fig. 1.6 Comparison of simulation (solid) and theoretical (dashed) mse curves as a function of the ratio between the analysis window length (N) and the impulse response length (N_h). (a) Comparison for several SNR values (input signal length is 3 seconds); (b) Comparison for several signal lengths (SNR is -10 dB).

to make the MTF approximation appropriate for system identification in the STFT domain.

1.5 The Cross-MTF Approximation

The discussion in the previous section indicates that the system-estimate accuracy achieved by the MTF approximation may be degraded as a consequence of using either a finite-support analysis window or a finite-length input signal. Furthermore, the exact STFT representation of linear systems (1.11) implies that the drawback of the MTF approximation may be related to ignoring cross-terms between subbands. Using data from adjacent frequency bins and including cross-multiplicative terms between distinct subbands, we may improve the system estimate accuracy without significantly increasing the computational cost. This has motivated the derivation of a new STFT model for linear systems, which is referred to as the CMTF approximation [28, 29]. According to this model, the system output is modeled using $2K+1$ cross-terms around each frequency bin as

$$\hat{d}_{p,k} = \sum_{k'=k-K}^{k+K} h_{k,k' \bmod N} x_{p,k' \bmod N} , \tag{1.38}$$

where $h_{k,k'}$ is a cross-term from frequency bin k' to frequency bin k. Note that for $K = 0$, (1.38) reduces to the MTF approximation (1.31).

1.5.1 Adaptive Estimation of Cross-Terms

Let us rewrite equation (1.38) in the form of equation (1.8) as

$$\hat{d}_{p,k} = \mathbf{x}_k^T(p)\theta_k , \tag{1.39}$$

where $\theta_k = \begin{bmatrix} h_{k,(k-K)\bmod N} & \cdots & h_{k,(k+K)\bmod N} \end{bmatrix}^T$ is the model parameter vector and $\mathbf{x}_k(p) = \begin{bmatrix} x_{p,(k-K)\bmod N} & \cdots & x_{p,(k+K)\bmod N} \end{bmatrix}^T$. A recursive estimate of θ_k can be found using the least-mean-square (LMS) algorithm [12]

$$\hat{\theta}_k(p+1) = \hat{\theta}_k(p) + \mu e_{p,k}\mathbf{x}_k^*(p) , \tag{1.40}$$

where $\hat{\theta}_k(p)$ is the estimate of θ_k at frame index p, $e_{p,k} = y_{p,k} - \mathbf{x}_k^T(p)\hat{\theta}_k(p)$ is the error signal in the kth frequency bin, $y_{p,k}$ is defined in (1.7), and μ is a step-size.

Let $\epsilon_k(p) = E\{|e_{p,k}|^2\}$ denote the transient mse in the kth frequency bin. Then, assuming that $x_{p,k}$ and $\xi_{p,k}$ are uncorrelated zero-mean white Gaussian signals, the mse can be expressed recursively as [28]

$$\epsilon_k(p+1) = \alpha(K)\,\epsilon_k(p) + \beta_k(K) , \tag{1.41}$$

where $\alpha(K)$ and $\beta_k(K)$ depend on the step-size μ and the number of cross-terms K. Accordingly, it can be shown [28] that the optimal step-size that results in the fastest convergence for each K is given by

$$\mu_{\mathrm{opt}} = \frac{1}{2\sigma_x^2(K+1)} , \tag{1.42}$$

where σ_x^2 is the variance of $x_{p,k}$. Equation (1.42) indicates that as the number of cross-terms increases (K increases), a smaller step-size has to be utilized. Consequently, the MTF approximation ($K = 0$) is associated with faster convergence, but suffers from higher steady-state mse $\epsilon_k(\infty)$. Estimation of additional cross-terms results in a slower convergence, but improves the steady-state mse. Since the number of cross-terms is fixed during the adaptation process, this approach may suffer from either slow convergence (typical to large K) or relatively high steady-state mse (typical to small K) [28].

1.5.2 Adaptive Control Algorithm

To improve both convergence rate and steady-state mse of the estimation algorithm in (1.40), we can adaptively control the number of cross-terms and find the optimal number that achieves the mmse at each iteration [29]. This strategy of controlling the number of cross-terms is related to filter-length control (e.g., [49, 50]). However, existing length-control algorithms operate in the time domain, focusing on linear FIR adaptive filters. Here, we extend the approach presented in [49] to construct an adaptive control procedure for CMTF adaptation implemented in the STFT domain.

Let

$$K_{\mathrm{opt}}(p) = \arg\min_{K} \epsilon_k(p) . \tag{1.43}$$

Then, $2K_{\mathrm{opt}}(p)+1$ denotes the optimal number of cross-terms at iteration p. At the beginning of the adaptation process, the proposed algorithm should initially select a small number of cross-terms (usually $K = 0$) to achieve initial fast convergence, and then, as the adaptation process proceeds, it should gradually increase this number to improve the steady-state performance. This is done by simultaneously updating three system models, each consists of different number of cross-terms, as illustrated in Fig. 1.7. The vectors θ_{1k}, θ_{2k} and θ_{3k} denote three model-parameter vectors of lengths $2K_1(p)+1$, $2K_2(p)+1$ and $2K_3(p) + 1$, respectively. These vectors are estimated simultaneously at each iteration using the normalized LMS (NLMS) algorithm

$$\hat{\theta}_{ik}(p + 1) = \hat{\theta}_{ik}(p) + \frac{\mu_i(p)}{\|\mathbf{x}_{ik}(p)\|^2} e_{p,k}^i \mathbf{x}_{ik}^*(p) , \tag{1.44}$$

where $i = 1, 2, 3$, $\mathbf{x}_{ik}(p) = [\, x_{p,(k-K_i(p))\bmod N} \cdots x_{p,(k+K_i(p))\bmod N} \,]^T$, $e_{p,k}^i = y_{p,k} - \mathbf{x}_{ik}^T(p)\hat{\theta}_{ik}(p)$ is the resulting error signal, and $\mu_i(p)$ is the relative step-size. The estimate of the second parameter vector $\hat{\theta}_{2k}(p)$ is the one of interest as it determines the desired signal estimate $\hat{d}_{p,k} = \mathbf{x}_{2k}^T(p)\hat{\theta}_{2k}(p)$. Therefore, the dimension of $\hat{\theta}_{2k}(p)$, $2K_2(p) + 1$, represents the optimal number of cross-terms in each iteration. Let

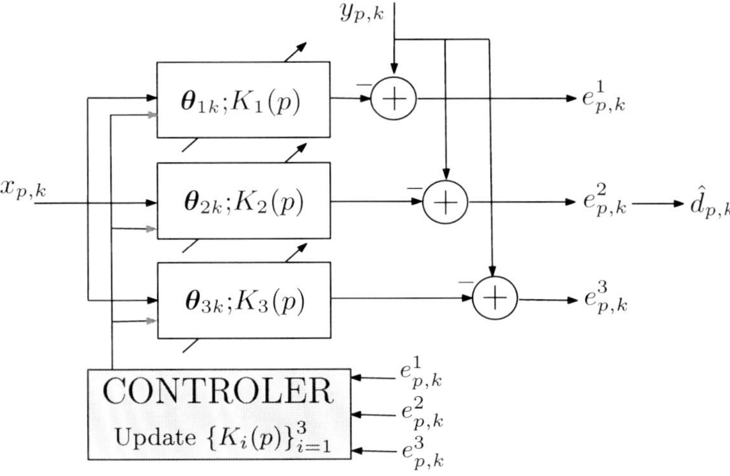

Fig. 1.7 Adaptive control scheme for CMTF adaptation in the STFT domain. The parameter vector of the second model θ_{2k} determines the model output $(\hat{d}_{p,k})$, and its dimension is updated by the controller which uses the decision rule (1.46).

$$\hat{\epsilon}_{ik}(p) = \frac{1}{Q} \sum_{q=p-Q+1}^{p} |e_{q,k}^i|^2 \,, \quad i = 1,2,3 \tag{1.45}$$

denote the estimate of the transient mse at the pth iteration, where Q is a constant parameter. These estimates are computed every Q frames, and the value of $K_2(p)$ is then determined by the following decision rule:

$$K_2(p+1) = \begin{cases} K_2(p) + 1 & \text{; if } \hat{\epsilon}_{1k}(p) > \hat{\epsilon}_{2k}(p) > \hat{\epsilon}_{3k}(p) \\ K_2(p) & \text{; if } \hat{\epsilon}_{1k}(p) > \hat{\epsilon}_{2k}(p) \leq \hat{\epsilon}_{3k}(p) \\ K_2(p) - 1 & \text{; otherwise} \end{cases} \tag{1.46}$$

Accordingly, $K_1(p+1)$ and $K_3(p+1)$ are updated by

$$K_1(p+1) = K_2(p+1) - 1 \,, \tag{1.47}$$
$$K_3(p+1) = K_2(p+1) + 1 \,,$$

and the adaptation proceeds by updating the resized vectors $\hat{\theta}_{ik}(p)$ using (1.44). Note that the parameter Q should be sufficiently small to enable tracking during variations in the optimal number of cross-terms, and sufficiently large to achieve an efficient approximation of the mse by (1.45).

The decision rule in (1.46) can be explained as follows. When the optimum number of cross-terms is equal or larger than $K_3(p)$, then $\hat{\epsilon}_{1k}(p) > \hat{\epsilon}_{2k}(p) > \hat{\epsilon}_{3k}(p)$ and all values are increased by one. In this case, the vectors are reinitialized by $\hat{\theta}_{1k}(p+1) = \hat{\theta}_{2k}(p)$, $\hat{\theta}_{2k}(p+1) = \hat{\theta}_{3k}(p)$, and $\hat{\theta}_{3k}(p+1) = \begin{bmatrix} 0 & \hat{\theta}_{3k}^T(p) & 0 \end{bmatrix}^T$. When $K_2(p)$ is the optimum number, then

$\hat{\epsilon}_{1k}(p) > \hat{\epsilon}_{2k}(p) \leq \hat{\epsilon}_{3k}(p)$ and the values remain unchanged. Finally, when the optimum number is equal or smaller than $K_1(p)$, we have $\hat{\epsilon}_{1k}(p) \leq \hat{\epsilon}_{2k}(p) < \hat{\epsilon}_{3k}(p)$ and all values are decreased by one. In this case, we reinitialize the vectors by $\hat{\theta}_{3k}(p+1) = \hat{\theta}_{2k}(p)$, $\hat{\theta}_{2k}(p+1) = \hat{\theta}_{1k}(p)$, and $\hat{\theta}_{1k}(p+1)$ is obtained by eliminating the first and last elements of $\hat{\theta}_{1k}(p)$. The decision rule is aimed at reaching the mmse for each frequency bin separately. That is, distinctive frequency bins may have different values of $K_2(p)$ at each frame index p. Clearly, this decision rule is unsuitable for applications where the error signal to be minimized is in the time domain. In such cases, the optimal number of cross-terms is the one that minimizes the time-domain mse $E\{|e(n)|^2\}$ [contrary to (1.43)]. Let

$$\hat{\epsilon}_i(n) = \frac{1}{\tilde{Q}} \sum_{m=n-\tilde{Q}+1}^{m} |e_i(m)|^2 \ , \quad i = 1, 2, 3 \tag{1.48}$$

denote the estimate of the time-domain mse, where $e_i(n)$ is the inverse STFT of $e^i_{p,k}$, and $\tilde{Q} \triangleq (Q-1)L + N$. Then, as in (1.45), these averages are computed every Q frames (corresponding to QL time-domain iterations), and $K_2(n)$ is determined similarly to (1.46) by substituting $\hat{\epsilon}_i(n)$ for $\hat{\epsilon}_{ik}(p)$ and n for p. Note that now all frequency bins have the same number of cross-terms $[2K_2(p) + 1]$ at each frame.

For completeness of discussion, let us evaluate the computational complexity of the proposed algorithm. Updating $2K + 1$ cross-terms with the NLMS adaptation formula (1.44), requires $8K + 6$ arithmetic operations for every L input samples [28]. Therefore, since three vectors of cross-terms are updated simultaneously in each frame, the adaptation process of the proposed approach requires $8[K_1(p) + K_2(p) + K_3(p)] + 6$ arithmetic operations. Using (1.47) and computing the desired signal estimate $\hat{d}_{p,k} = \mathbf{x}_{2k}^T(p)\hat{\theta}_{2k}(p)$, the overall complexity of the proposed approach is given by $28K_2(p) + 7$ arithmetic operation for every L input samples and each frequency bin. The computations required for updating $K_2(p)$ [see (1.45)–(1.47)] are relatively negligible, since they are carried out only once every Q iterations. When compared to the conventional MTF approach ($K = 0$), the proposed approach involves an increase of $28K_2(p) + 1$ arithmetic operations for every L input samples and every frequency bin.

1.6 Experimental Results

In this section, we present experimental results that demonstrate the performance of the approaches introduced in this chapter. The performance evaluation is carried out in terms of mse for both synthetic white Gaussian signals and real speech signals. In the following experiments, we use a Hamming synthesis window of length N with 50% overlap (i.e., $L = 0.5N$), and a corresponding minimum-energy analysis window that satisfies the completeness condition (1.6) [31]. The sample rate is 16 kHz.

1.6.1 Crossband Filters Estimation

In the first experiment, we examine the performance of the crossband filters approach under the assumptions made in Section 1.3.3. That is, the STFT of the input signal $x_{p,k}$ is a zero-mean white Gaussian process. Note that, $x_{p,k}$ is not necessarily a valid STFT signal, as not always a sequence whose STFT is given by $x_{p,k}$ may exist [51]. Similarly, the STFT of the noise signal $\xi_{p,k}$ is also a zero-mean white Gaussian process, which is uncorrelated with $x_{p,k}$. The impulse response $h(n)$ used in this experiment was measured in an office which exhibits a reverberation time of about 300 ms. The length of the STFT synthesis window is set to $N = 256$ (16 ms), and the crossband filters are estimated offline using (1.21). Figure 1.8 shows the resulting mse curves $\epsilon_k(K)$ [defined in (1.23)] as a function of the input SNR, obtained for frequency-bin $k = 1$ and for observable data length of $P = 200$ samples [Fig. 1.8(a)] and $P = 1000$ samples [Fig. 1.8(b)]. Results are averaged out over 200 independent runs (similar results are obtained for the other frequency bins). Clearly, as expected from the discussion in Section 1.3.3, as the SNR increases, a larger number of crossband filters should be utilized to achieve the mmse. We observe that the intersection-points of the mse curves are a monotonically increasing series. Furthermore, a comparison of Figs. 1.8(a) and (b) indicates that the intersection-points values decrease as we increase P [as expected from (1.29)]. This verifies that when the signal length increases (while the SNR remains constant), more crossband filters need to be used in order to attain the mmse.

1.6.2 Comparison of the Crossband Filters and MTF Approaches

In the second experiment, we compare the crossband filters approach to the MTF approach and investigate the influence of the STFT analysis window length (N) on their performances. The input signal $x(n)$ in this case is a speech signal of length 1.5 sec, where the additive noise $\xi(n)$ is a zero-mean white Gaussian process. We use the same impulse response $h(n)$ as in the previous experiment. The parameters of the crossband filters and MTF approaches are estimated offline using (1.21) and (1.33), respectively, and the resulting time-domain mse is computed by

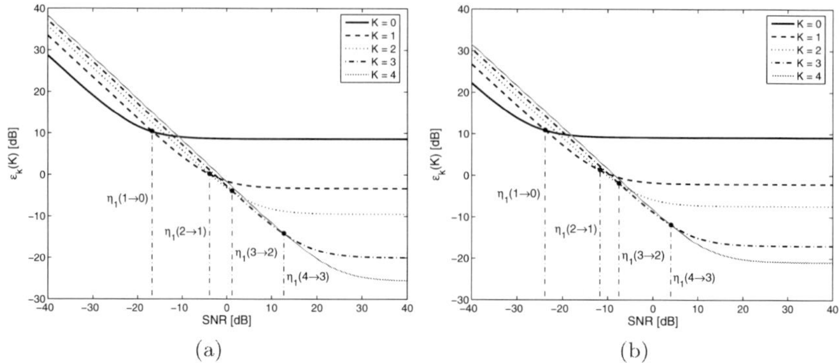

Fig. 1.8 MSE curves as a function of the input SNR using LS estimates of the crossband filters, for white Gaussian signals. (a) $P = 200$, (b) $P = 1000$.

$$\epsilon_{\text{time}} = 10 \log \frac{E\left\{\left[d(n) - \hat{d}(n)\right]^2\right\}}{E\left\{d^2(n)\right\}}, \tag{1.49}$$

where $\hat{d}(n)$ is the inverse STFT of the corresponding model output $\hat{d}_{p,k}$. Figure 1.9 shows the mse curves as a function of the input SNR obtained for an analysis window of length $N = 256$ [16 ms, Fig. 1.9(a)] and for a longer window of length $N = 2048$ [128 ms, Fig. 1.9(b)]. As expected, the performance of the MTF approach can be generally improved by using a longer analysis window. This is because the MTF approach heavily relies on the assumption that the support of the analysis window is sufficiently large compared with the duration of the system impulse response (see Section 1.4). As the SNR increases, a lower mse is attained by the crossband filters approach, even for long analysis window. For instance, Fig. 1.9(b) shows that for 20 dB SNR the MTF model achieves an mse value of -20 dB, whereas the crossband-filters model decreases the mse by approximately 10 dB by including three crossband filters ($K = 1$) in the model. Furthermore, it seems to be preferable to reduce the window length, as seen from Fig. 1.9(a), as it enables a decrease of approximately 7 dB in the mse (for a 20 dB SNR) by using the crossband filters approach. A short window is also essential for the analysis of nonstationary input signals, which is the case in acoustic echo cancellation applications. However, a short window support necessitates the estimation of more crossband filters for performance improvement, and correspondingly increases the computational complexity. It should also be noted that for low SNR values, a lower mse can be achieved by using the MTF approach, even when the large support assumption is not valid [Fig. 1.9(a)].

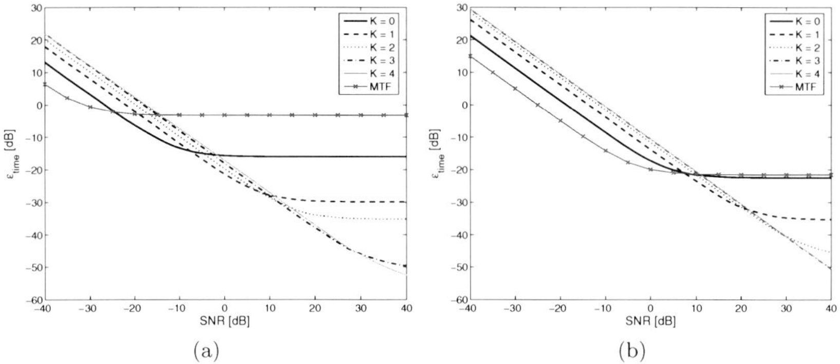

Fig. 1.9 MSE curves in the time domain as a function of the input SNR, obtained by the crossband filters approach and the MTF approach for a real speech input signal. (a) Length of analysis window is 16 ms ($N = 256$). (b) Length of analysis window is 128 ms ($N = 2048$).

1.6.3 CMTF Adaptation for Acoustic Echo Cancellation

In the third experiment, we demonstrate the CMTF approach (see Section 1.5) in an acoustic echo cancellation application [1, 2, 3] using real speech signals. The cross-terms are adaptively updated by the NLMS algorithm using a step-size $\mu = 1/(K + 1)$, where $2K + 1$ is the number of estimated cross-terms. The adaptive-control algorithm, introduced in Section 1.5.2, is employed and its performance is compared to that of an adaptive algorithm that utilizes a fixed number of cross-terms. The evaluation includes objective quality measures, a subjective study of temporal waveforms, and informal listening tests.

The experimental setup is depicted in Fig. 1.10. We use an ordinary office with a reverberation time T_{60} of about 100 ms. The measured acoustic signals are recorded by a DUET conference speakerphone, Phoenix Audio Technologies, which includes an omnidirectional microphone near the loudspeaker (more features of the DUET product are available at [60]). The farend signal is played through the speakerphone's built-in loudspeaker, and received together with the near-end signal by the speakerphone's built-in microphone. The small distance between the loudspeaker and the microphone yields relatively high SNR values, which may justify the estimation of more cross-terms. Employing the MTF approximation in this case, and ignoring all the cross-terms may result in insufficient echo reduction. It is worth noting that estimation of crossband filters, rather than CMTF, may be even more advantageous, but may also result in a significant increase in computational cost. In this experiment, the signals are sampled at 16 kHz. A far-end speech signal $x(n)$ is generated by the loudspeaker and received by the microphone

Fig. 1.10 Experimental setup. A speakerphone (Phoenix Audio DUET Executive Conference Speakerphone) is connected to a laptop using its USB interface. Another speakerphone without its cover shows the placement of the built-in microphone and loudspeaker.

as an echo signal $d(n)$ together with a near-end speech signal and local noise [collectively denoted by $\xi(n)$]. The distance between the near-end source and the microphone is 1 m. According to the room reverberation time, the effective length of the echo path is 100 ms, i.e., $N_h = 1600$. We use a synthesis window of length 200 ms (corresponding to $N = 3200$), which is twice the length of the echo path. A commonly-used quality measure for evaluating the performance of acoustic echo cancellers (AECs) is the echo-return loss enhancement (ERLE), defined in dB by

$$\text{ERLE}(K) = 10 \log_{10} \frac{E\left\{y^2(n)\right\}}{E\left\{e^2(n)\right\}}, \tag{1.50}$$

where $e(n) = y(n) - \hat{d}(n)$ is the error signal, and $\hat{d}(n)$ is the inverse STFT of the estimated echo signal.

Figures 1.11(a)–(b) show the far-end and microphone signals, respectively, where a double-talk situation (simultaneously active far-end and near-end speakers) occurs between 3.4 s and 4.4 s (indicated by two vertical dotted lines). Since such a situation may cause divergence of the adaptive algorithm, a double-talk detector (DTD) is usually employed to detect near-end signal and freeze the adaptation [52, 53]. Since the design of a DTD is beyond the scope of this chapter, we manually choose the periods where double-talk occurs and freeze the adaptation in these intervals. Figures 1.11(c)–(d) show the error signal $e(n)$ obtained by the CMTF approach with a fixed number of cross-terms [$K = 0$ (MTF) and $K = 2$, respectively], and Fig. 1.11(e) shows the error signal obtained by the adaptive-control algorithm. For the latter, the time-domain decision rule, based on the mse estimate in (1.48), is

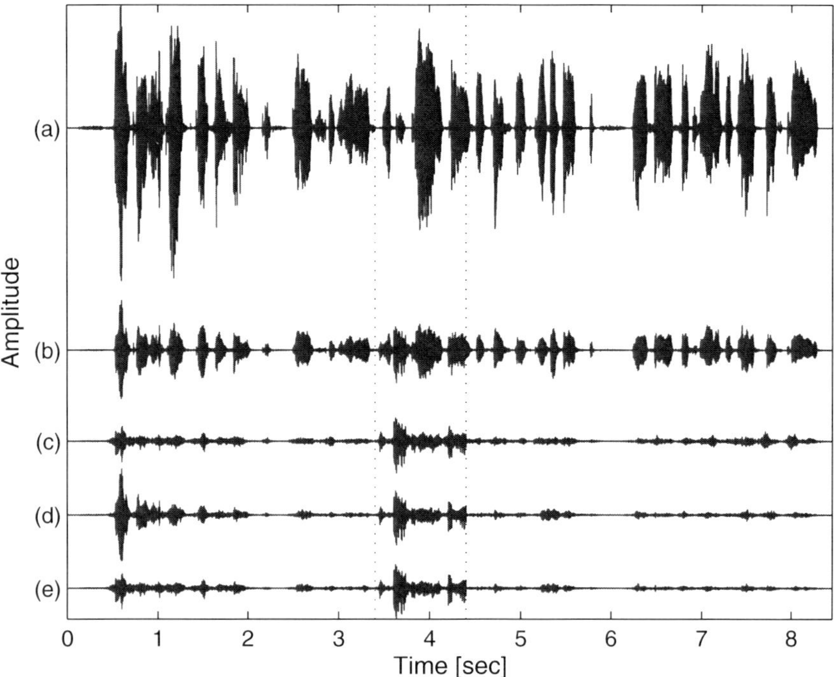

Fig. 1.11 Speech waveforms and error signals $e(n)$, obtained by adaptively updating the cross-terms using the NLMS algorithm. A double-talk situation is indicated by vertical dotted lines. (a) Far-end signal (b) Microphone signal. (c)–(d) Error signals obtained by using the CMTF approach with fixed number of cross-terms: $K = 0$ (MTF) and $K = 2$, respectively. (e) Error signal obtained by the adaptive-control algorithm described in Section 1.5.2.

employed using $Q = 5$. The ERLE values of the corresponding error signals were computed after convergence of the algorithms, and are given by 12.8 dB ($K = 0$), 17.2 dB ($K = 2$), and 18.6 dB (adaptive control). Clearly, the MTF approach ($K = 0$) achieves faster convergence than the CMTF approach (with $K = 2$), but suffers from higher steady-state ERLE. The slower convergence of the CMTF approach is attributable to the relatively small step-size forced by estimating more cross-terms [see (1.42)]. The adaptive control algorithm overcomes this problem by selecting the optimal number of cross-terms in each iteration. Figure 1.11(e) verifies that the adaptive-control algorithm achieves both fast convergence as the MTF approach and high ERLE as the CMTF approach. Subjective listening tests confirm that the CMTF approach in general, and the adaptive-control algorithm in particular, achieve a perceptual improvement in speech quality over the conventional MTF approach ($K = 0$).

It is worthwhile noting that the relatively small ERLE values obtained in this experiment, may be attributable to the nonlinearity introduced by the loudspeaker and its amplifier. Estimating the overall nonlinear system with a linear model yields a model mismatch that degrades the system estimate accuracy. Several techniques for nonlinear acoustic echo cancellation have been proposed (e.g., [54, 55, 56]). However, combining such techniques with the CMTF approximation is beyond the scope of this chapter.

1.7 Conclusions

We have considered the problem of linear system identification in the STFT domain and introduced three different approaches for that purpose. We have investigated the influence of crossband filers on a system identifier operating in the STFT domain, and derived important explicit expressions for the attainable mse in subbands. We show that in general, the number of crossband filters that should be utilized in the system identifier is larger for stronger and longer input signals.

The widely-used MTF approximation, which avoids the crossband filters by approximating the linear system as multiplicative in the STFT domain, was also considered. We have investigated the performance of a system identifier that utilizes this approximation and showed that the mse performance does not necessarily improve with increasing window length, mainly due to the finite length of the input signal. The optimal window length that achieves the mmse depends on the SNR and length of the input signal. We compared the performance of the MTF and crossband filters approaches and showed that for high SNR conditions, the crossband filters approach is considerably more advantageous, even for long analysis window.

Finally, motivated by the insufficient estimation accuracy of the MTF approach, we have introduced the CMTF model and proposed an adaptive-control algorithm for estimating its parameters. We have demonstrated the effectiveness of the resulting algorithm to an acoustic echo cancellation scenario, and showed a substantial improvement in both steady-state performance and speech quality, compared to the MTF approach.

Recently, a novel approach that extends the linear models described in this chapter has been introduced for improved *nonlinear* system identification in the STFT domain [56], and a detailed mean-square analysis of this approach is given in [57, 58]. The problem of nonlinear system identification in the STFT domain is also discussed in [59], and the reader is referred to there for further details.

Appendix: Derivation of the Crossband Filters

Substituting (1.4) into (1.10), we obtain

$$d_{p,k} = \sum_m \sum_\ell h(\ell) \sum_{k'=0}^{N-1} \sum_{p'} x_{p',k'} \psi_{p',k'}(m-\ell) \tilde{\psi}^*_{p,k}(m)$$

$$= \sum_{k'=0}^{N-1} \sum_{p'} x_{p',k'} h_{p,k,p',k'}, \tag{1.51}$$

where

$$h_{p,k,p',k'} = \sum_m \sum_\ell h(\ell) \psi_{p',k'}(m-\ell) \tilde{\psi}^*_{p,k}(m) \tag{1.52}$$

may be interpreted as the STFT of $h(n)$ using a composite analysis window

$$\sum_m \psi_{p',k'}(m-\ell) \tilde{\psi}^*_{p,k}(m) .$$

Substituting (1.3) and (1.5) into (1.52), we obtain

$$h_{p,k,p',k'} = \sum_m \sum_\ell h(\ell) \psi(m-\ell-p'L) e^{j\frac{2\pi}{N}k'(m-\ell-p'L)}$$

$$\times \tilde{\psi}(m-pL) e^{-j\frac{2\pi}{N}k(m-pL)}$$

$$= \sum_\ell h(\ell) \sum_m \tilde{\psi}(m) e^{-j\frac{2\pi}{N}km} \psi\left[(p-p')L - \ell + m\right]$$

$$\times e^{j\frac{2\pi}{N}k'((p-p')L-\ell+m)}$$

$$= \left\{ h(n) * \phi_{k,k'}(n) \right\} \big|_{n=(p-p')L} \triangleq h_{p-p',k,k'}, \tag{1.53}$$

where $*$ denotes convolution with respect to the time index n, and

$$\phi_{k,k'}(n) \triangleq e^{j\frac{2\pi}{N}k'n} \sum_m \tilde{\psi}(m) \psi(n+m) e^{-j\frac{2\pi}{N}m(k-k')} . \tag{1.54}$$

From (1.53), $h_{p,k,p',k'}$ depends on $(p-p')$ rather than on p and p' separately. Substituting (1.53) into (1.51), we obtain (1.11)–(1.13).

References

1. J. Benesty, T. Gänsler, D. R. Morgan, T. Gdnsler, M. M. Sondhi, and S. L. Gay, *Advances in Network and Acoustic Echo Cancellation.* New York: Springer, 2001.
2. E. Hänsler and G. Schmidt, *Acoustic Echo and Noise Control: A Practical Approach.* New Jersey: Wiley, 2004.
3. C. Breining, P. Dreiseitel, E. Hänsler, A. Mader, B. Nitsch, H. Puder, T. Schertler, G. Schmidt, and J. Tlip, "Acoustic echo control," *IEEE Signal Processing Mag.*, vol. 16, no. 4, pp. 42–69, Jul. 1999.
4. I. Cohen, "Relative transfer function identification using speech signals," *IEEE Trans. Speech Audio Processing*, vol. 12, no. 5, pp. 451–459, Sept. 2004.

5. Y. Huang, J. Benesty, and J. Chen, "A blind channel identification-based two-stage approach to separation and dereverberation of speech signals in a reverberant environment," *IEEE Trans. Speech Audio Processing*, vol. 13, no. 5, pp. 882–895, Sept. 2005.

6. M. Wu and D. Wang, "A two-stage algorithm for one-microphone reverberant speech enhancement," *IEEE Trans. Audio Speech Lang. Processing*, vol. 14, no. 3, pp. 774–784, May 2006.

7. S. Araki, R. Mukai, S. Makino, T. Nishikawa, and H. Saruwatari, "The fundamental limitation of frequency domain blind source separation for convolutive mixtures of speech," *IEEE Trans. Audio Speech Lang. Processing*, vol. 14, no. 3, pp. 774–784, May 2006.

8. F. Talantzis, D. B. Ward, and P. A. Naylor, "Performance analysis of dynamic acoustic source separation in reverberant rooms," *IEEE Trans. Audio Speech Lang. Processing*, vol. 14, no. 4, pp. 1378–1390, Jul. 2006.

9. S. Gannot, D. Burshtein, and E. Weinstein, "Signal enhancement using beamforming and nonstationarity with applications to speech," *IEEE Trans. Signal Processing*, vol. 49, no. 8, pp. 1614–1626, Aug. 2001.

10. S. Gannot and I. Cohen, "Speech enhancement based on the general transfer function GSC and postfiltering," *IEEE Trans. Speech Audio Processing*, vol. 12, no. 6, pp. 561–571, Nov. 2004.

11. L. Ljung, *System Identification: Theory for the User*. Upper Saddle River, New Jersey: Prentice-Hall, 1999.

12. S. Haykin, *Adaptive Filter Theory*. New Jersey: Prentice-Hall, 2002.

13. J. J. Shynk, "Frequncy-domain and multirate adaptive filtering," *IEEE Signal Processing Mag.*, vol. 9, no. 1, pp. 14–37, Jan. 1992.

14. P. P. Vaidyanathan, *Multirate Systems and Filters Banks*. New Jersey: Prentice-Hall, 1993.

15. H. Yasukawa, S. Shimada, and I. Furukawa, "Acoustic echo canceller with high speech quality," in *Proc. Int. Conf. on Acoustics, Speech and Signal Processing (ICASSP)*. Dallas, Texas: IEEE, Apr. 1987, pp. 2125–2128.

16. W. Kellermann, "Analysis and design of multirate systems for cancellation of acoustical echoes," in *Proc. Int. Conf. on Acoustics, Speech and Signal Processing (ICASSP)*, New-York City, Apr. 1988, pp. 2570–2573.

17. M. Harteneck, J. M. Páez-Borrallo, and R. W. Stewart, "An oversampled subband adaptive filter without cross adaptive filters," *Signal Processing*, vol. 64, no. 1, pp. 93–101, Mar. 1994.

18. V. S. Somayazulu, S. K. Mitra, and J. J. Shynk, "Adaptive line enhancement using multirate techniques," in *Proc. Int. Conf. on Acoustics, Speech and Signal Processing (ICASSP)*. Glasgow, Scotland: IEEE, May 1989, pp. 928–931.

19. A. Gilloire and M. Vetterli, "Adaptive filtering in subbands with critical sampling: Analysis, experiments, and application to acoustic echo cancellation," *IEEE Trans. Signal Processing*, vol. 40, no. 8, pp. 1862–1875, Aug. 1992.

20. S. S. Pradhan and V. U. Reddy, "A new approach to subband adaptive filtering," *IEEE Trans. Signal Processing*, vol. 47, no. 3, pp. 655–664, Mar. 1999.

21. B. E. Usevitch and M. T. Orchard, "Adaptive filtering using filter banks," *IEEE Trans. Circuits Syst. II*, vol. 43, no. 3, pp. 255–265, Mar. 1996.

22. C. Avendano, "Acoustic echo suppression in the STFT domain," in *Proc. IEEE Workshop on Application of Signal Processing to Audio and Acoustics*, New Paltz, NY, Oct. 2001, pp. 175–178.

23. Y. Lu and J. M. Morris, "Gabor expansion for adaptive echo cancellation," *IEEE Signal Processing Mag.*, vol. 16, pp. 68–80, Mar. 1999.

24. C. Faller and J. Chen, "Suppressing acoustic echo in a spectral envelope space," *IEEE Trans. Acoust., Speech, Signal Processing*, vol. 13, no. 5, pp. 1048–1062, Sept. 2005.

25. Y. Avargel and I. Cohen, "System identification in the short-time Fourier transform domain with crossband filtering," *IEEE Trans. Audio Speech Lang. Processing*, vol. 15, no. 4, pp. 1305–1319, May 2007.
26. ——, "On multiplicative transfer function approximation in the short-time Fourier transform domain," *IEEE Signal Processing Lett.*, vol. 14, no. 5, pp. 337–340, May 2007.
27. P. Smaragdis, "Blind separation of convolved mixtures in the frequency domain," *Neurocomputing*, vol. 22, no. 1-3, pp. 21–34, Nov. 1998.
28. Y. Avargel and I. Cohen, "Adaptive system identification in the short-time Fourier transform domain using cross-multiplicative transfer function approximation," *IEEE Trans. Audio Speech Lang. Processing*, vol. 16, no. 1, pp. 162–173, Jan. 2008.
29. ——, "Identification of linear systems with adaptive control of the cross-multiplicative transfer function approximation," in *Proc. Int. Conf. on Acoustics, Speech and Signal Processing (ICASSP)*, Las Vegas, Nevada, Apr. 2008, pp. 3789–3792.
30. M. R. Portnoff, "Time-frequency representation of digital signals and systems based on short-time Fourier analysis," *IEEE Trans. Signal Processing*, vol. ASSP-28, no. 1, pp. 55–69, Feb. 1980.
31. J. Wexler and S. Raz, "Discrete Gabor expansions," *Signal Processing*, vol. 21, pp. 207–220, Nov. 1990.
32. S. Qian and D. Chen, "Discrete Gabor transform," *IEEE Trans. Signal Processing*, vol. 41, no. 7, pp. 2429–2438, Jul. 1993.
33. Y. Avargel and I. Cohen, "Performance analysis of cross-band adaptation for subband acoustic echo cancellation," in *Proc. Int. Workshop Acoust. Echo Noise Control (IWAENC)*, Paris, France, Sept. 2006, pp. 1–4, paper no. 8.
34. A. Neumaier, "Solving ill-conditioned and singular linear systems: A tutorial on regularization," *SIAM Rev.*, vol. 40, no. 3, pp. 636–666, Sept. 1998.
35. G. H. Golub and C. F. Van Loan, *Matrix Computations*. Baltimore, MD: Johns Hopkins University Press, 1996.
36. A. Papoulis, *Probability, Random Variables, and Stochastic Processes*. Singapore: McGRAW-Hill, 1991.
37. D. R. Brillinger, *Time Series: Data Analysis and Theory*. Philadelphia: PA: SIAM, 2001.
38. Y. Ephraim and D. Malah, "Speech enhancement using a minimum mean-square error short-time spectral amplitude estimator," *IEEE Trans. Acoust., Speech, Signal Processing*, vol. 32, no. 6, pp. 1109–1121, Dec. 1984.
39. Y. Ephraim and I. Cohen, "Recent advancements in speech enhancement," in *The Electrical Engineering Handbook, Circuits, Signals, and Speech and Image Processing*, R. C. Dorf, Ed. Boca Raton, FL: CRC Press, 2006.
40. F. D. Ridder, R. Pintelon, J. Schoukens, and D. P. Gillikin, "Modified AIC and MDL model selection criteria for short data records," *IEEE Trans. Instrum. Meas.*, vol. 54, no. 1, pp. 144–150, Feb. 2005.
41. G. Schwarz, "Estimating the dimension of a model," *Ann. Stat.*, vol. 6, no. 2, pp. 461–464, 1978.
42. P. Stoica and Y. Selen, "Model order selection: a review of information criterion rules," *IEEE Signal Processing Mag.*, vol. 21, no. 4, pp. 36–47, Jul. 2004.
43. G. C. Goodwin, M. Gevers, and B. Ninness, "Quantifying the error in estimated transfer functions with application to model order selection," *IEEE Trans. Automat. Contr.*, vol. 37, no. 7, pp. 913–928, Jul. 1992.
44. H. Akaike, "A new look at the statistical model identification," *IEEE Trans. Automat. Contr.*, vol. AC-19, no. 6, pp. 716–723, Dec. 1974.
45. J. Rissanen, "Modeling by shortest data description," *Automatica*, vol. 14, no. 5, pp. 465–471, 1978.
46. M. Dentino, J. M. McCool, and B. Widrow, "Adaptive filtering in the frequency domain," *Proc. IEEE*, vol. 66, no. 12, pp. 1658–1659, Dec. 1978.

47. D. Mansour and J. A. H. Gray, "Unconstrained frequency-domain adaptive filter," *IEEE Trans. Acoust., Speech, Signal Processing*, vol. ASSP-30, no. 5, pp. 726–734, Oct. 1982.

48. P. C. W. Sommen, "Partitioned frequency domain adaptive filters," in *Proc. 23rd Asilomar Conf. Signals, Systems, Computers*, Pacific Grove, CA, Nov. 1989, pp. 677–681.

49. R. C. Bilcu, P. Kuosmanen, and K. Egiazarian, "On length adaptation for the least mean square adaptive filters," *Signal Processing*, vol. 86, pp. 3089–3094, Oct. 2006.

50. Y. Gong and C. F. N. Cowan, "An LMS style variable tap-length algorithm for structure adaptation," *IEEE Trans. Signal Processing*, vol. 53, no. 7, pp. 2400–2407, Jul. 2005.

51. D. W. Griffin and J. S. Lim, "Signal estimation from modified short-time Fourier transform," *IEEE Trans. Acoust., Speech, Signal Processing*, vol. ASSP-32, no. 2, pp. 236–243, Apr. 1984.

52. J. Benesty, D. R. Morgan, and J. H. Cho, "A new class of doubletalk detectors based on cross-correlation," *IEEE Trans. Speech Audio Processing*, vol. 8, no. 2, pp. 168–172, Mar. 2000.

53. J. H. Cho, D. R. Morgan, and J. Benesty, "An objective technique for evaluating doubletalk detectors in acoustic echo cancelers," *IEEE Trans. Speech Audio Processing*, vol. 7, no. 6, pp. 718–724, Nov. 1999.

54. A. Guérin, G. Faucon, and R. L. Bouquin-Jeannès, "Nonlinear acoustic echo cancellation based on Volterra filters," *IEEE Trans. Speech Audio Processing*, vol. 11, no. 6, pp. 672–683, Nov. 2003.

55. A. Stenger and W. Kellermann, "Adaptation of a memoryless preprocessor for nonlinear acoustic echo cancelling," *Signal Processing*, vol. 80, pp. 1747–1760, Sept. 2000.

56. Y. Avargel and I. Cohen, "Representation and identification of nonlinear systems in the short-time Fourier transform domain," *submitted to IEEE Trans. Signal Processing*.

57. ——, "Nonlinear systems in the short-time Fourier transform domain: Estimation error analysis," *in preperation*.

58. ——, "Adaptive nonlinear system identification in the short-time Fourier transform domain," *to appear in IEEE Trans. Signal Processing*.

59. ——, "Representation and identification of nonlinear systems in the short-time fourier transform domain," in *Speech Processing in Modern Communication: Challenges and Perspectives*, I. Cohen, J. Benesty, and S. Gannot, Eds.   Berlin, Germany: Springer, 2009.

60. [Online]. Available: http://phnxaudio.com.mytempweb.com/?tabid=62

Chapter 2
Identification of the Relative Transfer Function between Sensors in the Short-Time Fourier Transform Domain

Ronen Talmon, Israel Cohen, and Sharon Gannot

Abstract [1]In this chapter, we delve into the problem of relative transfer function (RTF) identification. First, we focus on identification algorithms that exploit specific properties of the input data. In particular, we exploit the non-stationarity of speech signals and the existence of segments where speech is absent in arbitrary utterances. Second, we explore approaches that aim at better modeling the signals and systems. We describe a common approach to represent a linear convolution in the short-time Fourier transform (STFT) domain as a multiplicative transfer function (MTF). Then, we present a new modeling approach for a linear convolution in the STFT domain as a convolution transfer function (CTF). The new approach is associated with larger model complexity and enables better representation of the signals and systems in the STFT domain. Then, we employ RTF identification algorithms based on the new model, and demonstrate improved results.

2.1 Introduction

Identification of the relative transfer function (RTF) between sensors take a significant role in various multichannel hands-free communication systems [1], [2]. This transfer function is often referred to as acoustic transfer function ratio since it represents the coupling between two sensors with respect to a desired source [3], [4]. In reverberant and noisy environments, estimates of the RTF are often used for constructing beamformers and noise cancelers [5].

Ronen Talmon and Israel Cohen
Technion–Israel Institute of Technology, Israel, e-mail: `{ronenta2@tx,icohen@ee}.technion.ac.il`

Sharon Gannot
Bar-Ilan University, Israel, e-mail: `gannot@eng.biu.ac.il`

[1] This work was supported by the Israel Science Foundation under Grant 1085/05 .

I. Cohen et al. (Eds.): Speech Processing in Modern Communication, STSP 3, pp. 33–47.
springerlink.com © Springer-Verlag Berlin Heidelberg 2010

For example, the RTF may be used in a blocking channel, where the desired signal is blocked in order to derive a reference noise-only signal, which can be exploited later in a noise canceler for attenuating interfering sources [3], [6], [7], [8], [9].

In the literature, there are many widely spread solutions for the general problem of system identification. A typical system identification problem is usually defined as a given input signal, which goes through a system of particular interest, whose noisy output is captured by an arbitrary sensor. Given the input and the noisy output signals, one aims at recovering this system of interest. However, identification of the RTF differs from a system identification problem. First, the RTF identification is considered a blind problem, since the RTF represents the coupling of two measured signal with respect to an unknown source. Second, in the general system identification problem, the estimated system is usually assumed to be independent of the input source and the additive noise, whereas, these assumptions cannot be made in the RTF identification task as will be shown later in this chapter.

In this chapter, we tackle the problem of RTF identification in two fronts. First, we focus on identification algorithms that exploit specific properties of the input data [1], [10]. In particular, we exploit the non-stationarity of speech signals and the existence of segments where speech is absent in arbitrary utterances. Second, we explore approaches that aim at better modeling the signals and systems [11]. We describe a common approach to represent a linear convolution in the short-time Fourier transform (STFT) domain as a multiplicative transfer function (MTF) [12]. Then, we present a new modeling approach for a linear convolution in the STFT domain as a convolution transfer function (CTF) [13]. The new approach is associated with larger model complexity and enables better representation of the signals and systems in the STFT domain.

This chapter is organized as follows. In Section 2.2, we present two RTF identification methods that exploit specific properties of the input data. In Section 2.3, we present a new modeling approach for the representation and identification of the RTF. Finally, in Section 2.4, we demonstrate the performance of the presented RTF identification methods and show an example for their use in a specific multichannel application.

2.2 Identification of the RTF Using Multiplicative Transfer Function Approximation

In typical rooms, the acoustic impulse response may be very long, and as a consequence the relative impulse response is modeled as a very long filter as well[2][14]. Thus, identification of such long filters in the time domain is inefficient. In addition, speech signals are better represented in the STFT domain than in the time domain. The spectrum of the STFT samples of speech signals is flat, yielding an improved convergence rate of adaptive algorithms. Hence, we focus on RTF identification methods which are carried out in the STFT domain.

In this section, we first tackle the problem of appropriately representing the signals and systems in the STFT domain. Then, we describe two RTF identification methods that exploit specific characteristics of the input signals. The first method exploits the non-stationarity property of speech signals with respect to the stationarity of the additive noise signals. The second method exploits the silent segments of the input data that consist of noise-only measurements, which usually appear in speech utterances.

2.2.1 Problem Formulation and the Multiplicative Transfer Function Approximation

A speech signal captured by microphones in a typical room is usually distorted by reverberation and corrupted by background noise. Suppose that $s(n)$ represents a speech signal and that $u(n)$ and $w(n)$ represent stationary noise signals uncorrelated with the speech source. Thus, the signals captured by a pair of primary and reference microphones are given by

$$x(n) = h_x(n) * s(n) + u(n), \qquad (2.1)$$

$$y(n) = h_y(n) * s(n) + w(n), \qquad (2.2)$$

where $*$ denotes convolution and $h_x(n)$ and $h_y(n)$ represent the acoustic room impulse responses of the primary and reference microphones to the speech source, respectively. Let $h(n)$ represent the *relative* impulse response between the microphones with respect to the speech source, which satisfies $h_y(n) = h(n) * h_x(n)$. Then (2.1) and (2.2) can be rewritten as

$$y(n) = h(n) * x(n) + v(n), \qquad (2.3)$$

$$v(n) = w(n) - h(n) * u(n), \qquad (2.4)$$

where in (2.3) we have a linear time-invariant (LTI) system with an input $x(n)$, output $y(n)$, and additive noise $v(n)$. In this work, our goal is to identify the response $h(n)$. This formulation, which indicates that the additive noise signal $v(n)$ depends on both $x(n)$ and $h(n)$, distinguishes the RTF identifi-

---

[2] Note that the relative impulse response is infinite since it represents the ratio between two room impulse responses. However, since the energy of the relative impulse response decays rapidly according to the room reverberation time, it can be modeled using a finite support without significant data loss.

cation problem from an ordinary system identification problem, where the additive noise is assumed uncorrelated with the input and the estimated system.

A common approach in speech processing is to divide the signals into overlapping time frames and analyze them using the STFT. Suppose the observation interval is divided into P time frames of length N (yielding N frequency bins). By assuming that the support of $h(n)$ is finite and small compared to the length of the time frame, (2.3) can be approximated in the STFT domain as

$$y_{p,k} = h_k x_{p,k} + v_{p,k}, \tag{2.5}$$

where p is the time frame index, k is the frequency subband index and h_k is the RTF. Modeling such an LTI system in the STFT domain is often referred to as a multiplicative transfer function (MTF) approximation. It is worthwhile noting that the relative impulse response is an infinite impulse response filter. Thus, the MTF approximation implies that modeling the relative impulse response as a short FIR filter conveys most of its energy.

2.2.2 RTF Identification Using Non-Stationarity

In the following we describe an RTF identification method using the non-stationarity of speech signals, assuming that the noise is stationary and additive, and that the RTF is not varying during the interval of interest [1].

According to (2.5), the cross power spectral density (PSD) between the microphone signals $x(n)$ and $y(n)$ can be written as

$$\phi_{yx}(p,k) = h_k \phi_{xx}(p,k) + \phi_{vx}(k). \tag{2.6}$$

Notice that writing the auto PSD of $x(n)$ implies that the speech is stationary in each time frame, which restricts the time frames to be relatively short (< 40 ms). In addition, the noise signals are assumed to be stationary, hence the cross PSD term $\phi_{vx}(k)$ is independent of the time frame index p. By rewriting (2.6) in terms of PSD estimates, we obtain

$$\hat{\phi}_{yx}(p,k) = \tilde{\phi}_{xx}^T(p,k)\,\theta(k) + \epsilon(p,k), \tag{2.7}$$

where $\theta(k) = [h_k \;\; \phi_{vx}(k)]^T$ represents the unknown variables given microphone measurements, $\tilde{\phi}_{xx}(p,k) = \left[\hat{\phi}_{xx}(p,k) \;\; 1\right]^T$, and $\epsilon(p,k)$ is the PSD estimation error. Since the speech signal is assumed to be non-stationary, the PSD $\phi_{xx}(p,k)$ may vary significantly from one time frame to another. Consequently, in (2.7) we obtain a set of $P \geq 2$ linearly independent equations that is used to estimate two variables.

By concatenating all the time frames, we obtain the following matrix form

$$\hat{\Phi}_{yx}(k) = \tilde{\Phi}_{xx}(k)\theta(k) + \mathbf{e}(k), \tag{2.8}$$

where $\tilde{\Phi}_{xx}(k) = \left[\tilde{\phi}_{xx}\left(1,k\right)\cdots\tilde{\phi}_{xx}\left(P,k\right)\right]^{T} = \left[\hat{\Phi}_{xx}(k)\ \ \mathbf{1}\right]$, $\mathbf{1}$ is a vector of ones, and $\hat{\Phi}_{xx}(k)$, $\hat{\Phi}_{yx}(k)$ and $\mathbf{e}(k)$ are column stack vectors of $\hat{\phi}_{xx}\left(p,k\right)$, $\hat{\phi}_{yx}\left(p,k\right)$ and $\epsilon\left(p,k\right)$, respectively. Now, the weighted least square (WLS) estimate of $\theta(k)$ is given by

$$\hat{\theta} = \arg\min_{\theta}\left\{\left(\hat{\Phi}_{yx} - \tilde{\Phi}_{xx}\theta\right)^{H} W\left(\hat{\Phi}_{yx} - \tilde{\Phi}_{xx}\theta\right)\right\}$$
$$= \left(\tilde{\Phi}_{xx}^{H} W \tilde{\Phi}_{xx}\right)^{-1}\tilde{\Phi}_{xx}^{H} W \hat{\Phi}_{yx}, \tag{2.9}$$

where W is the weight matrix and the frequency subband index k is omitted for notational simplicity. From (2.9), using uniform weights, we obtain the following estimator of the RTF

$$\hat{h} = \frac{\frac{1}{P}\hat{\Phi}_{xx}^{H}\hat{\Phi}_{yx} - \frac{1}{P^{2}}\mathbf{1}^{T}\hat{\Phi}_{xx}\hat{\Phi}_{yx}^{H}\mathbf{1}}{\frac{1}{P}\hat{\Phi}_{xx}^{H}\hat{\Phi}_{xx} - \left(\frac{1}{P}\mathbf{1}^{T}\hat{\Phi}_{xx}\right)^{2}}. \tag{2.10}$$

Thus, by relying on the diversity of the measured signals PSD across time frames, we obtain an unbiased estimate of the RTF.

2.2.3 RTF Identification Using Speech Signals

One limitation of the non-stationarity estimator (2.10) is that both the RTF and the noise cross PSD are estimated simultaneously through the same WLS criterion. The RTF identification requires large weights in high SNR subintervals and low weights in low SNR subintervals, whereas the noise cross PSD requires inversely distributed weights. In order to overcome this conflict, the RTF identification and the noise cross PSD estimation are decoupled by exploiting segments where the speech is absent [10].

Let $d(n)$ denote the reverberated speech component captured at the primary microphones, which satisfies $d(n) = h_x(n)*s(n)$. Since the speech signal is uncorrelated with the noise signals, from (2.1), (2.3) and (2.4), we have

$$\phi_{yx}\left(p,k\right) = h_k\phi_{dd}\left(p,k\right) + \phi_{wu}\left(k\right). \tag{2.11}$$

Now, writing (2.11) in term of PSD estimates yields

$$\hat{\phi}\left(p,k\right) = h_k\hat{\phi}_{dd}\left(p,k\right) + \epsilon\left(p,k\right), \tag{2.12}$$

where $\hat{\phi}(p,k) \triangleq \hat{\phi}_{yx}(p,k) - \hat{\phi}_{wu}(k)$, the reverberated speech PSD satisfies $\hat{\phi}_{dd}(p,k) = \hat{\phi}_{xx}(p,k) - \hat{\phi}_{uu}(k)$ and $\epsilon(p,k)$ denotes the PSD estimation error. The PSD terms $\hat{\phi}_{yx}(p,k)$ and $\hat{\phi}_{xx}(p,k)$ can be obtained directly from the measurements, while the noise PSD terms $\hat{\phi}_{wu}(k)$ and $\hat{\phi}_{uu}(k)$ can be estimated based on periods where the speech signal is absent and the measurements consist of noise only signals. It is worthwhile noting that we exploited here both speech and noise characteristics. First, we assume that an arbitrary speech utterance contains silent periods in addition to the non-stationarity assumption. Second, we assume that the noise signal statistics are slowly changing in time, and thus, the noise PSD terms can be calculated based on silent periods and applied during the whole observation interval.

Once again, by concatenating all the time frames we obtain the following matrix form of (2.12)

$$\hat{\boldsymbol{\Phi}}(k) = h_k \hat{\boldsymbol{\Phi}}_{dd}(k) + \mathbf{e}(k), \tag{2.13}$$

where $\hat{\boldsymbol{\Phi}}(k)$, $\hat{\boldsymbol{\Phi}}_{dd}(k)$ and $\mathbf{e}(k)$ are column stack vectors of $\hat{\phi}(p,k)$, $\hat{\phi}_{dd}(p,k)$ and $\epsilon(p,k)$, respectively.

Since the RTF represents the coupling between the primary and reference microphones with respect to the speech signal, it has to be identified based solely on time frames which contain the speech signal. Let $I(p,k)$ denote an indicator function for the speech signal presence and let $\mathbf{I}(k)$ be a diagonal matrix with the elements $[I(1,k), \cdots, I(P,k)]$ on its diagonal. Then, the WLS solution of (2.13) is given by

$$\begin{aligned}
\hat{h} &= \arg\min_{h} \left\{ (\mathbf{Ie})^{H} W (\mathbf{Ie}) \right\} \\
&= \arg\min_{h} \left\{ \left(\hat{\boldsymbol{\Phi}} - \hat{\boldsymbol{\Phi}}_{dd}h \right)^{H} \mathbf{I}W\mathbf{I} \left(\hat{\boldsymbol{\Phi}} - \hat{\boldsymbol{\Phi}}_{dd}h \right) \right\} \\
&= \left(\hat{\boldsymbol{\Phi}}_{dd}^{H} \mathbf{I}W\mathbf{I}\hat{\boldsymbol{\Phi}}_{dd} \right)^{-1} \hat{\boldsymbol{\Phi}}_{dd}^{H}\mathbf{I}W\mathbf{I}\hat{\boldsymbol{\Phi}},
\end{aligned} \tag{2.14}$$

where the subband index k is omitted for notational simplicity. It can be shown [10] that the variance of the estimator based on speech signals (2.14) is lower than the variance of the estimator based on non-stationarity (2.10).

2.3 Identification of the RTF Using Convolutive Transfer Function Approximation

In the previous section, we described RTF identification methods which mainly focused on exploiting unique characteristics of the speech signal. In this section, we aim at appropriately representing the problem in the STFT domain. The MTF approximation, which was described in Section 2.2.1, enables to replace a linear convolution in the time domain with a scalar multiplication in the STFT domain. This approximation becomes more accurate when the length of the time frame increases, relative to the length of the impulse response. Consequently, applying the MTF approximation to the RTF identification problem is inappropriate. The relative impulse response is of an infinite length, whereas the MTF approximation models the relative impulse response as a finite response of a much shorter length than the time frame. Moreover, since acoustic impulse responses in typical rooms are long, the RTF must be modeled as a long finite response, in order to convey most of its energy. Thus, controlling the time frame length has a significant influence. The use of short time frames restricts the length of the relative impulse response, yielding a significant loss of data. On the other hand, by using long time frames, fewer observations in each frequency subband are available[3], which may increase the estimation variance.

In the following, we present an alternative modeling of the problem, which enables a more accurate and flexible representation of the input signals in the STFT domain [13]. The new modeling employs an approximation of a linear convolution in the time domain as a linear convolution in the STFT domain. This approximation enables representation of long impulse responses in the STFT domain using short time frames. Based on the analysis of the system identification with cross-band filtering [11], we show that the new approach becomes especially advantageous as the SNR increases. In addition, unlike the MTF model, this so-called CTF approximation, enables flexibility in adjusting the estimated RTF length and the time frame length independently. Next, we formulate the CTF approximation and apply it on an RTF estimator.

2.3.1 The Convolutive Transfer Function Approximation

A filter convolution in the time domain can be represented as a sum of N cross-band convolutions in the STFT domain. Accordingly, (2.3) and (2.4) can be written as

$$y_{p,k} = \sum_{k'=0}^{N-1} \sum_{p'} x_{p-p',k'} h_{p',k',k} + v_{p,k}, \tag{2.15}$$

$$v_{p,k} = w_{p,k} - \sum_{k'=0}^{N-1} \sum_{p'} u_{p-p',k'} h_{p',k',k}, \tag{2.16}$$

---

[3] Since the observation interval is of finite length.

where $h_{p,k',k}$ are the cross-band filter coefficients between frequency bands k' and k.

Now, we approximate the convolution in the time domain as a convolution between the STFT samples of the input signal and the corresponding band-to-band filter (i.e. $k = k'$). Then, (2.15) and (2.16) reduce to

$$y_{p,k} = \sum_{p'} x_{p-p',k} h_{p',k,k} + v_{p,k}, \tag{2.17}$$

$$v_{p,k} = w_{p,k} - \sum_{p'} u_{p-p',k} h_{p',k,k}. \tag{2.18}$$

As previously mentioned, this convolutive transfer function (CTF) model enables to control the time frame length and the RTF length independently. It enables better representation of the input data by appropriately adjusting the length of the time frame along with better RTF modeling by appropriately adjusting the length of the RTF in each subband.

Let $\mathbf{h}_k$ denote a column stack vector of the band-to-band filter coefficients. Let $\mathbf{X}_k$ be a Toeplitz matrix constructed from the STFT coefficients of $x(n)$ in the kth subband. Similary, let $\mathbf{U}_k$ be a Toeplitz matrix constructed from the STFT coefficients of $u(n)$. Thus, (2.17) and (2.18) can be represented in matrix form as

$$\mathbf{y}_k = \mathbf{X}_k \mathbf{h}_k + \mathbf{v}_k, \tag{2.19}$$

$$\mathbf{v}_k = \mathbf{w}_k - \mathbf{U}_k \mathbf{h}_k, \tag{2.20}$$

where

$$\mathbf{y}_k = [y_{1,k}, \cdots, y_{P,k}]^T$$

and $\mathbf{v}_k$ and $\mathbf{w}_k$ are defined in a similar way to $\mathbf{y}_k$.

2.3.2 RTF Identification Using the Convolutive Transfer Function Approximation

Based on the CTF model and by taking the expectation of the cross multiplication of the two observed signals $y(n)$ and $x(n)$ in the STFT domain, we have from (2.19)

$$\Phi_{yx}(k) = \Psi_{xx}(k)\mathbf{h}_k + \Phi_{vx}(k), \tag{2.21}$$

where $\Psi_{xx}(k)$ terms are defined as

$$[\Psi_{xx}(k)]_{p,l} = E\left\{x_{p-l,k}x_{p,k}^*\right\} \triangleq \psi_{xx}(p,l,k).$$

Note that $\psi_{xx}(p,l,k)$ denotes the cross PSD between the signal $x(n)$ and its delayed version $x'(n) \triangleq x(n-lL)$ at time frame p and subband k. Since the speech signal $s(n)$ is assumed to be uncorrelated with the noise signal $u(n)$, by taking the expectation of the cross multiplication of $v(n)$ and $x(n)$ in the STFT domain, we get from (2.20)

$$\Phi_{vx}(k) = \Phi_{wu}(k) - \Psi_{uu}(k)\mathbf{h}_k, \tag{2.22}$$

where

$$[\Psi_{uu}(k)]_{p,l} = E\left\{u_{p-l,k}u^*_{p,k}\right\} \triangleq \psi_{uu}(p,l,k) := \psi_{uu}(l,k),$$

where $\psi_{uu}(p,l,k)$ denotes the cross PSD between the signal $u(n)$ and its delayed version $u'(n) \triangleq x(n-lL)$ at time frame p and subband k. Since the noise signals are stationary during the observation interval, the noise PSD term is independent of the time frame index p. From (2.21) and (2.22), we obtain

$$\Phi_{yx}(k) = (\Psi_{xx}(k) - \Psi_{uu}(k))\,\mathbf{h}_k + \Phi_{wu}(k), \tag{2.23}$$

and using PSD estimates yields

$$\hat{\Phi}(k) = \hat{\Psi}(k)\mathbf{h}_k + \mathbf{e}(k), \tag{2.24}$$

where $\mathbf{e}(k)$ denotes column stack vector of the PSD estimation errors and

$$\hat{\Phi}(k) \triangleq \hat{\Phi}_{yx}(k) - \hat{\Phi}_{wu}(k), \tag{2.25}$$

$$\hat{\Psi}(k) \triangleq \hat{\Psi}_{xx}(k) - \hat{\Psi}_{uu}(k). \tag{2.26}$$

Now, by taking into account only frames where speech is present, the WLS solution to (2.24) is given by

$$\hat{\mathbf{h}} = \left(\hat{\Psi}^H\mathbf{I}W\mathbf{I}\hat{\Psi}\right)^{-1}\hat{\Psi}^H\mathbf{I}W\mathbf{I}\hat{\Phi}, \tag{2.27}$$

where W is the weight matrix and the subband index k is omitted for notational simplicity.

Thus, in (2.27) we have an equivalent estimator to (2.14) based on the CTF model rather than the MTF model. In addition, it can be shown [13] that when the STFT samples of the signals are uncorrelated, or when each of the band-to-band filters contains a single tap, the estimator in (2.27) reduces exactly to (2.14).

2.4 Relative Transfer Function Identification in Speech Enhancement Applications

Identifying the relative transfer function between two microphones has an important role in multichannel hands-free communication systems. As previously mentioned, in common beamforming applications based on measurements captured in microphone arrays, the RTFs may be used to steer the beam towards a desired direction. Another common use of the RTF is in a so-called blocking matrix. A blocking matrix is a component that is aimed at blocking a particular signal from multichannel measurements. In this section, we demonstrate the blocking performance obtained by the previously presented RTF identification methods and describe a possible use in a specific multichannel application.

2.4.1 Blocking Matrix

In the following we evaluate the blocking ability of the RTF identification methods using speech signals based on both the MTF and CTF models [(2.14) and (2.27), respectively]. For evaluating the performance, we use a measure of the signal blocking factor (SBF) defined by

$$\text{SBF} = 10 \log_{10} \frac{E\left\{s^2(n)\right\}}{E\left\{r^2(n)\right\}}, \tag{2.28}$$

where $E\{s^2(n)\}$ is the energy contained in the speech received at the primary sensor, and $E\{r^2(n)\}$ is the energy contained in the leakage signal $r(n) = h(n) * s(n) - \hat{h}(n) * s(n)$. The leakage signal represents the difference between the reverberated speech at the reference sensor and its estimate given the speech at the primary sensor.

We simulate two microphones measuring noisy speech signals, placed inside an enclosure. The room acoustic impulse responses are generated according to Allen and Berkley's image method [15]. The speech source signal is a recorded speech from the TIMIT database [16] and the noise source signal is a computer generated white zero mean Gaussian noise with variance that varies to control the SNR level. The relative impulse response is infinite but under both models, it is approximated as a finite response filter. Under the MTF model, the RTF length is determined by the length of the time frame, whereas under the CTF model the RTF length can be set as desired. Thus, we set the estimated RTF length to be 1/8 of the room reverberation time T_{60}. This particular ratio was set since empirical tests produced satisfactory results.

Figure 2.1 shows the SBF curves obtained by the RTF identification methods based on the MTF and CTF models as a function of the SNR at the primary microphone. It may indicate two obvious trends. First, we observe that the RTF identification based on CTF approximation achieves higher SBF than the RTF identification based on MTF approximation in higher SNR conditions, whereas, the RTF identification that relies on MTF model achieves higher SBF in lower SNR conditions. Since the RTF identification

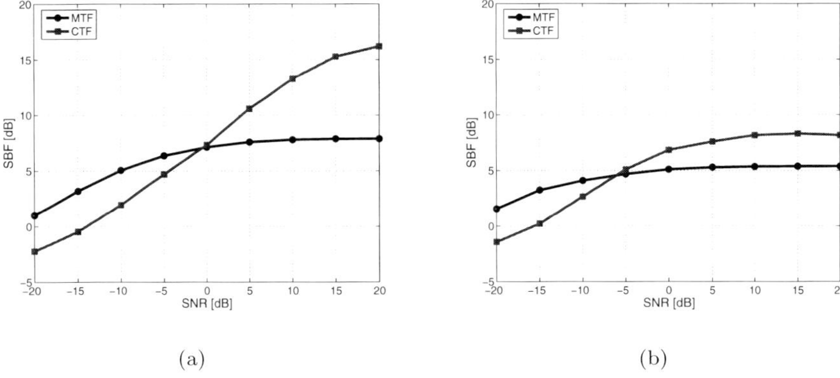

Fig. 2.1 SBF curves obtained by using the MTF and CTF approximations under various SNR conditions. The time frame length is $N = 512$ with 75% overlap. (a) Reverberation time $T_{60} = 0.2$ s. (b) Reverberation time $T_{60} = 0.4$ s.

using CTF model is associated with greater model complexity, it requires more reliable data, meaning, higher SNR values. Second, we observe that the gain for higher SNR levels is much higher in the case of $T_{60} = 0.2$ s than in the case of $T_{60} = 0.4$ s. In the latter case, where the impulse response is longer, the model mismatch using only a single band-to-band filter is larger than the model mismatch in the former case. Thus, in order to obtain larger gain when the reverberation time is longer, more cross-band filters should be employed to represent the system. More details and analytic analysis is presented in [11].

One of the significant advantages of the CTF model over the MTF model is the different influence of controlling the time frame length. Thus, we compare the methods that rely on the MTF and CTF approximations for various time frame lengths. Theoretically, under the MTF approximation, longer time frames enable identification of a longer RTF at the expense of fewer observations in each frequency bin. Thus, under the MTF model, controlling the time frame length controls both the representation of the data in the STFT domain and the estimated RTF. On the other hand, under the CTF model, the length of the estimated RTF can be set independently from the time frame length. Thus, under the CTF approximation, controlling the time frame length controls only the representation of the data in the STFT domain (whereas the RTF length is set independently). Figure 2.2 shows the SBF curves obtained by the RTF identification methods based on the MTF and CTF models as a function of the time frame length N with a fixed 75% overlap. It is worthwhile noting that this demonstration is most favorable to the method that relies on the MTF approximation since the number of variables under the MTF model increases as the time frame increases, whereas the number of estimated vari-

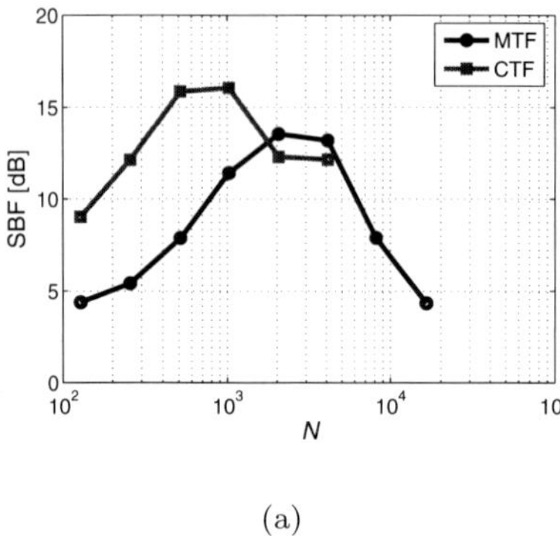 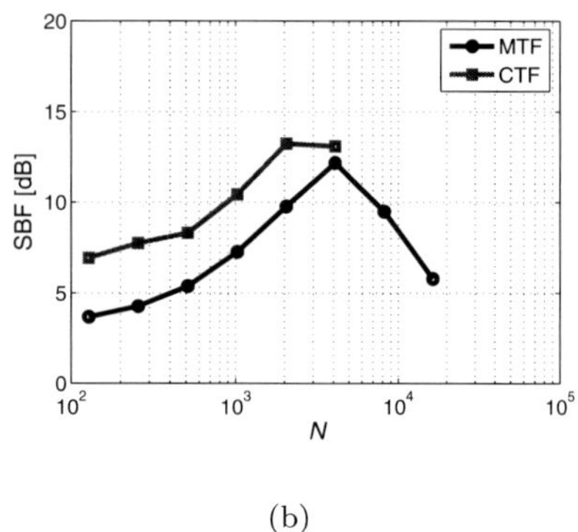

(a)   (b)

Fig. 2.2 SBF curves for the compared methods using various time frame lengths N. The SNR level is 20 dB. (a) Reverberation time $T_{60} = 0.2$ s. (b) Reverberation time $T_{60} = 0.4$ s.

ables under the CTF model is fixed (since the RTF length is fixed, longer time frame yields shorter band-to-band filters). The results demonstrate the trade-off between the increase of the estimated RTF length and the decrease of the estimation variance under the MTF model, which is further studied in [12]. In addition, we observe a trade-off under the CTF model. As the time frame length increases, the band-to-band filters become shorter and easier to identify, whereas less frames of observations are available. This trade-off between the length of the band-to-band filters and the number of data frames is studied for the general system identification case in [11]. We can also observe that the RTF identification method under the MTF approximation does not reach the maximal performance obtained by the RTF identification method under the CTF model. Since the model mismatch using the MTF approximation is too large, it cannot be compensated by taking longer time frames and estimating more variables. On the other hand, the CTF approximation enables better representation of the input data by appropriately adjusting the length of time frames, while the estimated RTF length is set independently according to the reverberation time.

2.4.2 The Transfer Function Generalized Sidelobe Canceler

In reverberant environments, the signals captured by the microphone array are distorted by the room impulse responses and corrupted by noise. Beamforming techniques, which aim at recovering the desired signal from the microphone array signals are among the most common speech enhancement applications. The basic idea behind beamformers is to exploit spatial and spectral information to form a beam and steer it to a desired look direction. Consequently, signals arriving from this look direction are reinforced, whereas signals arriving from all the other directions are attenuated. An example for such a beamformer, in which the RTF identification plays a critical role is the so-called transfer function generalized sidelobe canceler (TF-GSC) [3]. The TF-GSC consists of three blocks, organized in two branches. The first block is a fixed beamformer, which is designed to produce undistorted but noisy version of the desired signal and is built using estimates of the RTFs. The second block is a blocking matrix, which blocks the desired signal and produces noise-only reference signals. The blocking matrix is built using estimates of the RTFs. In Section 2.4.1 we demonstrated the blocking ability obtained in this case. The third block is an adaptive algorithm that aims at canceling the residual noise at the output of fixed beamformer (first block), given the noise-only reference signals obtained from the blocking matrix (second block). It is clear from the above description that the RTF identification quality has a significant influence on the results of such an application. For example, inaccurate RTF estimation results in inadequate steering of the fixed beamformer, which introduces distortions at the output of the upper branch. Moreover, inaccurate RTF estimation may also result in poor blocking ability, which has two-fold consequences. First, components of the desired signal may degrade the adaptive noise cancelation. Second, the leaked signal at the lower branch output is summed together with the output of the upper branch, yielding significant distortion at the beamformer output.

As previously explained, the duration of the relative impulse response in reverberant environments may reach several thousands taps. Thus, acquiring estimates of the RTF is a difficult task. In order to deal with this challenge, the TF-GSC presented in [3], is based entirely on the MTF model. This means that the TF-GSC suffers from the same limitations as the RTF identification method that relies on the MTF approximation, which were explored here. Therefore, an interesting direction for future research and a very promising lead is to incorporate the new approach for transfer function approximation (the CTF model) into the TF-GSC. Consequently, the RTF identification based on the CTF approximation may be used for building the TF-GSC blocks more accurately, enabling better representation of the desired source signal and stronger suppression of undesired interferences.

2.5 Conclusions

In this chapter we delved into the problem of RTF identification in the STFT domain. In particular, we described two identification methods that exploit specific properties of the desired signal. First, we showed an unbiased estimator of the RTF based on the non-stationarity of speech signals. However, this estimator requires simultaneous estimation of the noise statistics, which results in a significant difficulty distributing the weights of the time frames. Second, we showed an estimator that solves this conflict by decoupling the RTF estimator and the noise statistics estimation using silent fragments, enabling improved estimation variance. Next, we focused on the signals and systems representation in the STFT domain. We described an alternative representation of the time domain linear convolution. Instead of the commonly used approximation, which models the linear convolution in the time domain as a multiplicative transfer function in the STFT domain, we approximated the time domain convolution as a convolutive transfer function in the STFT domain. Then, we employed RTF identification methods on the new model. We showed that the RTF identification method based on the new representation obtains improved results. In addition, we demonstrated the trade-off emerging from setting the time frames length. As we described in the last section, this new modeling approach can be applied in various applications, enabling better representation of the data, which may lead to improved performances.

References

1. O. Shalvi and E. Weinstein, "System identification using nonstationary signals," *IEEE Trans. Signal Processing*, vol. 40, no. 8, pp. 2055–2063, Aug. 1996.
2. O. Hoshuyama, A. Sugiyama, and A. Hirano, "A robust adaptive beamformer for microphone arrays with a blocking matrix using constrained adaptive filters," *IEEE Trans. Signal Processing*, vol. 47, no. 10, pp. 2677–2684, Oct. 1999.
3. S. Gannot, D. Burshtein, and E. Weinstein, "Signal enhancement using beamforming and nonstationarity with applications to speech," *IEEE Trans. Signal Processing*, vol. 49, no. 8, pp. 1614–1626, Aug. 2001.
4. T. G. Dvorkind and S. Gannot, "Time difference of arrival estimation of speech source in a noisy and reverberant environment," *Signal Processing*, vol. 85, pp. 177–204, 2005.
5. S. Gannot, D. Burshtein, and E. Weinstein, "Theoretical performance analysis of the general transfer function GSC, " *Proc. Int. Workshop on Acoustic Echo and Noise Control (IWAENC)*, pp 103–106, 2001.
6. S. Gannot and I. Cohen, "Speech enhancement based on the general transfer function GSC and postfiltering," *IEEE Trans. Speech, Audio Processing*, vol. 12, no. 6, pp. 561–571, Nov. 2004.
7. J. Chen, J. Benesty, and Y. Huang, "A minimum distortion noise reduction algorithm with multiple microphones," *IEEE Trans. Audio, Speech, Language Processing*, vol. 16, pp. 481–493, Mar. 2008.
8. G. Reuven, S. Gannot, and I. Cohen, "Joint noise reduction and acoustic echo cancellation using the transfer-function generalized sidelobe canceller," *Special Issue of Speech Communication on Speech Enhancement*, vol. 49, pp. 623–635, Jul.-Aug. 2007.
9. G. Reuven, S. Gannot, and I. Cohen, "Dual source transfer-function generalized sidelobe canceller," *IEEE Trans. Audio, Speech, Language Processing*, vol. 16, pp. 711–727, May 2008.
10. I. Cohen, "Relative transfer function identification using speech signals," *IEEE Trans. Speech, Audio Processing*, vol. 12, no. 5, pp. 451–459, Sept. 2004.

11. Y. Avargel and I. Cohen, "System identification in the short time Fourier transform domain with crossband filtering," *IEEE Trans. Audio, Speech, Language Processing*, vol. 15, no. 4, pp. 1305–1319, May 2007.
12. Y. Avargel and I. Cohen, "On multiplicative transfer function approximation in the short time Fourier transform domain," *IEEE Signal Processing Letters*, vol. 14, pp. 337–340, 2007.
13. R. Talmon, I. Cohen, and S. Gannot, "Relative transfer function identification using convolutive transfer function approximation," *to appear in IEEE Trans. Audio, Speech, Language Processing*, 2009.
14. E. A. P. Habets, *Single- and Multi-Microphone Speech Dereverberation using Spectral Enhancement*, Ph.D. thesis, Technische Universiteit Eindhoven, The Netherlands, Jun 2007.
15. J. B. Allen and D. A. Berkley, "Image method for efficiently simulating small room acoustics," *Journal of the Acoustical Society of America*, vol. 65, no. 4, pp. 943–950, 1979.
16. J. S. Garofolo, "Getting started with the DARPA TIMIT CD-ROM: an acoustic-phonetic continous speech database," National Inst. of Standards and Technology (NIST), Gaithersburg, MD, Feb. 1993.

Chapter 3
Representation and Identification of Nonlinear Systems in the Short-Time Fourier Transform Domain

Yekutiel Avargel and Israel Cohen

Abstract [1] In this chapter, we introduce a novel approach for improved nonlinear system identification in the short-time Fourier transform (STFT) domain. We first derive explicit representations of discrete-time Volterra filters in the STFT domain. Based on these representations, approximate nonlinear STFT models, which consist of parallel combinations of linear and nonlinear components, are developed. The linear components are represented by crossband filters between subbands, while the nonlinear components are modeled by multiplicative cross-terms. We consider the identification of quadratically nonlinear systems and introduce batch and adaptive schemes for estimating the model parameters. Explicit expressions for the obtainable mean-square error (mse) in subbands are derived for both schemes. We show that estimation of the nonlinear component improves the mse only for high signal-to-noise ratio (SNR) conditions and long input signals. Furthermore, a significant reduction in computational cost as well as substantial improvement in estimation accuracy can be achieved over a time-domain Volterra model, particularly when long-memory systems are considered. Experimental results validate the theoretical derivations and demonstrate the effectiveness of the proposed approach.

3.1 Introduction

Identification of nonlinear systems has recently attracted great interest in many applications, including acoustic echo cancellation [1, 2, 3], channel

Yekutiel Avargel
Technion–Israel Institute of Technology, Israel, e-mail: `kutiav@tx.technion.ac.il`

Israel Cohen
Technion–Israel Institute of Technology, Israel, e-mail: `icohen@ee.technion.ac.il`

[1] This work was supported by the Israel Science Foundation under Grant 1085/05.

I. Cohen et al. (Eds.): Speech Processing in Modern Communication, STSP 3, pp. 49–87.
springerlink.com   © Springer-Verlag Berlin Heidelberg 2010

equalization [4, 5], biological system modeling [6], image processing [7], and loudspeaker linearization [8]. Volterra filters [9, 10, 11, 12, 13, 14] are widely used for modeling nonlinear physical systems, such as loudspeaker-enclosure-microphone (LEM) systems in nonlinear acoustic echo cancellation applications [2, 15, 16], and digital communication channels [4, 17], just to mention a few. An important property of Volterra filters, which makes them useful in nonlinear estimation problems, is the linear relation between the system output and the filter coefficients. Traditionally, Volterra-based approaches have been carried out in the time or frequency domains. Time-domain approaches employ conventional linear estimation methods in batch or adaptive forms in order to estimate the Volterra kernels. These approaches, however, often suffer from extremely high computational cost due to the large number of parameters of the Volterra model, especially for long-memory systems [18, 13]. Another major drawback of the Volterra model is its severe ill-conditioning [19], which leads to high estimation-error variance and to slow convergence of the adaptive Volterra filter. To overcome these problems, several approximations for the time-domain Volterra filter have been proposed, including orthogonalized power filters [20], Hammerstein models [21], parallel-cascade structures [22], and multi-memory decomposition [23].

Other algorithms, which operate in the frequency domain, have been proposed to ease the computational burden [24, 25, 26]. A discrete frequency-domain model, which approximates the Volterra filter using multiplicative terms, is defined in [24, 25]. A major limitation of this model is its underlying assumption that the observation data length is relatively large. When the data is of limited size (or when the nonlinear system is not time-invariant), this long duration assumption is very restrictive. Other frequency-domain approaches use cumulants and polyspectra information to estimate the Volterra transfer functions [26]. Although computationally efficient, these approaches often assume a Gaussian input signal, which limits their applicability.

In this chapter, we introduce a novel approach for improved nonlinear system identification in the short-time Fourier transform (STFT) domain, which is based on a time-frequency representation of the Volterra filter [27]. A typical nonlinear system identification scheme in the STFT domain is illustrated in Fig. 3.1. Similarly to STFT-based linear identification techniques [28, 29, 30], representing and identifying nonlinear systems in the STFT domain is motivated by the processing in distinct subbands, which may result in reduced computational cost and improved convergence rate, compared to time-domain methods. We show that a homogeneous time-domain Volterra filter [9] with a certain kernel can be perfectly represented in the STFT domain, at each frequency bin, by a sum of Volterra-like expansions with smaller-sized kernels. Based on this representation, an approximate nonlinear model, which simplifies the STFT representation of Volterra filters and significantly reduces the model complexity, is developed. The resulting model consists of a parallel combination of linear and nonlinear components. The linear component is represented by crossband filters between the subbands

[31, 28, 32], while the nonlinear component is modeled by multiplicative cross-terms, extending the so-called cross-multiplicative transfer function (CMTF) approximation [33]. It is shown that the proposed STFT model generalizes the conventional discrete frequency-domain model [24], and forms a much richer representation for nonlinear systems.

Concerning system identification, we employ the proposed model and introduce two schemes for estimating the model parameters: one is a batch scheme based on a least-squares (LS) criterion [27, 34], where in the second scheme, the parameters are adaptively estimated using the least-mean-square (LMS) algorithm [35]. For both schemes, the proposed approach is more advantageous in terms of computational complexity than the time-domain Volterra approach. We analyze the performance of both schemes and derive explicit expressions for the obtainable mean-squared error (mse) in each frequency bin. The analysis provides important insights into the influence of nonlinear undermodeling (i.e., employing a purely linear model in the estimation process) and the number of estimated crossband filters on the mse performance. We show that for low signal-to-noise ratio (SNR) conditions and short input signals, a lower mse is achieved by allowing for nonlinear undermodeling and utilizing a purely linear model. However, as the SNR or the input signal length increases, the performance can be generally improved by incorporating a nonlinear component into the model. When estimating long-memory systems, a substantial improvement in estimation accuracy over the Volterra model can be achieved by the proposed model, especially for high SNR conditions. Experimental results with white Gaussian signals and real speech signals demonstrate the advantages of the proposed approach and support the theoretical derivations.

The chapter is organized as follows. In Section 3.2, we introduce the time-domain Volterra model and review existing methods for Volterra system identification. In Section 3.3, we derive an explicit representation of discrete-time Volterra filters in the STFT domain. In Section 3.4, we introduce a simplified model for nonlinear systems in the STFT domain. In Section 3.5, we consider the identification of quadratically nonlinear systems in the STFT domain using batch and adaptive methods and derive explicit expressions for the mse in subbands. Finally, in Section 3.6, we present some experimental results.

3.2 Volterra System Identification

The Volterra filter is one of the most commonly-used models for nonlinear systems in the time domain [9, 10, 11, 12, 13, 14]. In this section, we introduce the Volterra model and briefly review existing methods for estimating its parameters. Throughout this chapter, scalar variables are written with lowercase letters and vectors are indicated with lowercase boldface letters. Capital boldface letters are used for matrices and norms are always ℓ_2 norms.

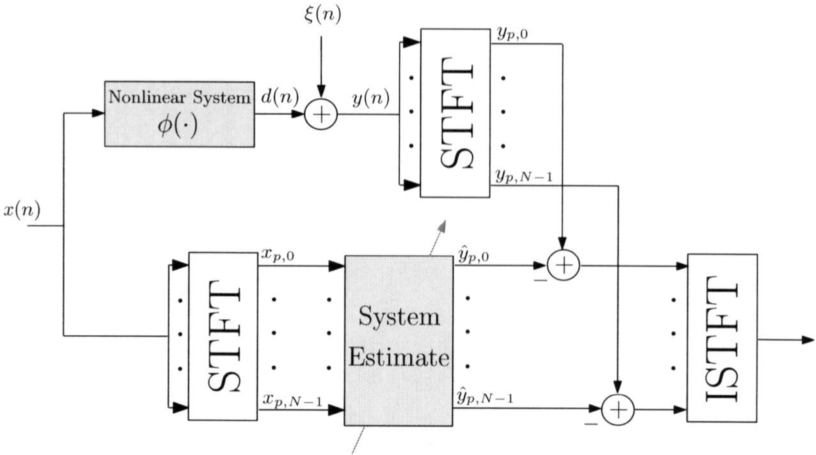

Fig. 3.1 Nonlinear system identification in the STFT domain. The unknown time-domain nonlinear system $\phi(\cdot)$ is estimated using a given model in the STFT domain.

Consider a generalized qth-order nonlinear system, with $x(n)$ and $d(n)$ being its input and output, respectively (see Fig. 3.1). A Volterra representation of this system is given by [9]

$$d(n) = \sum_{\ell=1}^{q} d_\ell(n), \qquad (3.1)$$

where $d_\ell(n)$ represents the output of the ℓth-order homogeneous Volterra filter, which is related to the input $x(n)$ by

$$d_\ell(n) = \sum_{m_1=0}^{N_\ell-1} \cdots \sum_{m_\ell=0}^{N_\ell-1} h_\ell(m_1, \ldots m_\ell) \prod_{i=1}^{\ell} x(n - m_i), \qquad (3.2)$$

where $h_\ell(m_1, \ldots m_\ell)$ is the ℓth-order Volterra kernel, and N_ℓ $(1 \leq \ell \leq q)$ represents its memory length. It is easy to verify that the representation in (3.2) consists of $(N_\ell)^\ell$ parameters, such that representing the system by the full model (3.1) requires $\sum_{\ell=1}^{q} (N_\ell)^\ell$ parameters. The representation in (3.2) is called symmetric if the Volterra kernels satisfy [9]

$$h_\ell(m_1, \ldots m_\ell) = h_\ell(m_{\pi(1)}, \ldots m_{\pi(\ell)}) \qquad (3.3)$$

for any permutation $\pi(1, \ldots, \ell)$. This representation, however, is redundant and often replaced by the triangular representation:

$$d_\ell(n) = \sum_{m_1=0}^{N_\ell-1} \sum_{m_2=m_1}^{N_\ell-1} \cdots \sum_{m_\ell=m_{\ell-1}}^{N_\ell-1} g_\ell(m_1, \ldots m_\ell) \prod_{i=1}^{\ell} x(n - m_i), \qquad (3.4)$$

where $g_\ell(m_1, \ldots m_\ell)$ is the ℓth-order triangular Volterra kernel. The representation in (3.4) consists of $\binom{N_\ell + \ell - 1}{\ell}$ parameters, such that the system's full model (3.1) requires $\sum_{\ell=1}^{q} \binom{N_\ell + \ell - 1}{\ell}$ parameters. The reduction in model complexity compared to the symmetric representation in (3.2) is obvious. Comparing (3.2) and (3.4), the triangular kernels can be expressed in terms of the symmetric kernels as [9]

$$g_\ell(m_1, \ldots m_\ell) = \ell! h_\ell(m_1, \ldots m_\ell) u(m_2 - m_1) \cdots u(m_\ell - m_{\ell-1}), \qquad (3.5)$$

where $u(n)$ is the unit step function [i.e., $u(n) = 1$ for $n \geq 0$, and $u(n) = 0$ otherwise]. Note that either of these representations (symmetric or triangular) is uniquely specified by the other.

The problem of nonlinear system identification in the time domain can be formulated as follows: given an input signal $x(n)$ and noisy observation $y(n)$, construct a model for describing the input-output relationship, and select its parameters so that the model output $\hat{y}(n)$ best estimates (or predicts) the measured output signal [36]. In Volterra-based approaches, the model parameters to be estimated are the Volterra kernels. An important property of Volterra filters is the linear relation between the system output and the filter coefficients (either in the symmetric or the triangular representation), which enables to employ algorithms from linear estimation theory for estimating the Volterra-model parameters. Specifically, let $\hat{y}(n)$ represent the output of an qth-order Volterra model, which attempts to estimate a measured output signal of an (unknown) nonlinear system. Then, it can be written in a vector form as

$$\hat{y}(n) = \mathbf{x}^T(n)\theta, \qquad (3.6)$$

where θ is the model parameter vector, and $\mathbf{x}(n)$ is the corresponding input data vector. An estimate of θ can now be derived using conventional linear estimation algorithms in batch or adaptive forms. Batch methods were introduced (e.g., [18, 13]), providing both LS and mse estimates. More specifically, denoting the observable data length by N_x, the LS estimate of the Volterra kernels is given by

$$\hat{\theta}_{\mathrm{LS}} = \left(\mathbf{X}^H \mathbf{X}\right)^{-1} \mathbf{X}^H \mathbf{y}, \qquad (3.7)$$

where $\mathbf{X}^T = \begin{bmatrix} \mathbf{x}(0) \ \mathbf{x}(1) \ \cdots \ \mathbf{x}(N_x - 1) \end{bmatrix}$ and $\mathbf{y}$ is the observable data vector. Similarly, the mse estimate is given by

$$\hat{\theta}_{\mathrm{MSE}} = \left[E\left\{ \mathbf{x}(n)\mathbf{x}^T(n) \right\} \right]^{-1} E\left\{ \mathbf{x}(n)y(n) \right\}. \qquad (3.8)$$

Adaptive methods were also proposed for estimating the Volterra kernels (e.g., [10, 16, 2]). These methods often employ the LMS algorithm [37] due to its robustness and simplicity. Accordingly, the estimate of the parameter vector θ at time n, denoted by $\hat{\theta}(n)$, is updated using the following recursion

$$\hat{\theta}(n + 1) = \hat{\theta}(n) + \mu e(n)\mathbf{x}(n), \qquad (3.9)$$

where μ is a step size, and $e(n) = y(n) - \mathbf{x}^T(n)\hat{\theta}(n)$ is the error signal. To speed-up convergence, the affine projection (AP) algorithm and the recursive least-squares (RLS) algorithm were employed for updating the adaptive Volterra filters [13, 15]. A common difficulty associated with the aforementioned time-domain approaches is their high computational cost, which is attributable to the large number of parameters of the Volterra model (i.e., the high dimensionality of the parameter vector θ). The complexity of the model, together with its severe ill-conditioning [19], leads to high estimation-error variance and to relatively slow convergence of the adaptive Volterra filter.

Alternatively, frequency-domain methods have been introduced for Volterra system identification, aiming at estimating the so-called Volterra transfer functions [24, 26, 25]. Statistical approaches based on higher order statistics (HOS) of the input signal use cumulants and polyspectra information [26]. Specifically, assuming Gaussian inputs, a closed form of the transfer function of an ℓth-order homogeneous Volterra filter is given by [38]

$$H_\ell(\omega_1, \ldots, \omega_\ell) = \frac{C_{yx\cdots x}(-\omega_1, \ldots, -\omega_\ell)}{\ell! C_{xx}(\omega_1) \cdots C_{xx}(\omega_\ell)}, \tag{3.10}$$

where $C_{xx}(\cdot)$ is the spectrum of $x(n)$, and $C_{yx\cdots x}(\cdot)$ is the $(\ell + 1)$th-order crosspolyspectrum between y and x [39]. The estimation of the transfer function $H_\ell(\omega_1, \ldots, \omega_\ell)$ is accomplished by deriving proper estimators for the cumulants and their spectra. However, a major drawback of cumulant estimators is their extremely-high variance, which necessitates enormous amount of data to achieve satisfactory performances. Moreover, the assumption of Gaussian inputs is very restrictive and limits the applicability of these approaches. In [24], a discrete frequency-domain model is defined, which approximates the Volterra filter in the frequency domain using multiplicative terms. Specifically for a second-order Volterra system, the frequency-domain model consists of a parallel combination of linear and quadratic components as follows:

$$\hat{Y}(k) = H_1(k)X(k) + \sum_{\substack{k',k''=0 \\ (k'+k'') \bmod N = k}}^{N-1} H_2(k', k'')X(k')X(k''), \tag{3.11}$$

where $X(k)$ and $\hat{Y}(k)$ are the Nth-length discrete Fourier transforms (DFT's) of the input $x(n)$ and the model output $\hat{y}(n)$, respectively, and $H_1(k)$ and $H_2(k', k'')$ are the linear and quadratic Volterra transfer functions (in the discrete Fourier domain), respectively. As in the time-domain Volterra representation, the output of the frequency-domain model depends linearly on its coefficients, and therefore can be written as

$$\hat{Y}(k) = \mathbf{x}_k^T \theta_k, \tag{3.12}$$

where θ_k is the model parameter vector at the kth frequency bin, and $\mathbf{x}_k$ is the corresponding transformed input-data vector. Using the formulation in (3.12), batch [24] and adaptive [40, 25] algorithms were proposed for estimating the model parameters. Although these approaches are computationally efficient and assume no particular statistics for the input signal, they require a long duration of the input signal to validate the multiplicative approximation. When the data is of limited size (or when the nonlinear system is not time-invariant), this long-duration assumption is very restrictive.

The drawbacks of the conventional time- and frequency-domain methods have recently motivated the use of subband (multirate) techniques for improved nonlinear system identification. In the following sections, a novel approach for improved nonlinear system identification in the STFT domain, which is based on a time-frequency representation of the time-domain Volterra filter, is introduced. Two identification schemes, in either batch or adaptive forms, are proposed.

3.3 Representation of Volterra Filters in the STFT Domain

In this section, we derive the representation of discrete-time Volterra filters in the STFT domain. We first consider the quadratic case, and subsequently generalize the results to higher orders of nonlinearity. We show that a time-domain Volterra kernel can be perfectly represented in the STFT domain by a sum of smaller-sized kernels in each frequency bin.

3.3.1 Second-Order Volterra Filters

Using the formulation in (3.1)–(3.2), the output of a second-order Volterra filter can be written as

$$d(n) = \sum_{m=0}^{N_1-1} h_1(m)x(n-m)$$

$$+ \sum_{m=0}^{N_2-1}\sum_{\ell=0}^{N_2-1} h_2(m,\ell)x(n-m)x(n-\ell)$$

$$\triangleq d_1(n) + d_2(n) , \qquad\qquad (3.13)$$

where $h_1(m)$ and $h_2(m,\ell)$ are the linear and quadratic Volterra kernels, respectively, and $d_1(n)$ and $d_2(n)$ denote the corresponding output signals of the linear and quadratic homogeneous components. The memory length N_1

of the linear kernel may be different in general from the memory length N_2 of the quadratic kernel. To find a representation of $d(n)$ in the STFT domain, let us first briefly review some definitions of the STFT representation of digital signals (for further details, see e.g., [41]).

The STFT representation of a signal $x(n)$ is given by

$$x_{p,k} = \sum_m x(m)\tilde{\psi}^*_{p,k}(m)\,, \tag{3.14}$$

where

$$\tilde{\psi}_{p,k}(n) \triangleq \tilde{\psi}(n - pL)e^{j\frac{2\pi}{N}k(n-pL)} \tag{3.15}$$

denotes a translated and modulated window function, $\tilde{\psi}(n)$ is an analysis window of length N, p is the frame index, k represents the frequency-bin index ($0 \leq k \leq N - 1$), L is the translation factor (or the decimation factor, in filter-bank interpretation) and * denotes complex conjugation. The inverse STFT, i.e., reconstruction of $x(n)$ from its STFT representation $x_{p,k}$, is given by

$$x(n) = \sum_p \sum_{k=0}^{N-1} x_{p,k}\psi_{p,k}(n)\,, \tag{3.16}$$

where

$$\psi_{p,k}(n) \triangleq \psi(n - pL)e^{j\frac{2\pi}{N}k(n-pL)}\,, \tag{3.17}$$

and $\psi(n)$ denotes a synthesis window of length N. To guarantee a perfect reconstruction of a signal $x(n)$ from its STFT coefficients $x_{p,k}$, we substitute (3.14) into (3.16) to obtain the so-called completeness condition:

$$\sum_p \psi(n - pL)\tilde{\psi}^*(n - pL) = \frac{1}{N} \qquad \text{for all } n\,. \tag{3.18}$$

Using the linearity of the STFT, $d(n)$ in (3.13) can be written in the time-frequency domain as

$$d_{p,k} = d_{1;p,k} + d_{2;p,k}\,, \tag{3.19}$$

where $d_{1;p,k}$ and $d_{2;p,k}$ are the STFT representations of $d_1(n)$ and $d_2(n)$, respectively. It is well known that in order to perfectly represent a linear system in the STFT domain, crossband filters between subbands are generally required [31, 28, 32]. Therefore, the output of the linear component can be expressed in the STFT domain as

$$d_{1;p,k} = \sum_{k'=0}^{N-1} \sum_{p'=0}^{\bar{N}_1-1} x_{p-p',k'} h_{p',k,k'}\,, \tag{3.20}$$

where $h_{p,k,k'}$ denotes a crossband filter of length $\bar{N}_1 = \lceil (N_1 + N - 1)/L \rceil + \lceil N/L \rceil - 1$ from frequency bin k' to frequency bin k. These filters are used for canceling the aliasing effects caused by the subsampling factor L. The crossband filter $h_{p,k,k'}$ is related to the linear kernel $h_1(n)$ by [28]

$$h_{p,k,k'} = \{h_1(n) * f_{k,k'}(n)\}|_{n=pL}, \tag{3.21}$$

where the discrete-time Fourier transform (DTFT) of $f_{k,k'}(n)$ with respect to the time index n is given by

$$F_{k,k'}(\omega) = \sum_n f_{k,k'}(n)e^{-jn\omega} = \tilde{\Psi}^*\left(\omega - \frac{2\pi}{N}k\right)\Psi\left(\omega - \frac{2\pi}{N}k'\right), \tag{3.22}$$

where $\tilde{\Psi}(\omega)$ and $\Psi(\omega)$ are the DTFT of $\tilde{\psi}(n)$ and $\psi(n)$, respectively. Note that the energy of the crossband filter from frequency bin k' to frequency bin k generally decreases as $|k - k'|$ increases, since the overlap between $\tilde{\Psi}\left(\omega - (2\pi/N)k\right)$ and $\Psi\left(\omega - (2\pi/N)k'\right)$ becomes smaller. Recently, the influence of crossband filters on a linear system identifier implemented in the STFT domain was investigated [28]. It is shown that increasing the number of crossband filters not necessarily implies a lower steady-state mse in subbands. In fact, the inclusion of more crossband filters in the identification process is preferable only when high SNR or long data are considered. As will be shown later, the same applies also when an additional nonlinear component is incorporated into the model.

The representation of the quadratic component's output $d_2(n)$ in the STFT domain can be derived in a similar manner to that of the linear component. Specifically, applying the STFT to $d_2(n)$ we may obtain after some manipulations (see Appendix)

$$\begin{aligned}
d_{2;p,k} &= \sum_{k',k''=0}^{N-1} \sum_{p',p''} x_{p',k'} x_{p'',k''} c_{p-p',p-p'',k,k',k''} \\
&= \sum_{k',k''=0}^{N-1} \sum_{p',p''} x_{p-p',k'} x_{p-p'',k''} c_{p',p'',k,k',k''},
\end{aligned} \tag{3.23}$$

where $c_{p-p',p-p'',k,k',k''}$ may be interpreted as a response of the quadratic system to a pair of impulses $\{\delta_{p-p',k-k'}, \delta_{p-p'',k-k''}\}$ in the time-frequency domain. Equation (3.23) indicates that for a given frequency-bin index k, the temporal signal $d_{2;p,k}$ consists of all possible interactions between pairs of input frequencies. The contribution of each frequency pair $\{k',k''|\,k',k'' \in \{0,\ldots,N-1\}\}$ to the output signal at frequency bin k is given as a Volterra-like expansion with $c_{p',p'',k,k',k''}$ being its quadratic kernel. The kernel $c_{p',p'',k,k',k''}$ in the time-frequency domain is related to the quadratic kernel $h_2(n,m)$ in the time domain by (see Appendix)

$$c_{p',p'',k,k',k''} = \{h_2(n,m) * f_{k,k',k''}(n,m)\}|_{n=p'L,\ m=p''L}, \qquad (3.24)$$

where $*$ denotes a 2D convolution and

$$f_{k,k',k''}(n,m) \triangleq \sum_{\ell} \tilde{\psi}^*(\ell) e^{-j\frac{2\pi}{N}k\ell} \psi(n+\ell) e^{j\frac{2\pi}{N}k'(n+\ell)} \psi(m+\ell) e^{j\frac{2\pi}{N}k''(m+\ell)}.$$

$$(3.25)$$

Equation (3.25) implies that for fixed k, k' and k'', the quadratic kernel $c_{p',p'',k,k',k''}$ is noncausal with $\lceil N/L \rceil - 1$ noncausal coefficients in each variable (p' and p''). Note that crossband filters are also noncausal with the same number of noncausal coefficients [28]. Hence, for system identification, an artificial delay of $(\lceil N/L \rceil - 1)L$ can be applied to the system output signal $d(n)$ in order to consider a noncausal response. It can also be seen from (3.25) that the memory length of each kernel is given by

$$\bar{N}_2 = \left\lceil \frac{N_2 + N - 1}{L} \right\rceil + \left\lceil \frac{N}{L} \right\rceil - 1, \qquad (3.26)$$

which is approximately L times lower than the memory length of the time-domain kernel $h_2(m,\ell)$. The support of $c_{p',p'',k,k',k''}$ is therefore given by $\mathcal{D} \times \mathcal{D}$ where

$$\mathcal{D} = \left[1 - \lceil N/L \rceil, \dots, \lceil (N_2 + N - 1)/L \rceil - 1 \right].$$

To give further insight into the basic properties of the quadratic STFT kernels $c_{p',p'',k,k',k''}$, we apply the 2D DTFT to $f_{k,k',k''}(n,m)$ with respect to the time indices n and m, and obtain

$$F_{k,k',k''}(\omega,\eta) = \tilde{\Psi}^*\left(\omega + \eta - \frac{2\pi}{N}k\right)\Psi\left(\omega - \frac{2\pi}{N}k'\right)\Psi\left(\omega - \frac{2\pi}{N}k''\right). \quad (3.27)$$

By taking $\Psi(\omega)$ and $\tilde{\Psi}(\omega)$ to be ideal low-pass filters with bandwidths π/N (i.e., $\Psi(\omega) = 0$ and $\tilde{\Psi}(\omega) = 0$ for $\omega \notin [-\pi/2N, \pi/2N]$), a perfect STFT representation of the quadratic time-domain kernel $h_2(n,m)$ can be achieved by utilizing only kernels of the form $c_{p',p'',k,k',(k-k') \bmod N}$, since in this case the product of $\Psi(\omega - (2\pi/N)k')$, $\Psi(\omega - (2\pi/N)k')$ and $\tilde{\Psi}^*(\omega + \eta - (2\pi/N)k)$ is identically zero for $k'' \neq (k - k') \bmod N$. Practically, the analysis and synthesis windows are not ideal and their bandwidths are greater than π/N, so $f_{k,k',(k-k') \bmod N}(n,m)$, and consequently $c_{p',p'',k,k',(k-k') \bmod N}$, are not zero. Nonetheless, one can observe from (3.27) that the energy of $f_{k,k',k''}(n,m)$ decreases as $|k'' - (k - k') \bmod N|$ increases, since the overlap between the translated window functions becomes smaller. As a result, not all kernels in the STFT domain should be considered in order to capture most of the energy of the STFT representation of $h_2(n,m)$. This is illustrated in Fig. 3.2, which shows the energy of $f_{k,k',k''}(n,m)$, defined as $E_{k,k'}(k'') \triangleq \sum_{n,m} |f_{k,k',k''}(n,m)|^2$, for $k = 1$, $k' = 0$ and $k'' \in \{(k - k' + i) \bmod N\}_{i=-10}^{10}$,

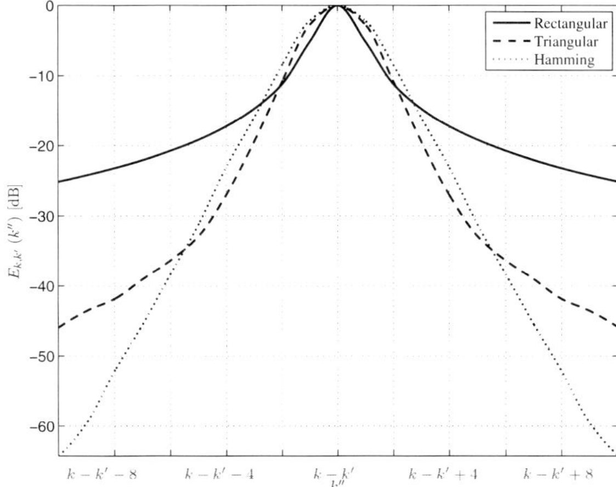

Fig. 3.2 Energy of $f_{k,k',k''}(n,m)$ [defined in (3.25)] for $k = 1$ and $k' = 0$, as obtained for different synthesis windows of length $N = 256$.

as obtained by using rectangular, triangular and Hamming synthesis windows of length $N = 256$. A corresponding minimum-energy analysis window that satisfies the completeness condition [42] for $L = 128$ (50% overlap) is also employed. The results confirm that the energy of $f_{k,k',k''}(n,m)$, for fixed k and k', is concentrated around the index $k'' = (k - k') \bmod N$.

As expected from (3.27), the number of useful quadratic kernels in each frequency bin is mainly determined by the spectral characteristics of the analysis and synthesis windows. That is, windows with a narrow mainlobe (e.g., a rectangular window) yield the sharpest decay, but suffer from wider energy distribution over k'' due to relatively high sidelobes energy. Smoother windows (e.g., Hamming window), on the other hand, enable better energy concentration. For instance, utilizing a Hamming window reduces the energy of $f_{k,k',k''}(n,m)$ for $k'' = (k - k' \pm 8) \bmod N$ by approximately 30 dB, when compared to using a rectangular window. These results will be used in the next section for deriving a useful model for nonlinear systems in the STFT domain.

3.3.2 High-Order Volterra Filters

Let us now consider a generalized ℓth-order homogeneous Volterra filter, whose input $x(n)$ and output $d_\ell(n)$ are related via (3.2). Applying the STFT to $d_\ell(n)$ and following a similar derivation to that made for the quadratic case [see (3.23)–(3.25), and the appendix at the end of this chapter], we obtain after some manipulations

$$d_{\ell;p,k} = \sum_{k_1,\ldots k_\ell=0}^{N-1} \sum_{p_1,\ldots p_\ell} c_{p_1,\ldots p_\ell,k,k_1,\ldots k_\ell} \prod_{i=1}^{\ell} x_{p-p_i,k_i} \,. \tag{3.28}$$

Equation (3.28) implies that the output of an ℓth-order homogeneous Volterra filter in the STFT domain, at a given frequency-bin index k, consists of all possible combinations of ℓ input frequencies. The contribution of each ℓ-fold frequency indices $\{k_1,\ldots k_\ell\}$ to the kth frequency bin is expressed in terms of an ℓth-order homogeneous Volterra expansion with the kernel $c_{p_1,\ldots p_\ell,k,k_1,\ldots k_\ell}$. Similarly to the quadratic case, it can be shown that the STFT kernel $c_{p_1,\ldots p_\ell,k,k_1,\ldots k_\ell}$ in the time-frequency domain is related to the kernel $h_\ell(m_1,\ldots m_\ell)$ in the time domain by

$$c_{p_1,\ldots p_\ell,k,k_1,\ldots k_\ell} = \{h_\ell(m_1,\ldots m_\ell) * f_{k,k_1,\ldots k_\ell}(m_1,\ldots m_\ell)\}\big|_{m_i=p_i L;\ i=1,\ldots\ell.} \, , \tag{3.29}$$

where $*$ denotes an ℓ-D convolution and

$$f_{k,k_1,\ldots k_\ell}(m_1,\ldots m_\ell) \triangleq \sum_n \tilde{\psi}^*(n)e^{-j\frac{2\pi}{N}kn} \prod_{i=1}^{\ell} \psi(m_i+n)e^{j\frac{2\pi}{N}k_i(m_i+n)} \,. \tag{3.30}$$

Equations (3.29)–(3.30) imply that for fixed indices $\{k_i\}_{i=1}^{\ell}$, the kernel $c_{p_1,\ldots p_\ell,k,k_1,\ldots k_\ell}$ is noncausal with $\lceil N/L \rceil - 1$ noncausal coefficients in each variable $\{p_i\}_{i=1}^{\ell}$, and its overall memory length is given by $\bar{N}_\ell = \lceil (N_\ell + N - 1)/L \rceil + \lceil N/L \rceil - 1$. Note that for $\ell = 1$ and $\ell = 2$, (3.28)–(3.30) reduce to the STFT representation of the linear kernel (3.20) and the quadratic kernel (3.23), respectively. Furthermore, applying the ℓ-D DTFT to $f_{k,k_1,\ldots k_\ell}(m_1,\ldots m_\ell)$ with respect to the time indices $m_1,\ldots m_\ell$, we obtain

$$F_{k,k_1,\ldots k_\ell}(\omega_1,\ldots \omega_\ell) = \tilde{\Psi}^*\left(\sum_{i=1}^{\ell} \omega_i - \frac{2\pi}{N}k\right) \prod_{m=1}^{\ell} \Psi\left(\omega_m - \frac{2\pi}{N}k_m\right) \,. \tag{3.31}$$

Then, had both $\tilde{\Psi}(\omega)$ and $\Psi(\omega)$ been ideal low-pass filters with bandwidths of $2\pi/\left(\lceil (\ell+1)/2 \rceil N\right)$, the overlap between the translated window functions in (3.31) would be identically zero for $k_\ell \neq \left(k - \sum_{i=1}^{\ell-1} k_i\right) \bmod N$, and thus only kernels of the form $c_{p_1,\ldots p_\ell,k,k_1,\ldots k_\ell}$ where $k_\ell = \left(k - \sum_{i=1}^{\ell-1} k_i\right) \bmod N$ would contribute to the output at frequency-bin index k. Practically, the energy is distributed over all kernels and particularly concentrated around the index $k_\ell = \left(k - \sum_{i=1}^{\ell-1} k_i\right) \bmod N$, as was demonstrated in Fig. 3.2 for the quadratic case ($\ell = 2$).

3.4 A New STFT Model For Nonlinear Systems

Representation of Volterra filters in the STFT domain involves a large number of parameters and high error variance, particularly when estimating the system from short and noisy data. In this section, we introduce an approximate model for improved nonlinear system identification in the STFT domain, which simplifies the STFT representation of Volterra filters and reduces the model complexity.

3.4.1 Quadratically Nonlinear Model

We start with an STFT representation of a second-order Volterra filter. Recall that modeling the linear kernel requires N crossband filters in each frequency bin [see (3.20)], where the length of each filter is approximately N_1/L. For system identification, however, only a few crossband filters need to be considered [28], which leads to a computationally efficient representation of the linear component. The quadratic Volterra kernel representation, on the other hand, consists of N^2 kernels in each frequency bin [see (3.23)], where the size of each kernel in the STFT domain is approximately $N_2/L \times N_2/L$. A perfect representation of the quadratic kernel is then achieved by employing $(NN_2/L)^2$ parameters in each frequency bin. Even though it may be reduced by considering the symmetric properties of the kernels, the complexity of such a model remains extremely large.

To reduce the complexity of the quadratic model in the STFT domain, let us assume that the analysis and synthesis filters are selective enough, such that according to Fig. 3.2, most of the energy of a quadratic kernel $c_{p',p'',k,k',k''}$ (for fixed k and k') is concentrated in a small region around the index $k'' = (k - k') \bmod N$. Accordingly, (3.23) can be efficiently approximated by

$$d_{2;p,k} \approx \sum_{\substack{k',k''=0 \\ (k'+k'') \bmod N = k}}^{N-1} \sum_{p',p''} x_{p-p',k'} x_{p-p'',k''} c_{p',p'',k,k',k''} \, . \tag{3.32}$$

A further simplification can be made by extending the so-called CMTF approximation, which was first introduced in [33, 43] for the representation of linear systems in the STFT domain. According to this model, a linear system is represented in the STFT domain by cross-multiplicative terms, rather than crossband filters, between distinct subbands. Following a similar reasoning, a kernel $c_{p',p'',k,k',k''}$ in (3.32) may be approximated as purely multiplicative in the STFT domain, so that (3.32) degenerates to

$$d_{2;p,k} \approx \sum_{\substack{k',k''=0 \\ (k'+k'') \bmod N = k}}^{N-1} x_{p,k'} x_{p,k''} c_{k',k''} . \tag{3.33}$$

We refer to $c_{k',k''}$ as a *quadratic cross-term*. The constraint $(k' + k'') \bmod N = k$ on the summation indices in (3.33) indicates that only frequency indices $\{k', k''\}$, whose sum is k or $k + N^2$, contribute to the output at frequency bin k. This concept is well illustrated in Fig. 3.3, which shows the (k', k'') two-dimensional plane. For calculating $d_{2;p,k}$ at frequency bin k, only points on the lines $k' + k'' = k$ and $k' + k'' = k + N$ need to be considered. Moreover, the quadratic cross-terms $c_{k',k''}$ have unique values only at the upper triangle ACH. Therefore, the intersection between this triangle and the lines $k' + k'' = k$ and $k' + k'' = k + N$ bounds the range of the summation indices in (3.33), such that $d_{2;p,k}$ can be compactly rewritten as

$$d_{2;p,k} \approx \sum_{k' \in \mathcal{F}} x_{p,k'} x_{p,(k-k') \bmod N} c_{k',(k-k') \bmod N} , \tag{3.34}$$

where $\mathcal{F} = \{0, 1, \ldots \lfloor k/2 \rfloor, k + 1, \ldots, k + 1 + \lfloor (N - k - 2)/2 \rfloor\} \subset [0, N - 1]$. Consequently, the number of cross-terms at the kth frequency bin has been reduced by a factor of two to $\lfloor k/2 \rfloor + \lfloor (N - k - 2)/2 \rfloor + 2$. Note that a further reduction in the model complexity can be achieved if the signals are assumed real-valued, since in this case $c_{k',k''}$ must satisfy $c_{k',k''} = c^*_{N-k',N-k''}$, and thus, only points in the grey area contribute to the model output (in this case, it is sufficient to consider only the first $\lfloor N/2 \rfloor + 1$ output frequency bins).

It is worthwhile noting the aliasing effects in the model output signal. Aliasing exists in the output as a consequence of sum and difference interactions that produce frequencies higher than one-half of the Nyquist frequency. The input frequencies causing these aliasing effects correspond to the points in the triangles BDO and FGO. To avoid aliasing, one must require that the value of $x_{p,k'} x_{p,k''} c_{k',k''}$ is zero for all indices k' and k'' inside these triangles.

Finally, using (3.20) and (3.34) for representing the linear and quadratic components of the system, respectively, we obtain

$$d_{p,k} = \sum_{k'=0}^{N-1} \sum_{p'=0}^{\bar{N}_1-1} x_{p-p',k'} h_{p',k,k'}$$
$$+ \sum_{k' \in \mathcal{F}} x_{p,k'} x_{p,(k-k') \bmod N} c_{k',(k-k') \bmod N} . \tag{3.35}$$

---

[2] Since k and k' range from 0 to $N - 1$, the contribution of the difference interaction of two frequencies to the kth frequency bin corresponds to the sum interaction of the same two frequencies to the $(k + N)$th frequency bin.

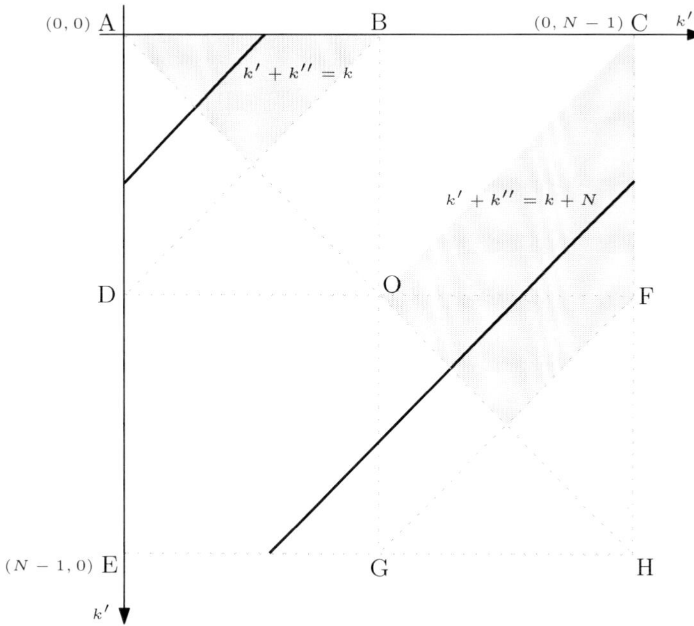

Fig. 3.3 Two-dimensional (k', k'') plane. Only points on the line $k' + k'' = k$ (corresponding to sum interactions) and the line $k' + k'' = k + N$ (corresponding to difference interactions) contribute to the output at the kth frequency bin.

Equation (3.35) represents an explicit model for quadratically nonlinear systems in the STFT domain. A block diagram of the proposed model is illustrated in Fig. 3.4. Analogously to the time-domain Volterra model, an important property of the proposed model is the fact that its output depends linearly on the coefficients, which means that conventional linear estimation algorithms can be applied for estimating its parameters (see Section 3.5).

The proposed STFT-domain model generalizes the conventional discrete frequency-domain Volterra model, described in (3.11). A major limitation of this model is its underlying assumption that the observation frame (N) is sufficiently large compared with the memory length of the linear kernel, which enables to approximate the linear convolution as multiplicative in the frequency domain. Similarly, under this large-frame assumption, the linear component in the proposed model (3.35) can be approximated as a multiplicative transfer function (MTF) [44, 45]. Accordingly, the STFT model in (3.35) reduces to

$$d_{p,k} = h_k x_{p,k} + \sum_{k' \in \mathcal{F}} x_{p,k'} x_{p,(k-k') \bmod N} c_{k',(k-k') \bmod N} , \qquad (3.36)$$

which is in one-to-one correspondence with the frequency-domain model (3.11). Therefore, the frequency-domain model can be regarded as a spe-

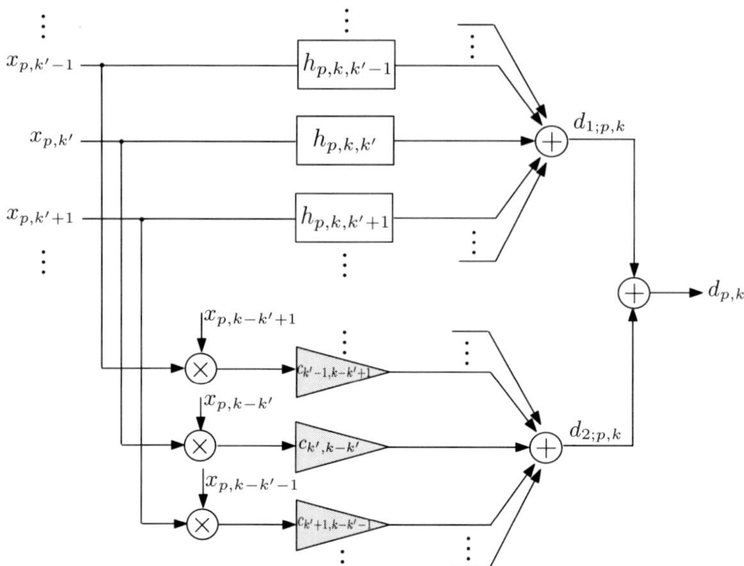

Fig. 3.4 Block diagram of the proposed model for quadratically nonlinear systems in the STFT domain. The upper branch represents the linear component of the system, which is modeled by the crossband filters $h_{p,k,k'}$. The quadratic component is modeled at the lower branch by using the quadratic cross-terms $c_{k,k'}$.

cial case of the proposed model for relatively large observation frames. In practice, a large observation frame may be very restrictive, especially when long and time-varying impulse responses are considered (as in acoustic echo cancellation applications [46]). A long frame restricts the capability to identify and track time variations in the system, since the system is assumed constant during the observation frame. Additionally, as indicated in [44], increasing the frame length (while retaining the relative overlap between consecutive frames), reduces the number of available observations in each frequency bin, which increases the variance of the system estimate. Attempting to identify the system using the models (3.11) or (3.36) yields a model mismatch that degrades the accuracy of the linear-component estimate. The crossband filters representation, on the other hand, outperforms the MTF approach and achieves a substantially lower mse value, even when relatively long frames are considered [28]. Clearly, the proposed model forms a much richer representation than that offered by the frequency-domain model, and may correspondingly be useful for a larger variety of applications.

In this context, it should be emphasized that the quadratic-component representation provided by the proposed time-frequency model (3.35) (and certainly by the frequency-domain model) may not exactly represent a second-order Volterra filter in the time domain, due to the approximations made in (3.32) and (3.33). Nevertheless, the proposed STFT model forms a new class of nonlinear models that may represent certain nonlinear systems more efficiently than the conventional time-domain Volterra model. In fact, as will be shown in Section 3.6, the proposed model may be more advantageous than the latter in representing nonlinear systems with relatively long memory due to its computational efficiency.

3.4.2 High-Order Nonlinear Models

For completeness of discussion, let us extend the STFT model to the general case of a qth-order nonlinear system. Following a similar derivation to that made for the quadratic case [see (3.32)–(3.33)], the output of a qth-order nonlinear system is modeled in the STFT domain as

$$d_{p,k} = d_{1;p,k} + \sum_{\ell=2}^{q} d_{\ell;p,k} \,, \tag{3.37}$$

where the linear component $d_{1;p,k}$ is given by (3.20), and the ℓth-order homogeneous component $d_{\ell;p,k}$ is given by

$$d_{\ell;p,k} = \sum_{\substack{k_1,\dots k_\ell=0 \\ \left(\sum_{i=1}^{\ell} k_i\right) \bmod N = k}}^{N-1} c_{k_1,\dots k_\ell} \prod_{i=1}^{\ell} x_{p,k_i} \,. \tag{3.38}$$

Clearly, only ℓ-fold frequencies $\{k_i\}_{i=1}^{\ell}$, whose sum is k or $k+N$, contribute to the output $d_{\ell;p,k}$ at frequency bin k. Consequently, the number of cross-terms $c_{k_1,\dots k_{\ell-1},k_\ell}$ $(\ell = 2,\dots,q)$ involved in representing a qth-order nonlinear system is given by $\sum_{\ell=2}^{q} N^{\ell-1} = (N^q - N)/(N-1)$. Note that this number can be further reduced by exploiting the symmetry property of the cross-terms, as was done for the quadratic case.

3.5 Quadratically Nonlinear System Identification

In this section, we consider the problem of identifying quadratically nonlinear systems using the proposed STFT model. The model parameters are estimated using either batch (Section 3.5.1) or adaptive (Section 3.5.2) methods, and a comparison to the time-domain Volterra model is carried out in terms of computational complexity. Without loss of generality, we consider here only the quadratic model due to its relatively simpler structure. The quadratic model is appropriate for representing the nonlinear behavior of many real world systems [47]. An extension to higher nonlinearity orders is straightforward.

Consider the STFT-based system identification scheme as illustrated in Fig. 3.1. The input signal $x(n)$ passes through an unknown quadratic time-invariant system $\phi(\cdot)$, yielding the clean output signal $d(n)$. Together with a corrupting noise signal $\xi(n)$, the system output signal is given by

$$y(n) = \{\phi x\}(n) + \xi(n) = d(n) + \xi(n) \, . \tag{3.39}$$

In the time-frequency domain, equation (3.39) may be written as

$$y_{p,k} = d_{p,k} + \xi_{p,k} \, . \tag{3.40}$$

To derive an estimator $\hat{y}_{p,k}$ for the system output in the STFT domain, we employ the proposed STFT model (3.35), but with the use of only $2K + 1$ crossband filters in each frequency bin. The value of K controls the under-modeling in the linear component of the model by restricting the number of crossband filters. Accordingly, the resulting estimate $\hat{y}_{p,k}$ can be written as

$$\begin{aligned}
\hat{y}_{p,k} = &\sum_{k'=k-K}^{k+K} \sum_{p'=0}^{\bar{N}_1-1} x_{p-p',k' \bmod N} h_{p',k,k' \bmod N} \\
&+ \sum_{k' \in \mathcal{F}} x_{p,k'} x_{p,(k-k') \bmod N} c_{k',(k-k') \bmod N} \, .
\end{aligned} \tag{3.41}$$

The influence of the number of estimated crossband filters $(2K + 1)$ on the system identifier performance is demonstrated in Section 3.6.

Let $\mathbf{h}_k$ be the $2K + 1$ filters at frequency bin k

$$\mathbf{h}_k = \left[\mathbf{h}_{k,(k-K) \bmod N}^T \ \mathbf{h}_{k,(k-K+1) \bmod N}^T \ \cdots \ \cdots \ \mathbf{h}_{k,(k+K) \bmod N}^T \right]^T , \tag{3.42}$$

where $\mathbf{h}_{k,k'} = \left[h_{0,k,k'} \ h_{1,k,k'} \ \cdots \ h_{\bar{N}_1-1,k,k'} \right]^T$ is the crossband filter from frequency bin k' to frequency bin k. Likewise, let

$$\bar{\mathbf{x}}_k(p) = \left[x_{p,k} \ x_{p-1,k} \ \cdots \ x_{p-M+1,k} \right]^T$$

and let

$$\mathbf{x}_{\mathrm{L},k}(p) = \left[\bar{\mathbf{x}}_{(k-K) \bmod N}^T(p) \ \bar{\mathbf{x}}_{(k-K+1) \bmod N}^T(p) \ \cdots \ \bar{\mathbf{x}}_{(k+K) \bmod N}^T(p) \right]^T \tag{3.43}$$

form the input data vector to the linear component of the model $\mathbf{h}_k(p)$. For notational simplicity, let us assume that k and N are both even, such that according to (3.34), the number of quadratic cross-terms in each frequency bin is $N/2 + 1$. Accordingly, let

$$\mathbf{c}_k = \begin{bmatrix} c_{0,k} & \cdots & c_{\frac{k}{2},\frac{k}{2}} & c_{k+1,N-1} & \cdots & c_{\frac{N+k}{2},\frac{N+k}{2}} \end{bmatrix}^T \tag{3.44}$$

denote the quadratic cross-terms at the kth frequency bin, and let

$$\mathbf{x}_{Q,k}(p) = \begin{bmatrix} x_{p,0}x_{p,k} & \cdots & x_{p,\frac{k}{2}}x_{p,\frac{k}{2}} & x_{p,k+1}x_{p,N-1} & \cdots & x_{p,\frac{N+k}{2}}x_{p,\frac{N+k}{2}} \end{bmatrix} \tag{3.45}$$

be the input data vector to the quadratic component of the model $\mathbf{c}_k(p)$. Then, the output signal estimate (3.41) can be written in a vector form as

$$\hat{y}_{p,k}(\theta_k) = \mathbf{x}_k^T(p)\theta_k , \tag{3.46}$$

where $\theta_k = [\mathbf{h}_k^T \; \mathbf{c}_k^T]^T$ is the model parameter vector in the kth frequency bin, and $\mathbf{x}_k(p) = [\mathbf{x}_{L,k}^T(p) \; \mathbf{x}_{Q,k}^T(p)]^T$ is the corresponding input data vector.

3.5.1 Batch Estimation Scheme

In the following, we estimate the model parameter vector θ_k in a batch form using an LS criterion and investigate the influence of nonlinear undermodeling on the system identifier performance.

Let N_x denote the time-domain observable data length and let $P \approx N_x/L$ be the number of samples given in a time-trajectory of the STFT representation (i.e., length of $x_{p,k}$ for a given k). Let

$$\mathbf{X}_k^T = \begin{bmatrix} \mathbf{x}_k(0) & \mathbf{x}_k(1) & \cdots & \mathbf{x}_k(P-1) \end{bmatrix} ,$$

where $\mathbf{x}_k(p)$ is defined in (3.46), and let $\mathbf{y}_k = \begin{bmatrix} y_{0,k} & y_{1,k} & \cdots & y_{P-1,k} \end{bmatrix}^T$ denote the observable data length. Then, the LS estimate of the model parameters at the kth frequency bin is given by

$$\hat{\theta}_k = \arg\min_{\theta_k} \|\mathbf{y}_k - \mathbf{X}_k\theta_k\|^2$$
$$= \left(\mathbf{X}_k^H\mathbf{X}_k\right)^{-1}\mathbf{X}_k^H\mathbf{y}_k , \tag{3.47}$$

where we assume that $\mathbf{X}_k^H\mathbf{X}_k$ is not singular[3]. Substituting $\hat{\theta}_k$ for θ_k in (3.46), we obtain the LS estimate of the system output in the STFT domain at the kth frequency bin.

---

[3] In the ill-conditioned case, when $\mathbf{X}_k^H\mathbf{X}_k$ is singular, matrix regularization is required [48].

3.5.1.1 Computational Complexity

Forming the LS normal equations $\left(\mathbf{X}_k^H \mathbf{X}_k\right) \hat{\theta}_k = \mathbf{X}_k^H \mathbf{y}_k$ in (3.47) and solving them using the Cholesky decomposition [49] require $Pd_{\theta_k}^2 + d_{\theta_k}^3/3$ arithmetic operations[4], where

$$d_{\theta_k} = (2K + 1)\,\bar{N}_1 + N/2 + 1$$

is the dimension of the parameter vector θ_k. Computation of the desired signal estimate (3.46) requires additional $2Pd_{\theta_k}$ arithmetic operations. Assuming P is sufficiently large and neglecting the computations required for the forward and inverse STFTs, the complexity associated with the proposed approach is

$$O_{\text{s,batch}} \sim O\left\{NP\left[(2K + 1)\,\bar{N}_1 + N/2 + 1\right]^2\right\}, \tag{3.48}$$

where the subscript s is for subband. Expectedly, we observe that the computational complexity increases as K increases. However, analogously to linear system identification [28], incorporating crossband filters into the model may yield lower mse for stronger and longer input signals, as demonstrated in Section 3.6.

In the time domain, the computational complexity of an LS-based estimation process using a second-order Volterra filter [see (3.7)] is given by [27]

$$O_{\text{f,batch}} \sim O\left(N_x\left[N_1 + N_2\left(N_2 + 1\right)/2\right]^2\right), \tag{3.49}$$

where the subscript f is for fullband. Rewriting the subband approach complexity (3.48) in terms of the fullband parameters (by using the relations $P \approx N_x/L$ and $\bar{N}_1 \approx N_1/L$), the ratio between the subband and fullband complexities can be written as

$$\frac{O_{\text{s,batch}}}{O_{\text{f,batch}}} \sim \frac{1}{r} \cdot \frac{\left(2N_1 \cdot \frac{2K+1}{rN} + N\right)^2}{\left(2N_1 + N_2^2\right)^2}, \tag{3.50}$$

where $r = L/N$ denote the relative overlap between consecutive analysis windows, and N_1 and N_2 are the memory lengths of the linear and quadratic Volterra kernels, respectively. Expectedly, we observe that the computational gain achieved by the proposed subband approach is mainly determined by the STFT analysis window length N, which represents the trade-off between the linear- and nonlinear-component complexities. Specifically, using a longer analysis window yields shorter crossband filters ($\sim N_1/N$), which reduces the computational cost of the linear component, but at the same time increases the nonlinear-component complexity by increasing the number of quadratic cross-terms ($\sim N$). Nonetheless, according to (3.50), the complexity of the proposed subband approach would typically be lower than that of the conven-

---

[4] An arithmetic operation is considered to be any complex multiplication, complex addition, complex subtraction, or complex division.

tional fullband approach. For instance, for $N = 256$, $r = 0.5$ (i.e., $L = 128$), $N_1 = 1024$, $N_2 = 80$, and $K = 2$ the proposed approach complexity is reduced by approximately 300, when compared to the fullband-approach complexity. The computational efficiency obtained by the proposed approach becomes even more significant when systems with relatively large second-order memory length are considered. This is because these systems necessitate an extremely large memory length N_2 for the quadratic kernel, when using the time-domain Volterra model, such that $N \ll N_2^2$ and consequently $O_{\mathrm{s,batch}} \ll O_{\mathrm{f,batch}}$.

3.5.1.2 Influence of Nonlinear Undermodeling

Nonlinear undermodeling, which is a consequence of employing a purely linear model for the estimation of nonlinear systems, has been examined recently in time and frequency domains [50, 51, 52]. The quantification of this error is of major importance since in many cases a purely linear model is fitted to the data, even though the system is nonlinear (e.g., employing a linear adaptive filter in acoustic echo cancellation applications [46]). Next, we examine the influence of nonlinear undermodeling in the STFT domain using the proposed model.

The (normalized) mse in the kth frequency bin is defined by[5]

$$\epsilon_k(K) = \frac{1}{E_d} E\left\{ \left\| \mathbf{d}_k - \mathbf{X}_k \hat{\theta}_k \right\|^2 \right\}, \tag{3.51}$$

where $E_d \triangleq E\left\{ \|\mathbf{d}_k\|^2 \right\}$, $\hat{\theta}_k$ is given in (3.47), and

$$\mathbf{d}_k = \begin{bmatrix} d_{0,k} & d_{1,k} & \cdots & d_{P-1,k} \end{bmatrix}^T .$$

To examine the influence of nonlinear undermodeling, let us define the mse achieved by estimating only the linear component of the model, i.e.,

$$\epsilon_{k,\mathrm{linear}}(K) = \frac{1}{E_d} E\left\{ \left\| \mathbf{d}_k - \mathbf{X}_{\mathrm{L},k} \hat{\theta}_{\mathrm{L},k} \right\|^2 \right\}, \tag{3.52}$$

where $\mathbf{X}_{\mathrm{L},k}^T = \begin{bmatrix} \mathbf{x}_{\mathrm{L},k}(0) & \mathbf{x}_{\mathrm{L},k}(1) & \cdots & \mathbf{x}_{\mathrm{L},k}(P-1) \end{bmatrix}$, $\mathbf{x}_{\mathrm{L},k}(p)$ is defined in (3.43), and

$$\hat{\theta}_{\mathrm{L},k} = \left(\mathbf{X}_{\mathrm{L},k}^H \mathbf{X}_{\mathrm{L},k} \right)^{-1} \mathbf{X}_{\mathrm{L},k}^H \mathbf{y}_k$$

---

[5] To avoid the well-known *overfitting* problem [36], the mse defined in (3.51) measures the fit of the optimal estimate $\mathbf{X}_k \hat{\theta}_k$ to the clean output signal $\mathbf{d}_k$, rather than to the measured (noisy) signal $\mathbf{y}_k$. Consequently, the growing model variability caused by increasing the number of model parameters is compensated, and a more reliable measure for the model estimation quality is achieved.

is the LS estimate of the $2K+1$ crossband filters of the model's linear component. To derive explicit expressions for the mse values in (3.51) and (3.52), let us first assume that the clean output of the true system $d_{p,k}$ satisfies the proposed quadratic model (3.35). We further assume that $x_{p,k}$ and $\xi_{p,k}$ are uncorrelated zero-mean white Gaussian signals, and that $x_{p,k}$ is ergodic [53]. Denoting the SNR by $\eta = E\{|d_{p,k}|^2\}/E\{|\xi_{p,k}|^2\}$ and using the above assumptions, the mse values can be expressed as [34]

$$\epsilon_k(K) = \frac{\alpha_k(K)}{\eta} + \beta_k(K), \tag{3.53}$$

$$\epsilon_{k,\text{linear}}(K) = \frac{\alpha_{k,\text{linear}}(K)}{\eta} + \beta_{k,\text{linear}}(K), \tag{3.54}$$

where $\alpha_k(K)$, $\beta_k(K)$, $\alpha_{k,\text{linear}}(K)$ and $\beta_{k,\text{linear}}(K)$ depend on the number of estimated crossband filters and the parameters of the *true* system. We observe from (3.53)–(3.54) that the mse, for either a linear or a nonlinear model, is a monotonically decreasing function of η, which expectedly indicates that a better estimation of the model parameters is enabled by increasing the SNR. Let

$$\nu = \sigma_{d_Q}^2/\sigma_{d_L}^2 \tag{3.55}$$

denote the nonlinear-to-linear ratio (NLR), where $\sigma_{d_L}^2$ and $\sigma_{d_Q}^2$ are the powers of the output signals of the linear and quadratic components of the system, respectively. Then, it can be shown that [34]

$$\alpha_k(K) - \alpha_{k,\text{linear}}(K) = \frac{\gamma}{P},$$

$$\beta_k(K) - \beta_{k,\text{linear}}(K) = -\frac{\delta/P + \nu}{1 + \nu}, \tag{3.56}$$

where γ and δ are positive constants. According to (3.56), we have $\alpha_k(K) > \alpha_{k,\text{linear}}(K)$ and $\beta_k(K) < \beta_{k,\text{linear}}(K)$, which implies that $\epsilon_k(K) > \epsilon_{k,\text{linear}}(K)$ for low SNR ($\eta << 1$), and $\epsilon_k(K) \le \epsilon_{k,\text{linear}}(K)$ for high SNR ($\eta >> 1$). As a result, since $\epsilon_k(K)$ and $\epsilon_{k,\text{linear}}(K)$ are monotonically decreasing functions of η, they must intersect at a certain SNR value, denoted by $\bar{\eta}$. This is well illustrated in Fig. 3.5, which shows the theoretical mse curves $\epsilon_k(K)$ and $\epsilon_{k,\text{linear}}(K)$ as a function of the SNR, obtained for a high NLR ν_1 [Fig. 3.5(a)] and a lower one $0.2\nu_1$ [Fig. 3.5(b)]. For SNR values lower than $\bar{\eta}$, we get $\epsilon_{k,\text{linear}}(K) < \epsilon_k(K)$, and correspondingly a lower mse is achieved by allowing for nonlinear undermodeling (i.e., employing only a linear model). On the other hand, as the SNR increases, the mse performance can be generally improved by incorporating also the nonlinear component into the model. A comparison of Figs. 3.5(a) and (b) indicates that this improvement in performance becomes larger as ν increases (i.e., $|\Delta\epsilon|$ increases). This stems from the fact that the error induced by the undermodeling in the linear component (i.e., by not considering all of the crossband filters) is less substantial as

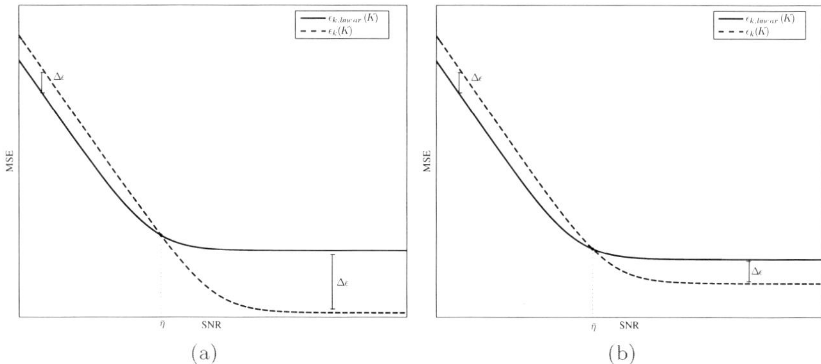

Fig. 3.5 Illustration of typical MSE curves as a function of the SNR, showing the relation between $\epsilon_{k,\text{linear}}(K)$ (solid) and $\epsilon_k(K)$ (dashed) for (a) high NLR ν_1 and (b) low NLR $0.2\nu_1$. $|\Delta\epsilon|$ denotes the nonlinear undermodeling error.

the nonlinearity strength increases, such that the true system can be more accurately estimated by the full model. Note that this can be theoretically verified from (3.56), which shows that $|\beta_k(K) - \beta_{k,\text{linear}}(K)|$ increases with increasing ν.

A comparison of Figs. 3.5(a) and (b) also indicates that the SNR intersection point $\bar{\eta}$ decreases as we increase ν. Consequently, as the nonlinearity becomes weaker (i.e., ν decreases), higher SNR values should be considered to justify the estimation of the nonlinear component. This can be verified from the theoretical value of $\bar{\eta}$, which can be obtained by requiring that $\epsilon_k(K) = \epsilon_{k,\text{linear}}(K)$, yielding

$$\bar{\eta} = \frac{1 + \nu}{\gamma^{-1}\delta + \gamma^{-1}P\nu}. \tag{3.57}$$

Expectedly, we observe that $\bar{\eta}$ is a monotonically decreasing function of ν (assuming $P > N/2 + 1$ [34], which holds in our case due to the ergodicity assumption of $x_{p,k}$). Equation (3.57) implies that $\bar{\eta}$ is a monotonically decreasing function of the observable data length in the STFT domain (P). Therefore, for a fixed SNR value, as more data is available in the identification process, a lower mse is achieved by estimating also the parameters of the nonlinear component. Recall that the system is assumed time invariant during P frames (its estimate is updated every P frames), in case the time variations in the system are relatively fast, we should decrease P and correspondingly allow for nonlinear undermodeling to achieve lower mse.

It is worthwhile noting that the discussion above assumes a fixed number of crossband filters (i.e., K is fixed). Nonetheless, as in linear system identification [28], the number of estimated crossband filters may significantly affect the system identifier performance. It can be shown that $\epsilon_k(K + 1) > \epsilon_k(K)$ for low SNR ($\eta << 1$), and $\epsilon_k(K + 1) \leq \epsilon_k(K)$ for high SNR ($\eta >> 1$) [34].

The same applies also for $\epsilon_{k,\text{linear}}(K)$. Accordingly, for every noise level there exists an optimal number of crossband filters, which increases as the SNR increases.

The results in this section are closely related to model-structure selection and model-order selection, which are fundamental problems in many system identification applications [36, 54, 55, 56, 57, 58, 59]. In our case, the model structure may be either linear or nonlinear, where a richer and larger structure is provided by the latter. The larger the model structure, the better the model fits to the data, at the expense of an increased variance of parametric estimates [36]. Generally, the structure to be chosen is affected by the level of noise in the data and the length of the observable data. As the SNR increases or as more data is employable, a richer structure can be used, and correspondingly a better estimation can be achieved by incorporating a nonlinear model rather than a linear one. Once a model structure has been chosen, its optimal order (i.e., the number of estimated parameters) should be selected, where in our case the model order is determined by the number of crossband filters. Accordingly, as the SNR increases, whether a linear or a nonlinear model is employed, more crossband filters should be utilized to achieve a lower mse.

3.5.2 Adaptive Estimation Scheme

Since practically many real-world systems are time-varying, the estimation process should be made adaptive in order to track these variations. Next, we introduce an LMS-based adaptive algorithm for the estimation of the model parameter vector θ_k from (3.46), and present explicit expressions for the transient and steady-state mse in subbands.

Let $\hat{\theta}_k(p)$ be the estimate of θ_k at frame index p. Using the LMS algorithm [37], $\hat{\theta}_k(p)$ can be recursively updated as

$$\hat{\theta}_k(p+1) = \hat{\theta}_k(p) + \mu e_{p,k} \mathbf{x}_k^*(p) \, , \tag{3.58}$$

where $e_{p,k} = y_{p,k} - \mathbf{x}_k^T(p)\hat{\theta}_k(p)$ is the error signal in the kth frequency bin, $y_{p,k}$ is defined in (3.40), and μ is a step-size. Note that since $\theta_k = [\,\mathbf{h}_k^T \ \mathbf{c}_k^T\,]^T$, the adaptive estimation (3.58) assumes a similar step-size μ for both linear ($\mathbf{h}_k$) and quadratic ($\mathbf{c}_k$) components of the model. However, in some cases, for instance, when one component varies slower than the other, it is necessary to use different step-sizes for each component in order to enhance the tracking capability of the algorithm. Specifically, let $\hat{\mathbf{h}}_k(p)$ and $\hat{\mathbf{c}}_k(p)$ be the estimates at frame index p of the $2K+1$ crossband filters $\mathbf{h}_k$ and the $N/2+1$ quadratic cross-terms $\mathbf{c}_k$, respectively. Then, the corresponding LMS update equations are given by

$$\hat{\mathbf{h}}_k(p+1) = \hat{\mathbf{h}}_k(p) + \mu_{\mathrm{L}} e_{p,k} \mathbf{x}_{\mathrm{L},k}^*(p) \, , \tag{3.59}$$

and

$$\hat{\mathbf{c}}_k(p+1) = \hat{\mathbf{c}}_k(p) + \mu_{\mathrm{Q}} e_{p,k} \mathbf{x}_{\mathrm{Q},k}^*(p) , \tag{3.60}$$

where

$$e_{p,k} = y_{p,k} - \mathbf{x}_{\mathrm{L},k}^T(p)\hat{\mathbf{h}}_k(p) - \mathbf{x}_{\mathrm{Q},k}^T(p)\hat{\mathbf{c}}_k(p) \tag{3.61}$$

is the error signal, $\mathbf{x}_{\mathrm{L},k}(p)$ is defined in (3.43), $\mathbf{x}_{\mathrm{Q},k}(p)$ is defined in (3.45), and μ_{L} and μ_{Q} are the step sizes of the linear and quadratic components of the model, respectively. Substituting $\hat{\theta}_k(p) = [\hat{\mathbf{h}}_k^T(p) \; \hat{\mathbf{c}}_k^T(p)]^T$ for θ_k in (3.46), we obtain the LMS estimate of the system output in the STFT domain at the pth frame and the kth frequency bin.

3.5.2.1 Computational Complexity

The adaptation formulas given in (3.59) and (3.60) require $(2K+1)\bar{N}_1 + N/2 + 3$ complex multiplications, $(2K+1)\bar{N}_1 + N/2 + 1$ complex additions, and one complex substraction to compute the error signal. Moreover, computing the desired signal estimate (3.46) results in an additional $2(2K+1)\bar{N}_1 + N + 1$ arithmetic operations. Note that each arithmetic operation is not carried out every input sample, but only once for every L input samples, where L denotes the decimation factor of the STFT representation. Thus, the adaptation process requires $4(2K+1)\bar{N}_1 + 2N + 6$ arithmetic operations for every L input samples and each frequency bin. Finally, repeating the process for each frequency bin and neglecting the computations required for the forward and inverse STFTs, the complexity associated with the proposed subband approach is given by

$$O_{\mathrm{s,adaptive}} \sim O\left(\frac{N}{L} \left\{ 4\left[(2K+1)\bar{N}_1 + N/2 \right] + 6 \right\} \right) . \tag{3.62}$$

Expectedly, we observe that the computational complexity increases as K increases.

In the time domain, the computational complexity of an LMS-based estimation process using a second-order Volterra filter [see (3.9] is given by [35]

$$O_{\mathrm{f,adaptive}} \sim O\left\{ 4\left(N_1 + N_2\left(N_2 + 1 \right)/2 \right) + 1 \right\} . \tag{3.63}$$

Rewriting the subband approach complexity (3.62) in terms of the fullband parameters (by using the relation $\bar{N}_1 \approx N_1/L$), the ratio between the subband and fullband complexities can be written as

$$\frac{O_{\mathrm{s,adaptive}}}{O_{\mathrm{f,adaptive}}} \sim \frac{N}{L} \cdot \frac{2N_1(2K+1)/L + N}{2N_1 + N_2^2} . \tag{3.64}$$

According to (3.64), the complexity of the proposed subband approach would be typically lower than that of the conventional fullband approach. For instance, with $N = 128$, $L = 64$ (50% overlap), $N_1 = 1024$, $N_2 = 80$, and $K = 2$, computational complexity of the proposed approach is smaller by a factor of 15 compared to that of the fullband approach. The computational efficiency of the proposed model was also demonstrated for a batch estimation scheme (see Section 3.5.1.1).

3.5.2.2 Convergence Analysis

In the following, we provide a convergence analysis of the proposed adaptive scheme and derive expressions for both transient and steady-state mse. The transient mse is defined by

$$\epsilon_k(p) = E\left\{|e_{p,k}|^2\right\}. \tag{3.65}$$

As in the batch estimation scheme (Section 3.5.1.2), in order to make the following analysis mathematically tractable, we assume that the clean output signal $d_{p,k}$ satisfies the proposed quadratic model (3.35), and that $x_{p,k}$ and $\xi_{p,k}$ are statistically independent zero-mean white complex Gaussian signals with variances σ_x^2 and σ_ξ^2, respectively. The common independence assumption, which states that the current input data vector is statistically independent of the currently updated parameters vector (e.g., [60, 61]), is also used; that is, the vector $[\mathbf{x}_{\mathrm{L},k}^T(p)\ \mathbf{x}_{\mathrm{Q},k}^T(p)]^T$ is independent of $[\hat{\mathbf{h}}_k^T(p)\ \hat{\mathbf{c}}_k^T(p)]^T$. Let us define the misalignment vectors of the linear and quadratic components, respectively, as

$$\mathbf{g}_{\mathrm{L},k}(p) = \hat{\mathbf{h}}_k(p) - \bar{\mathbf{h}}_k, \tag{3.66}$$

and

$$\mathbf{g}_{\mathrm{Q},k}(p) = \hat{\mathbf{c}}_k(p) - \bar{\mathbf{c}}_k, \tag{3.67}$$

where $\bar{\mathbf{h}}_k$ and $\bar{\mathbf{c}}_k$ are respectively the $2K+1$ crossband filters and the $N/2+1$ cross-terms of the true system [defined similarly to (3.42) and (3.44)]. Then, substituting (3.61) into (3.65), and using the definition of $y_{p,k}$ from (3.40) and (3.35), the mse can be expressed as [35, Appendix I]

$$\epsilon_k(p) = \epsilon_k^{\min} + \sigma_x^2 E\left\{\|\mathbf{g}_{\mathrm{L},k}(p)\|^2\right\} + \sigma_x^4 E\left\{\|\mathbf{g}_{\mathrm{Q},k}(p)\|^2\right\}, \tag{3.68}$$

where

$$\epsilon_k^{\min} = \sigma_\xi^2 + \sigma_x^2\left(\|\bar{\mathbf{h}}_{\mathrm{full},k}\|^2 - \|\bar{\mathbf{h}}_k\|^2\right) \tag{3.69}$$

is the minimum mse obtainable in the kth frequency bin, and $\bar{\mathbf{h}}_{\mathrm{full},k}$ is a vector consisting of all the crossband filters of the true system at the kth frequency bin. To accomplish the mse transient behavior, recursive formulas for $E\{\|\mathbf{g}_{\mathrm{L},k}(p)\|^2\}$ and $E\{\|\mathbf{g}_{\mathrm{Q},k}(p)\|^2\}$ are required. Defining $\mathbf{q}(p) \triangleq \left[E\{\|\mathbf{g}_{\mathrm{L},k}(p)\|^2\}\ E\{\|\mathbf{g}_{\mathrm{Q},k}(p)\|^2\}\right]^T$, using the above assumptions, and substituting (3.59) and (3.60) into (3.66) and (3.67), respectively, we obtain after some manipulations [35]:

$$\mathbf{q}(p+1) = \mathbf{A}\mathbf{q}(p) + \gamma, \tag{3.70}$$

where the elements of the 2×2 matrix $\mathbf{A}$ and the vector γ depend on $\epsilon_k^{\min}$ and the step-sizes μ_{L} and μ_{Q}. Equation (3.70) is convergent if and only if the eigenvalues of $\mathbf{A}$ are all within the unit circle. Finding explicit conditions on the step sizes μ_{L} and μ_{Q} that imposed by this demand is tedious and not straightforward. However, simplified expressions may be derived by assuming that the adaptive vectors $\hat{\mathbf{h}}_k(p)$ and $\hat{\mathbf{c}}_k(p)$ are not updated simultaneously. More specifically, assuming that $\hat{\mathbf{c}}_k(p)$ is constant during the adaptation of $\hat{\mathbf{h}}_k(p)$ (i.e., $\mu_{\mathrm{Q}} \ll \mu_{\mathrm{L}}$), the optimal step size that results in the fastest convergence of the linear component is given by [35]

$$\mu_{\mathrm{L,opt}} = \left[\sigma_x^2(2K+1)\bar{N}_1\right]^{-1}. \tag{3.71}$$

Note that since $\mu_{\mathrm{L,opt}}$ is inversely proportional to K, a lower step-size value should be utilized with increasing the number of crossband filters, which results in a slower convergence. Similarly, the optimal step-size for the quadratic component is given by [35]

$$\mu_{\mathrm{Q,opt}} = \left[\sigma_x^4 N/2\right]^{-1}. \tag{3.72}$$

It should be noted that when the assumption of the separated adaptation of the adaptive vectors does not hold [that is, $\hat{\mathbf{h}}_k(p)$ and $\hat{\mathbf{c}}_k(p)$ are updated simultaneously], the convergence of the algorithm is no longer guaranteed by using the derived optimal step sizes (they result in an eigenvalue on the unit circle). Practically, though, the stability of the algorithm can be guaranteed by using the so-called normalized LMS (NLMS) algorithm [37], which also leads to faster convergence.

Provided that μ_{L} and μ_{Q} satisfy the convergence conditions of the LMS algorithm, the steady-state solution of (3.70) is given by $\mathbf{q}(\infty) = [\mathbf{I} - \mathbf{A}]^{-1}\gamma$, which can be substituted into (3.68) to find an explicit expression for the steady-state mse [35]:

$$\epsilon_k(\infty) = f\left(\mu_{\mathrm{L}}, \mu_{\mathrm{Q}}\right)\epsilon_k^{\min}, \tag{3.73}$$

where

$$f\left(\mu_{\mathrm{L}}, \mu_{\mathrm{Q}}\right) = \frac{2}{2 - \mu_{\mathrm{L}}\sigma_x^2(2K+1)M - \mu_{\mathrm{Q}}\sigma_x^4 N/2}. \tag{3.74}$$

Note that since μ_{L} is inversely proportional to K [see (3.71)], we expect $f\left(\mu_{\mathrm{L}}, \mu_{\mathrm{Q}}\right)$ to be independent of K. Consequently, based on the definition of $\epsilon_k^{\min}$ from (3.69), a lower steady-state mse is expected by increasing the number of estimated crossband filters, as will be further demonstrated in Section 3.6.

It should be noted that the transient and steady-state performance of a purely linear model can be obtained as a special case of the above analysis by substituting $\mu_Q = 0$ and $E\{\|\mathbf{g}_{Q,k}(p)\|^2\} = \|\bar{\mathbf{c}}_k\|^2$. Accordingly, the resulting steady-state mse, denoted by $\epsilon_{k,\mathrm{linear}}(p)$, can be expressed as [35]

$$\epsilon_{k,\mathrm{linear}}(\infty) = f(\mu_{\mathrm{L}}, 0)\, \epsilon_{k,\mathrm{linear}}^{\min}, \tag{3.75}$$

where

$$\epsilon_{k,\mathrm{linear}}^{\min} = \epsilon_k^{\min} + \sigma_x^4 \|\bar{\mathbf{c}}_k\|^2 \tag{3.76}$$

represents the minimum mse that can be obtained by employing a linear model in the estimation process. It can be verified from (3.69), (3.74) and (3.76) that $\epsilon_k^{\min} \leq \epsilon_{k,\mathrm{linear}}^{\min}$ and $f(\mu_{\mathrm{L}}, \mu_{\mathrm{Q}}) \geq f(\mu_{\mathrm{L}}, 0)$, which implies that in some cases, a lower steady-state mse might be achieved by using a linear model, rather than a nonlinear one. A similar phenomenon was also indicated in the context of off-line system identification, where it was shown that the nonlinear undermodeling error is mainly influenced by the NLR (see Section 3.5.1.2). Specifically in our case, the ratio between $\epsilon_{k,\mathrm{linear}}^{\min}$ and $\epsilon_k^{\min}$ can be written as

$$\frac{\epsilon_{k,\mathrm{linear}}^{\min}}{\epsilon_k^{\min}} = 1 + \frac{\left\|\bar{\mathbf{h}}_{\mathrm{full},k}\right\|^2}{\sigma_\xi^2/\sigma_x^2 + \left\|\bar{\mathbf{h}}_{\mathrm{full},k}\right\|^2 - \left\|\bar{\mathbf{h}}_k\right\|^2} \cdot \nu, \tag{3.77}$$

where ν represents the NLR [defined in (3.55)]. Equation (3.77) indicates that as the nonlinearity becomes stronger (i.e., ν increases), the minimum mse attainable by the full nonlinear model ($\epsilon_k^{\min}$) would be much lower than that obtained by the purely linear model ($\epsilon_{k,\mathrm{linear}}^{\min}$), such that $\epsilon_k(\infty) < \epsilon_{k,\mathrm{linear}}(\infty)$. On the other hand, the purely linear model may achieve a lower steady-state mse when low NLR values are considered. In the limit, for $\nu \to 0$, we get $\epsilon_k^{\min} = \epsilon_{k,\mathrm{linear}}^{\min}$, and consequently $\epsilon_{k,\mathrm{linear}}(\infty) < \epsilon_k(\infty)$. Note, however, that since more parameters need to be estimated in the nonlinear model, we expect to obtain (for any NLR value) slower convergence than that of a linear model. These points will be demonstrated in the next section.

3.6 Experimental Results

In this section, we present experimental results which support the theoretical derivations and demonstrate the effectiveness of the proposed approach in estimating quadratically nonlinear systems. A comparison to the conventional time-domain Volterra approach is carried out in terms of mse performance for both synthetic white Gaussian signals and real speech signals. For the STFT, we use a half overlapping Hamming analysis window of $N = 256$ samples length (i.e., $L = 0.5N$). The inverse STFT is implemented with a minimum-energy synthesis window that satisfies the completeness condition [42]. The sample rate is 16 kHz.

3.6.1 Performance Evaluation for White Gaussian Inputs

In the first experiment, we examine the performances of the Volterra and proposed models under the assumption of white Gaussian signals. The system to be identified is formed as a parallel combination of linear and quadratic components as follows:

$$y(n) = \sum_{m=0}^{N_1^*-1} h(m)x(n-m) + \{\mathcal{L}x\}(n) + \xi(n), \qquad (3.78)$$

where $h(n)$ is the impulse response of the linear component, and $\{\mathcal{L}x\}(n)$ denotes the output of the quadratic component. The latter is generated according to the quadratic model (3.34), i.e.,

$$\{\mathcal{L}x\}(n) = S^{-1} \sum_{k'\in\mathcal{F}} x_{p,k'} x_{p,(k-k')\bmod N} c_{k',(k-k')\bmod N}, \qquad (3.79)$$

where S^{-1} denotes the inverse STFT operator and $\left\{ c_{k',(k-k')\bmod N} \middle| k' \in \mathcal{F} \right\}$ are the quadratic cross-terms of the true system. These terms are modeled here as a unit-variance zero-mean white Gaussian process. In addition, we model the linear impulse response as a nonstationary stochastic process with an exponential decay envelope, i.e., $h(n) = u(n)\beta(n)e^{-\alpha n}$, where $u(n)$ is the unit step function, $\beta(n)$ is a unit-variance zero-mean white Gaussian noise, and α is the decay exponent. In the following, we use $N_1^* = 768$, $\alpha = 0.009$, and an observable data length of $N_x = 24000$ samples. The input signal $x(n)$ and the additive noise signal $\xi(n)$ are uncorrelated zero-mean white Gaussian processes.

We employ the batch estimation scheme (see Section 3.5.1) and estimate the parameters of the Volterra and proposed models using the LS criterion. The resulting mse is computed in the time domain by

$$\epsilon_{\text{time}} = 10\log \frac{E\left\{[d(n)-\hat{y}(n)]^2\right\}}{E\left\{d^2(n)\right\}}, \qquad (3.80)$$

where $d(n) = y(n) - \xi(n)$ is the clean output signal and $\hat{y}(n)$ is the inverse STFT of the corresponding model output $\hat{y}_{p,k}$. For both models, a memory length of $N_1 = 768$ is employed for the linear kernel, where the memory length N_2 of the quadratic kernel in the Volterra model is set to 30. Figure 3.6

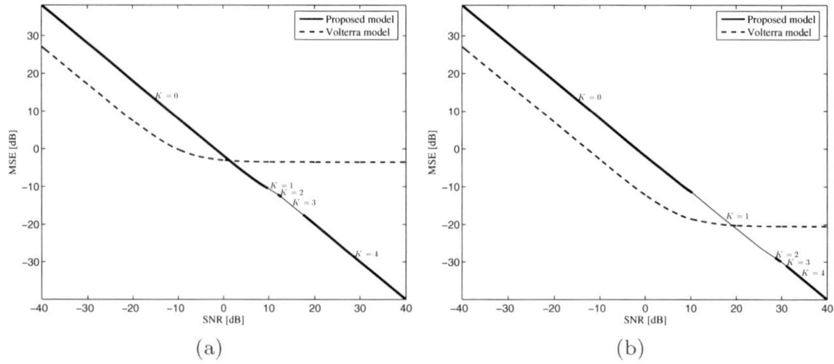

Fig. 3.6 MSE curves as a function of the SNR using LS estimates of the proposed STFT model [via (3.47)] and the conventional time-domain Volterra model [via (3.7)], for white Gaussian signals. The optimal value of K is indicated above the corresponding mse curve. (a) Nonlinear-to-linear ratio (NLR) of 0 dB, (b) NLR of -20 dB.

shows the resulting mse curves as a function of the SNR, as obtained for an NLR of 0 dB [Fig. 3.6(a)] and -20 dB [Fig. 3.6(b)]. For the proposed model, several values of K are employed in order to determine the influence of the number of estimated crossband filters on the mse performance, and the optimal value that achieves the minimal mse (mmse) is indicated above the mse curve. Note that a transition in the value of K is indicated by a variation in the width of the curve. Figure 3.6(a) implies that for relatively low SNR values, a lower mse is achieved by the conventional Volterra model. For instance, for an SNR of -20 dB, employing the Volterra model reduces the mse by approximately 10 dB, when compared to that achieved by the proposed model. However, for higher SNR conditions, the proposed model is considerably more advantageous. For an SNR of 20 dB, for instance, the proposed model enables a decrease of 17 dB in the mse using $K = 4$ (i.e., by incorporating 9 crossband filters into the model). Table 3.1 specifies the mse values obtained by each value of K for various SNR conditions. We observe that for high SNR values a significant improvement over the Volterra model can also be attained by using only the band-to-band filters (i.e., $K = 0$), which further reduces the computational cost of the proposed model. Clearly, as the SNR increases, a larger number of crossband filters should be utilized to attain the mmse, which is similar to what has been shown in the identification of purely linear systems [28]. Note that similar results are obtained for a smaller NLR value [Fig. 3.6(b)], with the only difference is that the two curves intersect at a higher SNR value.

The complexity of the fullband and subband approaches (for each value of K) is evaluated by computing the central processing unit (CPU) running

Table 3.1 MSE Obtained by the proposed model for several K values and by the Volterra model, using the batch scheme (Section 3.5.1) and under various SNR conditions. The nonlinear-to-linear ratio (NLR) is 0 dB.

K	MSE [dB]		
	SNR$= -10$ dB	SNR$= 20$ dB	SNR$= 35$ dB
0	8.08	-15.12	-16.05
1	8.75	-16.91	-18.8
2	9.31	-18.17	-21.55
3	9.82	-19.67	-28.67
4	10.04	-19.97	-34.97
Volterra	0.42	-3.25	-3.58

Table 3.2 Average running time in terms of CPU of the proposed approach (for several K values) and the Volterra approach. The length of the observable data is 24000 samples.

K	Running Time [sec]
0	5.15
1	6.79
2	8.64
3	10.78
4	13.23
Volterra	61.31

time[6] of the LS estimation process. The running time in terms of CPU seconds is averaged over several SNR conditions and summarized in Table 3.2. We observe, as expected from (3.50), that the running time of the proposed approach, for any value of K, is substantially lower than that of the Volterra approach. Specifically, the estimation process of the Volterra model is approximately 12 and 4.5 times slower than that of the proposed model with $K = 0$ and $K = 4$, respectively. Moreover, Table 3.2 indicates that the running time of the proposed approach increases as more crossband filters are estimated, as expected from (3.48).

3.6.2 Nonlinear Undermodeling in Adaptive System Identification

In the second experiment, the proposed-model parameters are adaptively estimated using the LMS algorithm (see Section 3.5.2), and the influence of nonlinear undermodeling is demonstrated by fitting both linear and nonlinear models to the observable data. The input signal $x_{p,k}$ and the additive noise signal $\xi_{p,k}$ are uncorrelated zero-mean white complex Gaussian processes in the STFT domain, which is in contrast with the previous experiment, for which the corresponding signals were assumed white in the time domain.

---

[6] The simulations were all performed under MATLAB; v.7.0, on a Core(TM)2 Duo P8400 2.27 GHz PC with 4 GB of RAM, running Windows Vista, Service Pack 1.

Although $x_{p,k}$ may not necessarily be a valid STFT signal (i.e., a time-domain sequence whose STFT is given by $x_{p,k}$ may not always exist [62]), we use this assumption to verify the mean-square theoretical derivations made in Section 3.5.2.2. The system to be identified is similar to that used in the previous experiment. A purely linear model is fitted to the data by setting the step size of the quadratic component to zero (i.e., $\mu_Q = 0$); whereas, a full nonlinear model is employed by updating the quadratic component with a step size of $\mu_Q = 0.25/\left(\sigma_x^4 N/2\right)$. For both cases, the linear kernel is updated with step size $\mu_L = 0.25/\left[\sigma_x^2(2K+1)\bar{N}_1\right]$ for two different values of K ($K = 1$ and 3).

Figure 3.7 shows the resulting mse curves $\epsilon_k(p)$ [defined in (3.65)] and $\epsilon_{k,\text{linear}}(p)$, as obtained from simulation results and from the theoretical derivations made in Section 3.5.2.2, for frequency bin $k = 11$, an SNR of 40 dB, and an NLR of -10 dB [Fig. 3.7(a)] and -30 dB [Fig. 3.7(b)]. It can be seen that the experimental results are accurately described by the theoretical mse curves. We observe from 3.7(a) that for a -10 dB NLR, a lower steady-state mse is achieved by using the nonlinear model. Specifically for $K = 3$, a significant improvement of 12 dB can be achieved over a purely linear model. On the contrary, Fig. 3.7(b) shows that for a lower NLR value (-30 dB), the inclusion of the nonlinear component in the model is not necessarily preferable. For example when $K = 1$, the linear model achieves the lowest steady-state mse, while for $K = 3$, the improvement achieved by the nonlinear model is insignificant, and apparently does not justify the substantial increase in model complexity. In general, by further decreasing the NLR, the steady-state mse associated with the linear model decreases, while the relative improvement achieved by the nonlinear model becomes smaller. These results, which were accurately described by the theoretical error analysis in Section 3.5.2.2 [see (3.73)–(3.77)], are attributable to the fact the linear model becomes more accurate as the nonlinearity strength decreases. As a result, the advantage of the nonlinear model due to its improved modeling capability becomes insignificant (i.e., $\epsilon_k^{\min} \approx \epsilon_{k,\text{linear}}^{\min}$), and therefore cannot compensate for the additional adaptation noise caused by also updating the nonlinear component of the model. Another interesting point that can be concluded from the comparison of Figs. 3.7(a) and (b) is the strategy of controlling the model structure and the model order. Specifically, for high NLR conditions [Fig. 3.7(a)], a linear model with a small K should be used at the beginning of the adaptation. Then, the model structure should be changed to nonlinear at an intermediate stage of the adaptation, and the number of estimated crossband filters should increase as the adaptation process proceeds in order to achieve the minimum mse at each iteration. On the other hand, for low NLR conditions [Fig. 3.7(b)], one would prefer to initially update a purely linear model in order to achieve faster convergence, and then to gradually increase the number of crossband filters. In this case, switching to a different model structure and also incorporating the nonlinear component into the model would be preferable only at an advanced stage of the adaptation process.

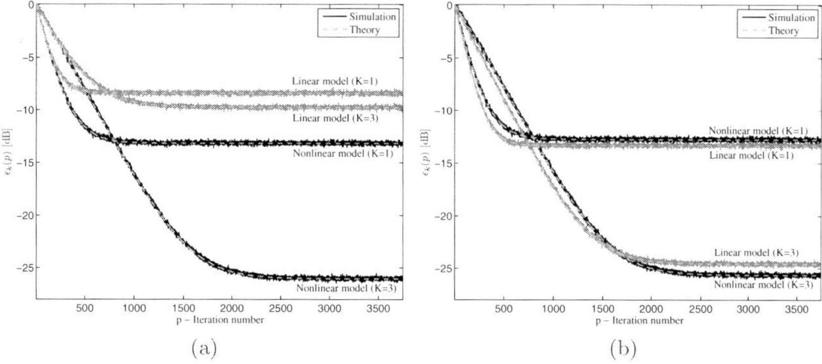

Fig. 3.7 Comparison of simulation and theoretical curves of the transient mse (3.65) for frequency bin $k = 11$ and white Gaussian signals, as obtained by adaptively updating a purely linear model (light) and a nonlinear model (dark) via (3.59)–(3.60). (a) Nonlinear-to-linear ratio (NLR) of -10 dB, (b) NLR of -30 dB.

3.6.3 Nonlinear Acoustic Echo Cancellation Application

In the third experiment, we demonstrate the application of the proposed approach to nonlinear acoustic echo cancellation [3, 2, 1] using the batch estimation scheme introduced in Section 3.5.1. The nonlinear behavior of the LEM system in acoustic echo cancellation applications is mainly introduced by the loudspeakers and their amplifiers, especially when small loudspeakers are driven at high volume. In this experiment, we use an ordinary office with a reverberation time T_{60} of about 100 ms. A far-end speech signal $x(n)$ is fed into a loudspeaker at high volume, thus introducing non-negligible nonlinear distortion. The signal $x(n)$ propagates through the enclosure and received by a microphone as an echo signal together with a local noise $\xi(n)$. The resulting noisy signal is denoted by $y(n)$. In this experiment, the signals are sampled at 16 kHz. Note that the acoustic echo canceller (AEC) performance is evaluated in the absence of near-end speech, since a double-talk detector (DTD) is usually employed for detecting the near-end signal and freezing the estimation process [63, 64]. A commonly-used quality measure for evaluating the performance of AECs is the echo-return loss enhancement (ERLE), defined in dB by

$$\text{ERLE} = 10 \log_{10} \frac{E\left\{y^2(n)\right\}}{E\left\{e^2(n)\right\}}, \tag{3.81}$$

where $e(n) = y(n) - \hat{y}(n)$ is the error signal, and $\hat{y}(n)$ is the inverse STFT of the estimated echo signal.

Figures 3.8(a) and (b) show the far-end signal and the microphone signal, respectively. Figures 3.8(c)–(e) show the error signals as obtained by using

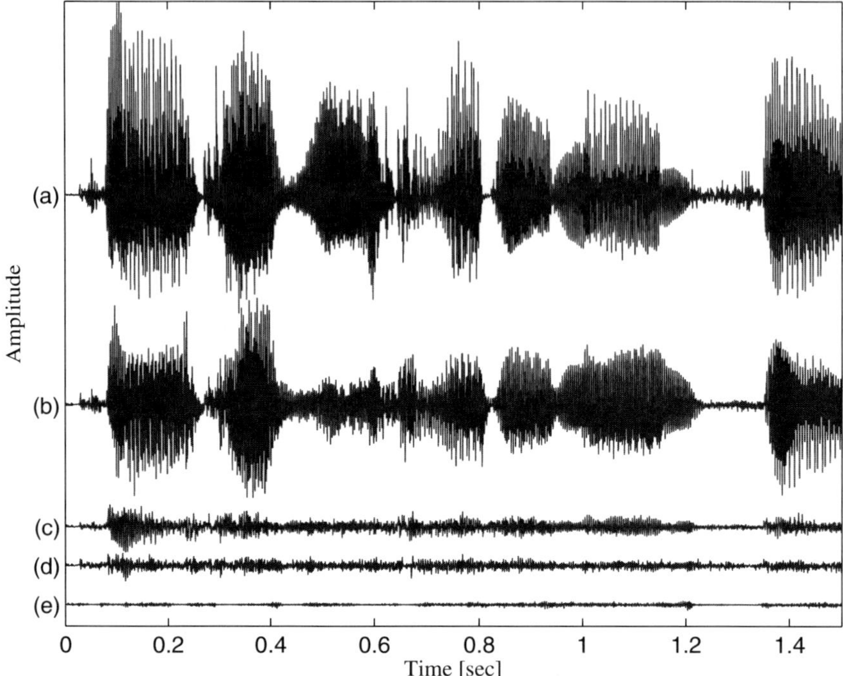

Fig. 3.8 Speech waveforms and residual echo signals, obtained by LS estimates of the proposed STFT model [via (3.47)] and the conventional time-domain Volterra model [via (3.7)]. (a) Far-end signal, (b) microphone signal. (c)–(e) Error signals obtained by a purely linear model in the time domain, the Volterra model with $N_2 = 90$, and the proposed model with $K = 1$, respectively.

a purely linear model in the time domain, a Volterra model with $N_2 = 90$, and the proposed model with $K = 1$, respectively. For all models, a length of $N_1 = 768$ is employed for the linear kernel. The ERLE values of the corresponding error signals were computed by (3.81), and are given by 14.56 dB (linear), 19.14 dB (Volterra), and 29.54 dB (proposed). Clearly, the proposed approach achieves a significant improvement over a time domain approach. This may be attributable to the long memory of the system's nonlinear components which necessitate long kernels for sufficient modeling of the acoustic path. Expectedly, a purely linear model does not provide a sufficient echo attenuation due to nonlinear undermodeling. Subjective listening tests confirm that the proposed approach achieves a perceptual improvement in speech quality over the conventional Volterra approach (audio files are available online [65]).

3.7 Conclusions

Motivated by the common drawbacks of conventional time- and frequency-domain methods, we have introduced a novel approach for identifying nonlinear systems in the STFT domain. We have derived an explicit nonlinear model, based on an efficient approximation of Volterra-filters representation in the time-frequency domain. The proposed model consists of a parallel combination of a linear component, which is represented by crossband filters between subbands, and a nonlinear component, modeled by multiplicative cross-terms. We showed that the conventional discrete frequency-domain model is a special case of the proposed model for relatively long observation frames. Furthermore, we considered the identification of quadratically nonlinear systems and introduced batch and adaptive schemes for the estimation of the model parameters. We provided an explicit estimation-error analysis for both schemes and showed that incorporating the nonlinear component into the model may not necessarily imply a lower steady-state mse in subbands. In fact, the estimation of the nonlinear component improves the mse performance only for stronger and longer input signals. This improvement in performance becomes larger as the nonlinearity becomes stronger. Moreover, as the SNR increases or as more data is employed in the estimation process, whether a purely-linear or a nonlinear model is utilized, additional crossband filters should be estimated to achieve a lower mse. We further showed that a significant reduction in computational cost can be achieved over the time-domain Volterra model by the proposed approach.

Experimental results have demonstrated the advantage of the proposed STFT model in estimating nonlinear systems with relatively large memory length. The time-domain Volterra model fails to estimate such systems due to its high complexity. The proposed model, on the other hand, achieves a significant improvement in mse performance, particularly for high SNR conditions. Overall, the results have met the expectations originally put into STFT-based estimation techniques. The proposed approach in the STFT domain offers both structural generality and computational efficiency, and consequently facilitates a practical alternative for conventional methods.

Appendix: The STFT Representation of the Quadratic Volterra Component

Using (3.13) and (3.14), the STFT of $d_2(n)$ can be written as

$$d_{2;p,k} = \sum_{n,m,\ell} h_2(m,\ell)x(n-m)x(n-\ell)\,\tilde{\psi}^*_{p,k}(n)\,. \qquad (3.82)$$

Substituting (3.16) into (3.82), we obtain

$$d_{2;p,k} = \sum_{n,m,\ell} h_2(m,\ell) \sum_{k'=0}^{N-1} \sum_{p'} x_{p',k'} \psi_{p',k'}(n-m)$$

$$\times \sum_{k''=0}^{N-1} \sum_{p''} x_{p'',k''} \psi_{p'',k''}(n-\ell) \tilde{\psi}_{p,k}^*(n)$$

$$= \sum_{k',k''=0}^{N-1} \sum_{p',p''} x_{p',k'} x_{p'',k''} c_{p,p',p'',k,k',k''}, \tag{3.83}$$

where

$$c_{p,p',p'',k,k',k''} = \sum_{n,m,\ell} h_2(m,\ell) \psi_{p',k'}(n-m) \psi_{p'',k''}(n-\ell) \tilde{\psi}_{p,k}^*(n). \tag{3.84}$$

Substituting (3.15) and (3.17) into (3.84), we obtain

$$c_{p,p',p'',k,k',k''} = \sum_{n,m,\ell} h_2(m,\ell) \psi(n-m-p'L) e^{j\frac{2\pi}{N}k'(n-m-p'L)} \tilde{\psi}^*(n-pL)$$

$$\times \psi(n-\ell-p''L) e^{j\frac{2\pi}{N}k''(n-\ell-p''L)} e^{-j\frac{2\pi}{N}k(n-pL)}$$

$$= \sum_{n,m,\ell} h_2(m,\ell) \psi\left[(p-p')L + n - m\right] e^{j\frac{2\pi}{N}k'\left[(p-p')L+n-m\right]}$$

$$\times \psi\left[(p-p'')L + n - \ell\right] e^{j\frac{2\pi}{N}k''\left[(p-p'')L+n-\ell\right]}$$

$$\times \tilde{\psi}^*(n) e^{-j\frac{2\pi}{N}kn}$$

$$= \left. \{h_2(n,m) * f_{k,k',k''}(n,m)\} \right|_{n=(p-p')L,\ m=(p-p'')L}$$

$$\triangleq c_{p-p',p-p'',k,k',k''},$$

$$\tag{3.85}$$

where $*$ denotes a 2D convolution with respect to the time indices n and m, and

$$f_{k,k',k''}(n,m) \triangleq \sum_{\ell} \tilde{\psi}^*(\ell) e^{-j\frac{2\pi}{N}k\ell} \psi(n+\ell) e^{j\frac{2\pi}{N}k'(n+\ell)} \psi(m+\ell) e^{j\frac{2\pi}{N}k''(m+\ell)}.$$

$$\tag{3.86}$$

From (3.85), $c_{p,p',p'',k,k',k''}$ depends on $(p-p')$ and $(p-p'')$ rather than on p, p' and p'' separately. Substituting (3.85) into (3.83), we obtain (3.23)–(3.25).

References

1. A. Stenger and W. Kellermann, "Adaptation of a memoryless preprocessor for nonlinear acoustic echo cancelling," *Signal Processing*, vol. 80, no. 9, pp. 1747–1760, 2000.

2. A. Guérin, G. Faucon, and R. L. Bouquin-Jeannès, "Nonlinear acoustic echo cancellation based on Volterra filters," *IEEE Trans. Speech Audio Processing*, vol. 11, no. 6, pp. 672–683, Nov. 2003.

3. H. Dai and W. P. Zhu, "Compensation of loudspeaker nonlinearity in acoustic echo cancellation using raised-cosine function," *IEEE Trans. Circuits Syst. II*, vol. 53, no. 11, pp. 1190–1194, Nov. 2006.

4. S. Benedetto and E. Biglieri, "Nonlinear equalization of digital satellite channels," *IEEE J. Select. Areas Commun.*, vol. SAC-1, pp. 57–62, Jan. 1983.

5. D. G. Lainiotis and P. Papaparaskeva, "A partitioned adaptive approach to nonlinear channel equalization," *IEEE Trans. Commun.*, vol. 46, no. 10, pp. 1325–1336, Oct. 1998.

6. D. T. Westwick and R. E. Kearney, "Separable least squares identification of nonlinear Hammerstein models: Application to stretch reflex dynamics," *Ann. Biomed. Eng.*, vol. 29, no. 8, pp. 707–718, Aug. 2001.

7. G. Ramponi and G. L. Sicuranza, "Quadratic digital filters for image processing," *IEEE Trans. Acoust., Speech, Signal Processing*, vol. 36, no. 6, pp. 937–939, Jun. 1988.

8. F. Gao and W. M. Snelgrove, "Adaptive linearization of a loudpseaker," in *Proc. IEEE Int. Conf. Acoust., Speech, Signal Processing*, Toronto, Canada, May 1991, pp. 3589–3592.

9. W. J. Rugh, *Nonlinear System Theory: The Volterra-Wiener Approach*. John Hopkins Univ. Press, 1981.

10. T. Koh and E. J. Powers, "Second-order Volterra filtering and its application to nonlinear system identification," *IEEE Trans. Acoust., Speech, Signal Processing*, vol. ASSP-33, no. 6, pp. 1445–1455, Dec. 1985.

11. M. Schetzen, *The Volterra and Wiener Theories of Nonlinear Systems*. New York: Krieger, 1989.

12. V. J. Mathews, "Adaptive polynomial filters," *IEEE Signal Processing Mag.*, vol. 8, no. 3, pp. 10–26, Jul. 1991.

13. G. O. Glentis, P. Koukoulas, and N. Kalouptsidis, "Efficient algorithms for Volterra system identification," *IEEE Trans. Signal Processing*, vol. 47, no. 11, pp. 3042–3057, Nov. 1999.

14. V. J. Mathews and G. L. Sicuranza, *Polynomial Signal Processing*. New York: Wiley, 2000.

15. A. Fermo, A. Carini, and G. L. Sicuranza, "Simplified Volterra filters for acoustic echo cancellation in GSM receivers," in *Proc. European Signal Processing Conf.*, Tampere, Finland, 2000, pp. 2413–2416.

16. A. Stenger., L. Trautmann, and R. Rabenstein, "Nonlinear acoustic echo cancellation with 2nd order adaptive Volterra filters," in *Proc. IEEE Int. Conf. Acoust., Speech, Signal Processing*, Phoenix, USA, Mar. 1999, pp. 877–880.

17. E. Biglieri, A. Gersho, R. D. Gitlin, and T. L. Lim, "Adaptive cancellation of nonlinear intersymbol interference for voiceband data transmission," *IEEE J. Select. Areas Commun.*, vol. 2, no. 5, pp. 765–777, Sept. 1984.

18. R. D. Nowak, "Penalized least squares estimation of Volterra filters and higher order statistics," *IEEE Trans. Signal Processing*, vol. 46, no. 2, pp. 419–428, Feb. 1998.

19. R. D. Nowak and B. D. V. Veen, "Random and pseudorandom inputs for Volterra filter identification," *IEEE Trans. Signal Processing*, vol. 42, no. 8, pp. 2124–2135, Aug. 1994.

20. F. Kuech and W. Kellermann, "Orthogonalized power filters for nonlinear acoustic echo cancellation," *Signal Processing*, vol. 86, no. 6, pp. 1168–1181, Jun. 2006.

21. E. W. Bai and M. Fu, "A blind approach to Hammerstein model identification," *IEEE Trans. Signal Processing*, vol. 50, no. 7, pp. 1610–1619, Jul. 2002.

22. T. M. Panicker, "Parallel-cascade realization and approximation of truncated Volterra systems," *IEEE Trans. Signal Processing*, vol. 46, no. 10, pp. 2829–2832, Oct. 1998.

23. W. A. Frank, "An efficient approximation to the quadratic Volterra filter and its application in real-time loudspeaker linearization," *Signal Processing*, vol. 45, no. 1, pp. 97–113, Jul. 1995.
24. K. I. Kim and E. J. Powers, "A digital method of modeling quadratically nonlinear systems with a general random input," *IEEE Trans. Acoust., Speech, Signal Processing*, vol. 36, no. 11, pp. 1758–1769, Nov. 1988.
25. C. H. Tseng and E. J. Powers, "Batch and adaptive Volterra filtering of cubically nonlinear systems with a Gaussian input," in *IEEE Int. Symp. Circuits and Systems (ISCAS)*, vol. 1, 1993, pp. 40–43.
26. P. Koukoulas and N. Kalouptsidis, "Nonlinear system identification using Gaussian inputs," *IEEE Trans. Signal Processing*, vol. 43, no. 8, pp. 1831–1841, Aug. 1995.
27. Y. Avargel and I. Cohen, "Representation and identification of nonlinear systems in the short-time Fourier transform domain," *submitted to IEEE Trans. Signal Processing*.
28. ——, "System identification in the short-time Fourier transform domain with cross-band filtering," *IEEE Trans. Audio Speech Lang. Processing*, vol. 15, no. 4, pp. 1305–1319, May 2007.
29. C. Faller and J. Chen, "Suppressing acoustic echo in a spectral envelope space," *IEEE Trans. Acoust., Speech, Signal Processing*, vol. 13, no. 5, pp. 1048–1062, Sept. 2005.
30. Y. Lu and J. M. Morris, "Gabor expansion for adaptive echo cancellation," *IEEE Signal Processing Mag.*, vol. 16, pp. 68–80, Mar. 1999.
31. Y. Avargel and I. Cohen, "Linear system identification in the short-time fourier transform domain," in *Speech Processing in Modern Communication: Challenges and Perspectives*, I. Cohen, J. Benesty, and S. Gannot, Eds. Berlin, Germany: Springer, 2009.
32. A. Gilloire and M. Vetterli, "Adaptive filtering in subbands with critical sampling: Analysis, experiments, and application to acoustic echo cancellation," *IEEE Trans. Signal Processing*, vol. 40, no. 8, pp. 1862–1875, Aug. 1992.
33. Y. Avargel and I. Cohen, "Adaptive system identification in the short-time Fourier transform domain using cross-multiplicative transfer function approximation," *IEEE Trans. Audio Speech Lang. Processing*, vol. 16, no. 1, pp. 162–173, Jan. 2008.
34. ——, "Nonlinear systems in the short-time Fourier transform domain: Estimation error analysis," *submitted*.
35. ——, "Adaptive nonlinear system identification in the short-time Fourier transform domain," *to appear in IEEE Trans. Signal Processing*.
36. L. Ljung, *System Identification: Theory for the User*. Upper Saddle River, New Jersey: Prentice-Hall, 1999.
37. S. Haykin, *Adaptive Filter Theory*. New Jersey: Prentice-Hall, 2002.
38. M. B. Priestley, *Spectral Analysis and Time Series*. New York: Academic, 1981.
39. J. M. Mendel, "Tutorial on higher-order statistics (spectra) in signal processing and system theory: Theoretical results and some applications," *Proc. IEEE*, vol. 79, no. 3, pp. 278–305, Mar. 1991.
40. S. W. Nam, S. B. Kim, and E. J. Powers, "On the identification of a third-order Volterra nonlinear systems using a freuqnecy-domain block RLS adaptive algorithm," in *Proc. IEEE Int. Conf. Acoust. Speech, Signal Processing*, Albuquerque, New Mexico, Apr. 1990, pp. 2407 – 2410.
41. M. R. Portnoff, "Time-frequency representation of digital signals and systems based on short-time Fourier analysis," *IEEE Trans. Signal Processing*, vol. ASSP-28, no. 1, pp. 55–69, Feb. 1980.
42. J. Wexler and S. Raz, "Discrete Gabor expansions," *Signal Processing*, vol. 21, pp. 207–220, Nov. 1990.
43. Y. Avargel and I. Cohen, "Identification of linear systems with adaptive control of the cross-multiplicative transfer function approximation," in *Proc. IEEE Int. Conf. Acoust. Speech, Signal Processing*, Las Vegas, Nevada, Apr. 2008, pp. 3789–3792.

44. ——, "On multiplicative transfer function approximation in the short-time Fourier transform domain," *IEEE Signal Processing Lett.*, vol. 14, no. 5, pp. 337–340, May 2007.
45. ——, "Nonlinear acoustic echo cancellation based on a multiplicative transfer function approximation," in *Proc. Int. Workshop Acoust. Echo Noise Control (IWAENC)*, Seattle, WA, USA, Sep. 2008, pp. 1 – 4, paper no. 9035.
46. C. Breining, P. Dreiseitel, E. Hänsler, A. Mader, B. Nitsch, H. Puder, T. Schertler, G. Schmidt, and J. Tlip, "Acoustic echo control," *IEEE Signal Processing Mag.*, vol. 16, no. 4, pp. 42–69, Jul. 1999.
47. G. L. Sicuranza, "Quadratic filters for signal processing," *Proc. IEEE*, vol. 80, no. 8, pp. 1263–1285, Aug. 1992.
48. A. Neumaier, "Solving ill-conditioned and singular linear systems: A tutorial on regularization," *SIAM Rev.*, vol. 40, no. 3, pp. 636–666, Sep. 1998.
49. G. H. Golub and C. F. V. Loan, *Matrix Computations*. Baltimore, MD: The Johns Hopkins University Press, 1996.
50. A. E. Nordsjo, B. M. Ninness, and T. Wigren, "Quantifying model errors caused by nonlinear undermodeling in linear system identification," in *Preprints 13th IFAC World Congr.*, San Francisco, CA, Jul. 1996, pp. 145–149.
51. B. Ninness and S. Gibson, "Quantifying the accuracy of Hammerstein model estimation," *Automatica*, vol. 38, no. 12, pp. 2037–2051, 2002.
52. J. Schoukens, R. Pintelon, T. Dobrowiecki, and Y. Rolain, "Identification of linear systems with nonlinear distortions," *Automatica*, vol. 41, no. 3, pp. 491–504, 2005.
53. A. Papoulis, *Probability, Random Variables, and Stochastic Processes*. Singapore: McGraw-Hill, 1991.
54. F. D. Ridder, R. Pintelon, J. Schoukens, and D. P. Gillikin, "Modified AIC and MDL model selection criteria for short data records," *IEEE Trans. Instrum. Meas.*, vol. 54, no. 1, pp. 144–150, Feb. 2005.
55. G. Schwarz, "Estimating the dimension of a model," *Ann. Stat.*, vol. 6, no. 2, pp. 461–464, 1978.
56. P. Stoica and Y. Selen, "Model order selection: a review of information criterion rules," *IEEE Signal Processing Mag.*, vol. 21, no. 4, pp. 36–47, Jul. 2004.
57. G. C. Goodwin, M. Gevers, and B. Ninness, "Quantifying the error in estimated transfer functions with application to model order selection," *IEEE Trans. Automat. Contr.*, vol. 37, no. 7, pp. 913–928, Jul. 1992.
58. H. Akaike, "A new look at the statistical model identification," *IEEE Trans. Automat. Contr.*, vol. AC-19, no. 6, pp. 716–723, Dec. 1974.
59. J. Rissanen, "Modeling by shortest data description," *Automatica*, vol. 14, no. 5, pp. 465–471, 1978.
60. L. L. Horowitz and K. D. Senne, "Perforamce advantage of complex LMS for controlling narrow-band adaptive arrays," *IEEE Trans. Acoust., Speech, Signal Processing*, vol. ASSP-29, no. 3, pp. 722–736, Jun. 1981.
61. K. Mayyas, "Performance analysis of the deficient length LMS adaptive algorithm," *IEEE Trans. Signal Processing*, vol. 53, no. 8, pp. 2727–2734, Aug. 2005.
62. D. W. Griffin and J. S. Lim, "Signal estimation from modified short-time Fourier transform," *IEEE Trans. Acoust., Speech, Signal Processing*, vol. ASSP-32, no. 2, pp. 236–243, Apr. 1984.
63. J. Benesty, D. R. Morgan, and J. H. Cho, "A new class of doubletalk detectors based on cross-correlation," *IEEE Trans. Speech Audio Processing*, vol. 8, no. 2, pp. 168–172, Mar. 2000.
64. J. H. Cho, D. R. Morgan, and J. Benesty, "An objective technique for evaluating doubletalk detectors in acoustic echo cancelers," *IEEE Trans. Speech Audio Processing*, vol. 7, no. 6, pp. 718–724, Nov. 1999.
65. Y. Avargel Homepage. [Online]. Available: http://sipl.technion.ac.il/~yekutiel

Chapter 4
Variable Step-Size Adaptive Filters for Echo Cancellation

Constantin Paleologu, Jacob Benesty, and Silviu Ciochină

Abstract The principal issue in acoustic echo cancellation (AEC) is to estimate the impulse response between the loudspeaker and the microphone of a hands-free communication device. Basically, we deal with a system identification problem, which can be solved by using an adaptive filter. The most common adaptive filters for AEC are the normalized least-mean-square (NLMS) algorithm and the affine projection algorithm (APA). These two algorithms have to compromise between fast convergence rate and tracking on the one hand and low misadjustment and robustness (against the near-end signal) on the other hand. In order to meet these conflicting requirements, the step-size parameter of the algorithms needs to be controlled. This is the motivation behind the development of variable step-size (VSS) algorithms. Unfortunately, most of these algorithms depend on some parameters that are not always easy to tune in practice. The goal of this chapter is to present a family of non-parametric VSS algorithms (including both VSS-NLMS and VSS-APA), which are very suitable for realistic AEC scenarios. They are developed based on another objective of AEC application, i.e., to recover the near-end signal from the error signal of the adaptive filter. As a consequence, these VSS algorithms are equipped with good robustness features against near-end signal variations, like double-talk. This idea can be extended even in the case of the recursive least-squares (RLS) algorithm, where the overall performance is controlled by a forgetting factor. Therefore, a variable forgetting factor RLS (VFF-RLS) algorithm is also presented. The simulation results indicate that these algorithms are reliable candidates for real-world applications.

Constantin Paleologu
University Politehnica of Bucharest, Romania, e-mail: `pale@comm.pub.ro`

Jacob Benesty
INRS-EMT, QC, Canada, e-mail: `benesty@emt.inrs.ca`

Silviu Ciochină
University Politehnica of Bucharest, Romania, e-mail: `silviu@comm.pub.ro`

I. Cohen et al. (Eds.): Speech Processing in Modern Communication, STSP 3, pp. 89–125.
springerlink.com   © Springer-Verlag Berlin Heidelberg 2010

4.1 Introduction

Nowadays, hands-free communication devices are involved in many popular applications, such as mobile telephony and teleconferencing systems. Due to their specific features, they can be used in a wide range of environments with different acoustic characteristics. In this context, an important issue that has to be addressed when dealing with such devices is the acoustic coupling between the loudspeaker and microphone. In other words, besides the voice of the near-end speaker and the background noise, the microphone of the hands-free equipment captures another signal (coming from its own loudspeaker), known as the acoustic echo. Depending of the environment's characteristics, this phenomenon can be very disturbing for the far-end speaker, which hears a replica of her/his own voice. From this point of view, there is a need to enhance the quality of the microphone signal by cancelling the unwanted acoustic echo.

The most reliable solution to this problem is the use of an adaptive filter that generates at its output a replica of the echo, which is further subtracted from the microphone signal [1], [2], [3]. Basically, the adaptive filter has to model an unknown system, i.e., the acoustic echo path between the loudspeaker and microphone, like in a "system identification" problem [4], [5], [6]. Even if this formulation is straightforward, the specific features of acoustic echo cancellation (AEC) represent a challenge for any adaptive algorithm.

First, the acoustic echo paths have excessive lengths in time (up to hundreds of milliseconds), due to the slow speed of sound in the air, together with multiple reflections caused by the environment; consequently, long length adaptive filters are required (hundreds or even thousands of coefficients), influencing the convergence rate of the algorithm. Also, the acoustic echo paths are time-variant systems (depending on temperature, pressure, humidity, and movement of objects or bodies), requiring good tracking capabilities for the echo canceller. As a consequence of these aspects related to the acoustic echo path characteristics, the adaptive filter works most likely in an under-modeling situation, i.e., its length is smaller than the length of the acoustic impulse response. Hence, the residual echo caused by the part of the system that can not be modeled acts like an additional noise and disturbs the overall performance. Second, the echo signal is combined with the near-end signal; ideally, the adaptive filter should separate this mixture and provide an estimate of the echo at its output and an estimate of the near-end from the error signal (from this point of view, the adaptive filter works as in an "interference cancelling" configuration [4], [5], [6]). This is not an easy task, since the near-end signal can contain both the background noise and the near-end speech; the background noise can be non-stationary and strong (it is also amplified by the microphone of the hands-free device), while the near-end speech acts like a large level disturbance. Last but not least, the input of the adaptive filter (i.e., the far-end signal) is mainly speech, which is a non-stationary and highly correlated signal that can influence the overall performance of the adaptive algorithm.

Maybe the most challenging situation in echo cancellation is the double-talk case, i.e., the talkers on both sides speak simultaneously. The behavior of the adaptive filter can be seriously affected in this case, up to divergence. For this reason, the echo canceller is usually equipped with a double-talk detector (DTD), in order to slow down or completely halt the adaptation process during double-talk periods [1], [2]. Nevertheless, there is some inherent delay in the decision of any DTD; during this small period, a few undetected large amplitude samples can perturb the echo path estimate considerably. Consequently, it is highly desirable to improve the robustness of the adaptive algorithm in order to handle a certain amount of double-talk without diverging.

There is not a perfect algorithm suitable for echo cancellation. Different types of algorithms are involved in the context of this application [1], [2], [3], [7]. One of the most popular is the normalized least-mean-square (NLMS) algorithm [4], [5], [6]. The main reasons behind this popularity are its moderate computational complexity, together with its good numerical stability. Also, the affine projection algorithm (APA) (originally proposed in [8]) and some of its versions, e.g., [9], [10], [11], were found to be very attractive choices for AEC applications. The main advantage of the APA over the NLMS algorithm consists of a superior convergence rate, especially for speech signals.

The classical NLMS algorithm and APA use a step-size parameter to control their performances. Nevertheless, a compromise should be made when choosing this parameter; a large value implies fast convergence rate and tracking, while a small value leads to low misadjustment and good robustness features. Since in echo cancellation there is a need for all these performance criteria, the step-size should be controlled. Hence, a variable step-size (VSS) algorithm represents a more proper choice. Different types of VSS-NLMS algorithms and VSS-APAs were developed (e.g., see [12], [13], [14], [15] and references therein). In general, most of them require the tuning of some parameters which are difficult to control in practice. For real-world AEC applications, it is highly desirable to use non-parametric algorithms, in the sense that no information about the acoustic environment is required.

One of the most interesting VSS adaptive filters is the non-parametric VSS-NLMS (NPVSS-NLMS) algorithm proposed in [14]. It was developed in a system identification context, aiming to recover the system noise (i.e., the noise that corrupts the output of the unknown system) from the error of the adaptive filter. In the context of echo cancellation, this system noise is the near-end signal. If only the background noise is considered, its power estimate (which is needed in the step-size formula of the NPVSS-NLMS algorithm) is easily obtained during silences of the near-end talker. This was the framework of the experimental results reported in [14], showing that the NPVSS-NLMS algorithm is very efficient and easy to control in practice. Nevertheless, the main challenge is the case when the near-end signal contains not only the background noise but also the near-end speech (i.e., double-talk scenario). Inspired by the original idea of the NPVSS-NLMS algorithm, several approaches have focused on finding other practical solutions for this problem. Consequently, different VSS-NLMS algorithms and also VSS-APA have been developed [16], [17], [18], [19]. Furthermore, the basic idea can be adapted to the recursive least-squares (RLS) algorithm, where the overall performance is controlled by the forgetting factor [4], [5], [6]; consequently, a variable forgetting factor RLS (VFF-RLS) algorithm was proposed in [20].

The main objective of this chapter is to present and study in a unified way the VSS algorithms. Their capabilities and performances are analyzed in terms of the classical criteria (e.g., convergence rate, tracking, robustness), but also from other practical points of view (e.g., available parameters, computational complexity). The rest of this chapter is organized as follows. Section 4.2 describes the derivation of the original NPVSS-NLMS algorithm. Several VSS-NLMS algorithms from the same family are presented in Section 4.3. A generalization to the VSS-APA is given in Section 4.4. An extension to the VFF-RLS algorithm is developed in Section 4.5. Simulation results are shown in Section 4.6. Finally, Section 4.7 concludes this work.

4.2 Non-Parametric VSS-NLMS Algorithm

Let us consider the framework of a system identification problem [4], [5], [6], where we try to model an unknown system using an adaptive filter, both driven by the same zero-mean input signal $x(n)$, where n is the time index. These two systems are assumed to be finite impulse response (FIR) filters of length L, defined by the real-valued vectors

$$\mathbf{h} = \begin{bmatrix} h_0 & h_1 & \cdots & h_{L-1} \end{bmatrix}^T,$$

$$\hat{\mathbf{h}}(n) = \begin{bmatrix} \hat{h}_0(n) & \hat{h}_1(n) & \cdots & \hat{h}_{L-1}(n) \end{bmatrix}^T,$$

where superscript ^{T} denotes transposition. The desired signal for the adaptive filter is

$$d(n) = \mathbf{x}^T(n)\mathbf{h} + \nu(n), \tag{4.1}$$

where

$$\mathbf{x}(n) = \begin{bmatrix} x(n) & x(n-1) & \cdots & x(n-L+1) \end{bmatrix}^T$$

is a real-valued vector containing the L most recent samples of the input signal $x(n)$ and $\nu(n)$ is the system noise [assumed to be quasi-stationary, zero mean, and independent of the input signal $x(n)$] that corrupts the output of the unknown system.

Using the previous notation we may define the a priori and a posteriori error signals as

$$e(n) = d(n) - \mathbf{x}^T(n)\hat{\mathbf{h}}(n-1) \tag{4.2}$$
$$= \mathbf{x}^T(n)\left[\mathbf{h} - \hat{\mathbf{h}}(n-1)\right] + \nu(n),$$
$$\varepsilon(n) = d(n) - \mathbf{x}^T(n)\hat{\mathbf{h}}(n) \tag{4.3}$$
$$= \mathbf{x}^T(n)\left[\mathbf{h} - \hat{\mathbf{h}}(n)\right] + \nu(n),$$

where the vectors $\hat{\mathbf{h}}(n-1)$ and $\hat{\mathbf{h}}(n)$ contain the adaptive filter coefficients at time $n-1$ and n, respectively. The well-known update equation for NLMS-type algorithms is

$$\hat{\mathbf{h}}(n) = \hat{\mathbf{h}}(n-1) + \mu(n)\mathbf{x}(n)e(n), \tag{4.4}$$

where $\mu(n)$ is a positive factor known as the step-size, which governs the stability, the convergence rate, and the misadjustment of the algorithm. A reasonable way to derive $\mu(n)$, taking into account the stability conditions, is to cancel the a posteriori error signal [21]. Replacing (4.4) in (4.3) with the requirement $\varepsilon(n) = 0$, it results that

$$\varepsilon(n) = e(n)\left[1 - \mu(n)\mathbf{x}^T(n)\mathbf{x}(n)\right] = 0, \tag{4.5}$$

and assuming that $e(n) \neq 0$, we find

$$\mu_{\mathrm{NLMS}}(n) = \frac{1}{\mathbf{x}^T(n)\mathbf{x}(n)}, \tag{4.6}$$

which represents the step-size of the classical NLMS algorithm; in practice, a positive constant (usually smaller than 1) multiplies this step-size to achieve a proper compromise between the convergence rate and the misadjustment [4], [5], [6]. According to (4.4) and (4.6), the update equation is given by

$$\hat{\mathbf{h}}(n) = \hat{\mathbf{h}}(n-1) + \frac{\mathbf{x}(n)}{\mathbf{x}^T(n)\mathbf{x}(n)}e(n). \tag{4.7}$$

We should note that the above procedure makes sense in the absence of noise [i.e., $\nu(n) = 0$], where the condition $\varepsilon(n) = 0$ implies that $\mathbf{x}^T(n)\left[\mathbf{h} - \hat{\mathbf{h}}(n)\right] = 0$. Finding the parameter $\mu(n)$ in the presence of noise will introduce noise in $\hat{\mathbf{h}}(n)$, since the condition $\varepsilon(n) = 0$ leads to $\mathbf{x}^T(n)\left[\mathbf{h} - \hat{\mathbf{h}}(n)\right] = -\nu(n) \neq 0$. In fact, we would like to have $\mathbf{x}^T(n)\left[\mathbf{h} - \hat{\mathbf{h}}(n)\right] = 0$, which implies that $\varepsilon(n) = \nu(n)$. Consequently, the step-size parameter $\mu(n)$ is found imposing the condition

$$E\left[\varepsilon^2(n)\right] = \sigma_\nu^2, \tag{4.8}$$

where $E[\cdot]$ denotes mathematical expectation and $\sigma_\nu^2 = E\left[\nu^2(n)\right]$ is the power of the system noise. Taking into account the fact that $\mu(n)$ is deterministic in nature, we have to follow condition (4.8) in order to develop an explicit relation for the step-size parameter. Following this requirement, we rewrite (4.5) as

$$\varepsilon(n) = e(n)\left[1 - \mu(n)\mathbf{x}^T(n)\mathbf{x}(n)\right] = \nu(n). \tag{4.9}$$

Squaring the previous equation, then taking the mathematical expectation of both sides and using the approximation

$$\mathbf{x}^T(n)\mathbf{x}(n) = L\sigma_x^2 = LE\left[x^2(n)\right], \text{ for } L \gg 1, \tag{4.10}$$

where σ_x^2 is the power of the input signal, it results

$$\left[1 - \mu(n)L\sigma_x^2\right]^2 \sigma_e^2(n) = \sigma_\nu^2, \tag{4.11}$$

where $\sigma_e^2(n) = E\left[e^2(n)\right]$ is the power of the error signal. Thus, developing (4.11), we obtain a quadratic equation

$$\mu^2(n) - \frac{2}{L\sigma_x^2}\mu(n) + \frac{1}{(L\sigma_x^2)^2}\left[1 - \frac{\sigma_\nu^2}{\sigma_e^2(n)}\right] = 0, \tag{4.12}$$

for which the obvious solution is

$$\begin{aligned}
\mu_{\mathrm{NPVSS}}(n) &= \frac{1}{\mathbf{x}^T(n)\mathbf{x}(n)}\left[1 - \frac{\sigma_\nu}{\sigma_e(n)}\right] \\
&= \mu_{\mathrm{NLMS}}(n)\alpha(n),
\end{aligned} \tag{4.13}$$

where $\alpha(n)$ [with $0 \le \alpha(n) \le 1$] is the normalized step-size. Therefore, the non-parametric variable step-size NLMS (NPVSS-NLMS) algorithm [14] is

$$\hat{\mathbf{h}}(n) = \hat{\mathbf{h}}(n-1) + \mu_{\mathrm{NPVSS}}(n)\mathbf{x}(n)e(n). \tag{4.14}$$

Looking at (4.13) it is obvious that before the algorithm converges, $\sigma_e(n)$ is large compared to σ_ν and consequently $\mu_{\mathrm{NPVSS}}(n) \approx \mu_{\mathrm{NLMS}}(n)$. When the algorithm starts to converge to the true solution, $\sigma_e(n) \approx \sigma_\nu$ and $\mu_{\mathrm{NPVSS}}(n) \approx 0$. In fact, this is the desired behaviour for the adaptive algorithm, leading to both fast convergence and low misadjustment. Moreover, the NPVSS-NLMS algorithm was derived with almost no assumptions compared to most of the members belonging to the family of variable step-size NLMS algorithms.

Similar at first glance with the NPVSS-NLMS, a so-called set membership NLMS (SM-NLMS) algorithm was proposed earlier in [22]. The step-size of this algorithm is

$$\mu_{\mathrm{SM}}(n) = \begin{cases} \mu_{\mathrm{NLMS}}(n)\left[1 - \frac{\eta}{|e(n)|}\right], & \text{if } |e(n)| > \eta \\ \\ 0, & \text{otherwise} \end{cases}, \qquad (4.15)$$

where the parameter η represents a bound on the noise. Nevertheless, since there is no averaging on $|e(n)|$, we cannot expect a low misadjustment as for NPVSS-NLMS algorithm. Simulations performed in [14] show that the NPVSS-NLMS outperforms, and by far, the SM-NLMS algorithm, which in fact achieves only a slight performance improvement over the classical NLMS.

In order to analyze the convergence of the misalignment for the NPVSS-NLMS algorithm, we suppose that the system is stationary. Defining the misalignment vector at time n as

$$\mathbf{m}(n) = \mathbf{h} - \hat{\mathbf{h}}(n), \qquad (4.16)$$

the update equation of the algorithm (4.14) can be rewritten in terms of the misalignment as

$$\mathbf{m}(n) = \mathbf{m}(n-1) - \mu_{\mathrm{NPVSS}}(n)\mathbf{x}(n)e(n). \qquad (4.17)$$

Taking the l_2 norm in (4.17), then the mathematical expectation of both sides and assuming that

$$E\left[\nu(n)\mathbf{x}^T(n)\mathbf{m}(n-1)\right] = 0, \qquad (4.18)$$

which is true if $\nu(n)$ is white, we obtain

$$E\left[\|\mathbf{m}(n)\|_2^2\right] - E\left[\|\mathbf{m}(n-1)\|_2^2\right] = -\mu_{\mathrm{NPVSS}}(n)\left[\sigma_e(n) - \sigma_\nu\right]\left[\sigma_e(n) + 2\sigma_\nu\right]$$
$$\leq 0. \qquad (4.19)$$

The previous expression proves that the length of the misalignment vector for the NPVSS-NLMS algorithm is nonincreasing, which implies that

$$\lim_{n\to\infty} \sigma_e^2(n) = \sigma_\nu^2. \qquad (4.20)$$

It should be noticed that the previous relation does not imply that $E\left[\|\mathbf{m}(\infty)\|_2^2\right] = 0$. However, under the independence assumption, we can show the equivalence. Indeed, from

$$e(n) = \mathbf{x}^T(n)\mathbf{m}(n-1) + \nu(n), \qquad (4.21)$$

it can be shown that

$$E\left[e^2(n)\right] = \sigma_\nu^2 + \mathrm{tr}\left[\mathbf{R}\mathbf{K}(n-1)\right] \qquad (4.22)$$

if $\mathbf{x}(n)$ are independent, where $\mathrm{tr}[\cdot]$ is the trace of a matrix, $\mathbf{R} = E\left[\mathbf{x}(n)\mathbf{x}^T(n)\right]$, and $\mathbf{K}(n-1) = E\left[\mathbf{m}(n-1)\mathbf{m}^T(n-1)\right]$. Taking (4.20) into account, (4.22) becomes

$$\mathrm{tr}\left[\mathbf{R}\mathbf{K}(\infty)\right] = 0. \tag{4.23}$$

Assuming that $\mathbf{R} > 0$, it results that $\mathbf{K}(\infty) = \mathbf{0}$ and consequently

$$E\left[\|\mathbf{m}(\infty)\|_2^2\right] = 0. \tag{4.24}$$

Finally, some practical considerations have to be stated. First, in order to avoid division by small numbers, all NLMS-based algorithms need to be regularized by adding a positive constant δ to the denominator of the step-size parameter. In general, this regularization parameter is chosen proportional to the input signal variance as $\delta = \mathrm{constant} \cdot \sigma_x^2$. A second consideration is with regards to the estimation of the parameter $\sigma_e(n)$. In practice, the power of the error signal is estimated as follows:

$$\hat{\sigma}_e^2(n) = \lambda\hat{\sigma}_e^2(n-1) + (1-\lambda)e^2(n), \tag{4.25}$$

where λ is a weighting factor. Its value is chosen as $\lambda = 1 - 1/(KL)$, where $K > 1$. The initial value for (4.25) is $\hat{\sigma}_e^2(0) = 0$. Theoretically, it is clear that $\sigma_e^2(n) \geq \sigma_\nu^2$, which implies that $\mu_{\mathrm{NPVSS}}(n) \geq 0$. Nevertheless, the estimation from (4.25) could result in a lower magnitude than the noise power estimate, $\hat{\sigma}_\nu^2$, which would make $\mu_{\mathrm{NPVSS}}(n)$ negative. In this situation, the problem is solved by setting $\mu_{\mathrm{NPVSS}}(n) = 0$.

The NPVSS-NLMS algorithm is summarizes in Table 4.1. Taking into account (4.13), (4.15), and (4.25), it is clear that if we set $\lambda = 0$ (i.e., no averaging of the error signal) and $\eta = \sigma_\nu$, it will result the SM-NLMS algorithm [22] mentioned before.

4.3 VSS-NLMS Algorithms for Echo Cancellation

Let us reformulate the system identification problem from the previous section in the context of the AEC application, as depicted in Fig. 4.1. The far-end signal $x(n)$ goes through the echo path $\mathbf{h}$, providing the echo signal $y(n)$. This signal is added to the near-end signal $\nu(n)$ [which can contain both the background noise, $w(n)$, and the near-end speech, $u(n)$], resulting the microphone signal $d(n)$. The adaptive filter, defined by the vector $\hat{\mathbf{h}}(n)$, aims to produce at its output an estimate of the echo, $\hat{y}(n)$, while the error signal $e(n)$ should contain an estimate of the near-end signal.

It can be noticed that the AEC scheme from Fig. 4.1 can be interpreted as a combination of two classes of adaptive system configurations (according to the adaptive filter theory [4], [5], [6]). First, it represents a "system identification" configuration, because the goal is to identify an unknown system (i.e., the acoustic echo path, $\mathbf{h}$) with its output corrupted by an apparently "un-

Table 4.1 NPVSS-NLMS algorithm.

Initialization:

$\hat{\mathbf{h}}(0) = \mathbf{0}_{L \times 1}$

$\hat{\sigma}_e^2(0) = 0$

Parameters:

$\lambda = 1 - \dfrac{1}{KL}$, weighting factor with $K > 1$

$\hat{\sigma}_\nu^2$, noise power known or estimated

$\delta > 0$, regularization

$\zeta > 0$, very small number to avoid division by zero

For time index $n = 1, 2, \ldots$:

$e(n) = d(n) - \mathbf{x}^T(n)\hat{\mathbf{h}}(n-1)$

$\hat{\sigma}_e^2(n) = \lambda \hat{\sigma}_e^2(n-1) + (1-\lambda)e^2(n)$

$\alpha(n) = 1 - \dfrac{\hat{\sigma}_\nu}{\zeta + \hat{\sigma}_e(n)}$

$$\mu_{\mathrm{NPVSS}}(n) = \begin{cases} \alpha(n)\left[\delta + \mathbf{x}^T(n)\mathbf{x}(n)\right]^{-1}, & \text{if } \alpha(n) > 0 \\ \\ 0, & \text{otherwise} \end{cases}$$

$\hat{\mathbf{h}}(n) = \hat{\mathbf{h}}(n-1) + \mu_{\mathrm{NPVSS}}(n)\mathbf{x}(n)e(n)$

---

desired" signal [i.e., the near-end signal, $\nu(n)$]. But it also can be viewed as an "interference cancelling" configuration, aiming to recover a "useful" signal [i.e., the near-end signal, $\nu(n)$] corrupted by an undesired perturbation [i.e., the acoustic echo, $y(n)$]; consequently, the "useful" signal should be recovered from the error signal of the adaptive filter. Therefore, since the existence of the near-end signal can not be omitted in AEC, the condition (4.8) is very reasonable. This indicates that the NPVSS-NLMS algorithm should perform very well in AEC applications (especially in terms of robustness against near-end signal variations, like double-talk).

According to the development from the previous section, the only parameter that is needed in the step-size formula of the NPVSS-NLMS algorithm is the power estimate of the system noise, $\hat{\sigma}_\nu^2$. In the case of AEC, this system noise is represented by the near-end signal. Nevertheless, the estimation of the near-end signal power is not always straightforward in real-world AEC applications. Several scenarios should be considered, as follows.

1. *Single-talk scenario.* In the single-talk case, the near-end signal consists only of the background noise, i.e., $\nu(n) = w(n)$. Its power could be es-

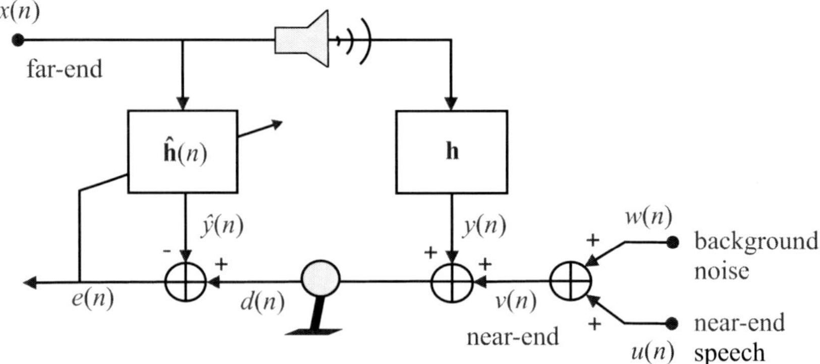

Fig. 4.1 Acoustic echo cancellation configuration.

timated (during silences of the near-end talker) and it can be assumed constant. This was the framework of the experimental results reported in [14]. Nevertheless, the background noise can be time-variant, so that the power of the background noise should be periodically estimated. Moreover, when the background noise changes between two consecutive estimations or during the near-end speech, its new power estimate will not be available immediately; consequently, until the next estimation period of the background noise, the algorithm behavior will be disturbed.

2. *Double-talk scenario.* In the double-talk case, the near-end signal consists of both the background noise and the near-end speech; so that $v(n) = w(n)+u(n)$. The main issue is to obtain an accurate estimate for the power of this combined signal, taking into account especially the non-stationary character of the speech signal.

3. *Under-modeling scenario.* In both previous cases the adaptive filter might work in an under-modeling situation, i.e., the length of $\hat{\mathbf{h}}(n)$ is smaller than the length of $\mathbf{h}$, so that an under-modeling noise appears (i.e., the residual echo caused by the part of the echo path that is not modeled). It can be interpreted as an additional noise that corrupts the near-end signal. Since it is unavailable in practice, the power of the under-modeling noise cannot be estimated in a direct manner, and consequently, it is difficult to evaluate its contribution to the near-end signal power.

Summarizing, there are several situations when the estimation of the near-end signal power is not an easy task. In order to deal with these aspects, different solutions have been developed.

A very simple way to roughly estimate the near-end signal power is to use the error signal $e(n)$, but with a larger value of the weighting factor [as compared to (4.25)], i.e.,

$$\hat{\sigma}_v^2(n) = \gamma\hat{\sigma}_v^2(n-1) + (1-\gamma)e^2(n), \tag{4.26}$$

with $\gamma > \lambda$ [23]. Let us name this algorithm the simple VSS-NLMS (SVSS-NLMS) algorithm. Since $\hat{\sigma}_\nu^2(n)$ from (4.26) can vary "around" $\hat{\sigma}_e^2(n)$, the step-size of the algorithm is computed using the absolute value, i.e.,

$$\mu_{\text{SVSS}}(n) = \frac{1}{\delta + \mathbf{x}^T(n)\mathbf{x}(n)}\left|1 - \frac{\hat{\sigma}_\nu}{\zeta + \hat{\sigma}_e(n)}\right|. \tag{4.27}$$

The value of γ influences the overall behavior of the algorithm. When γ is close to λ, the normalized step-size of the algorithm is small, which affects the convergence rate and tracking capabilities, but offers a good robustness against near-end signal variations. When γ is large as compared to λ, the convergence rate is improved, but the robustness issues are degraded.

A more elaborated method was proposed in [17]. It was demonstrated that the power estimate of the near-end signal can be evaluated as

$$\hat{\sigma}_\nu^2(n) = \hat{\sigma}_e^2(n) - \frac{1}{\hat{\sigma}_x^2(n)}\hat{\mathbf{r}}_{ex}^T(n)\hat{\mathbf{r}}_{ex}(n), \tag{4.28}$$

where the variance of $e(n)$ is estimated based on (4.25) and the other terms are evaluated in a similar manner, i.e.,

$$\hat{\sigma}_x^2(n) = \lambda\hat{\sigma}_x^2(n-1) + (1-\lambda)x^2(n), \tag{4.29}$$

$$\hat{\mathbf{r}}_{ex}(n) = \lambda\hat{\mathbf{r}}_{ex}(n-1) + (1-\lambda)\mathbf{x}(n)e(n). \tag{4.30}$$

Using (4.28), the NEW-NPVSS-NLMS algorithm proposed in [17] computes its step-size as

$$\mu_{\text{NEW}-\text{NPVSS}}(n) = \begin{cases} \left[\delta + \mathbf{x}^T(n)\mathbf{x}(n)\right]^{-1}\left[1 - \frac{\hat{\sigma}_\nu(n)}{\zeta + \hat{\sigma}_e(n)}\right], & \text{if } \xi(n) < \varsigma \\ \\ \left[\delta + \mathbf{x}^T(n)\mathbf{x}(n)\right]^{-1}, & \text{otherwise} \end{cases}, \tag{4.31}$$

where $\xi(n)$ is the convergence statistic and ς is a small positive quantity. The convergence statistic is evaluated as

$$\xi(n) = \left|\frac{\hat{r}_{ed}(n) - \hat{\sigma}_e^2(n)}{\hat{\sigma}_d^2(n) - \hat{r}_{ed}(n)}\right|, \tag{4.32}$$

where $\hat{\sigma}_e^2(n)$ is evaluated based on (4.25), while the other parameters are estimated as

$$\hat{\sigma}_d^2(n) = \lambda\hat{\sigma}_d^2(n-1) + (1-\lambda)d^2(n), \tag{4.33}$$

$$\hat{r}_{ed}(n) = \lambda\hat{r}_{ed}(n-1) + (1-\lambda)e(n)d(n). \tag{4.34}$$

It can be noticed that the step-size of the NEW-NPVSS-NLMS algorithm is computed using signals that are available, except for the constant ς. This parameter is related to the convergence state. In the initial phase of the algorithm, or when there is an echo path change, the condition $\xi(n) < \varsigma$ is not fulfilled; consequently, according to (4.31), the normalized step-size is equal to 1, forcing a fast adaptation. Otherwise, in the steady-state of the algorithm or when the near-end signal fluctuates (e.g., double-talk case), the condition $\xi(n) < \varsigma$ is fulfilled and the algorithm uses the first line of (4.31). The main problem is how to choose this convergence threshold, in terms of the value of ς. The experimental results will show that this parameter is important for the overall performance of the algorithm.

Both previous algorithms require the tuning of some parameters that influence their overall performances. In practice, it is not always easy to control such parameters. A more practical solution was proposed in [18]. It is known that the desired signal of the adaptive filter is expressed as $d(n) = y(n)+\nu(n)$. Since the echo signal and the near-end signal can be considered uncorrelated, the previous relation can be rewritten in terms of variances as

$$E\left[d^2(n)\right] = E\left[y^2(n)\right] + E\left[\nu^2(n)\right]. \tag{4.35}$$

Assuming that the adaptive filter has converged to a certain degree, we can use the approximation

$$E\left[y^2(n)\right] \approx E\left[\hat{y}^2(n)\right]. \tag{4.36}$$

Consequently, using power estimates, we may compute

$$\hat{\sigma}_\nu^2(n) \approx \hat{\sigma}_d^2(n) - \hat{\sigma}_{\hat{y}}^2(n), \tag{4.37}$$

where $\hat{\sigma}_d^2(n)$ is computed as in (4.33) and $\hat{\sigma}_{\hat{y}}^2(n)$ is evaluated in a similar manner, i.e.,

$$\hat{\sigma}_{\hat{y}}^2(n) = \lambda\hat{\sigma}_{\hat{y}}^2(n-1) + (1-\lambda)\hat{y}^2(n). \tag{4.38}$$

For the case 1., when only the background noise is present, i.e., $\nu(n) = w(n)$, an estimate of its power is obtained using the right-hand term in (4.37). This expression holds even if the level of the background noise changes, so that there is no need for the estimation of this parameter during silences of the near-end talker. For the case 2., when the near-end speech is present (assuming that it is uncorrelated with the background noise), the near-end signal variance can be expressed as $E\left[\nu^2(n)\right] = E\left[w^2(n)\right] + E\left[u^2(n)\right]$. Accordingly, the right-hand term in (4.37) still provides a power estimate of the near-end signal. Most importantly, this term depends only on the signals that are available within the AEC application, i.e., the microphone signal, $d(n)$, and the output of the adaptive filter, $\hat{y}(n)$. Consequently, the step-size of the practical VSS-NLMS (PVSS-NLMS) algorithm is evaluated as

$$\mu_{\text{PVSS}}(n) = \left[\delta + \mathbf{x}^T(n)\mathbf{x}(n)\right]^{-1}\left|1 - \frac{\sqrt{\left|\hat{\sigma}_d^2(n) - \hat{\sigma}_{\hat{y}}^2(n)\right|}}{\zeta + \hat{\sigma}_e(n)}\right|. \tag{4.39}$$

The absolute values in (4.39) prevent any minor deviations (due to the use of power estimates) from the true values, which can make the normalized step-size negative or complex. It is a non-parametric algorithm, since all the parameters in (4.39) are available. Also, good robustness against near-end signal variations is expected. The main drawback is due to the approximation in (4.36). This assumption will be biased in the initial convergence phase or when there is a change of the echo path. Concerning the first problem, we can use a regular NLMS algorithm in the first steps (e.g., in the first L iterations).

It is interesting to demonstrate that the step-size formula from (4.39) also covers the under-modeling scenario (i.e., case 3.) [16], [18]. Let us consider this situation, when the length of the adaptive filter L is smaller than the length of the echo path, denoted now by N. In this situation, the echo signal at time index n can be decomposed as

$$y(n) = y_L(n) + q(n). \tag{4.40}$$

The first term from the right-hand side of (4.40) represents the part of the acoustic echo that can be modeled by the adaptive filter. It can be written as

$$y_L(n) = \mathbf{x}^T(n)\mathbf{h}_L, \tag{4.41}$$

where the vector

$$\mathbf{h}_L = \begin{bmatrix} h_0\ h_1 \cdots h_{L-1} \end{bmatrix}^T$$

contains the first L coefficients of the echo path vector $\mathbf{h}$ of length N. The second term from the right-hand side of (4.40) is the under-modeling noise. This residual echo (which can not be modeled by the adaptive filter) can be expressed as

$$q(n) = \mathbf{x}_{N-L}^T(n)\mathbf{h}_{N-L}, \tag{4.42}$$

where

$$\mathbf{x}_{N-L}(n) = \begin{bmatrix} x(n-L)\ x(n-L-1) \cdots x(n-N+1) \end{bmatrix}^T,$$
$$\mathbf{h}_{N-L} = \begin{bmatrix} h_L\ h_{L+1} \cdots h_{N-1} \end{bmatrix}^T.$$

The term from (4.42) acts like an additional noise for the adaptive process, so that (4.9) should be rewritten as

$$e(n) \left[1 - \mu(n)\mathbf{x}^T(n)\mathbf{x}(n)\right] = \nu(n) + q(n). \tag{4.43}$$

Following the same procedure as in the case of the NPVSS-NLMS algorithm, the normalized step-size results as

$$\alpha(n) = 1 - \frac{\sqrt{\sigma_\nu^2(n) + \sigma_q^2(n)}}{\sigma_e(n)}, \tag{4.44}$$

where $\sigma_q^2(n) = E\left[q^2(n)\right]$ is the power of the under-modeling noise. In (4.44), it was considered that the near-end signal and the under-modeling noise are uncorrelated.

Unfortunately, expression (4.44) is useless in a real-world AEC application since it depends on some sequences that are unavailable, i.e., the near-end signal and the under-modeling noise. In order to solve this issue, the desired signal can be rewritten as

$$d(n) = y_L(n) + q(n) + \nu(n). \tag{4.45}$$

Next, let us assume that $y_L(n)$ and $q(n)$ are uncorrelated. This holds for a white signal. When the input signal is speech, it is difficult to analytically state this assumption. Nevertheless, we can extend it based on the fact that $L \gg 1$ in AEC scenario, and that for usual cases the correlation function has a decreasing trend with the time lag. Moreover, in general the first part of the acoustic impulse response $\mathbf{h}_L$ is more significant as compared to the tail $\mathbf{h}_{N-L}$. Squaring, then taking the expectation of both sides of (4.45), it results that

$$E\left[d^2(n)\right] = E\left[y_L^2(n)\right] + E\left[q^2(n)\right] + E\left[\nu^2(n)\right]. \tag{4.46}$$

Also, let us assume that the adaptive filter coefficients have converged to a certain degree [similar to (4.36)], so that

$$E\left[y_L^2(n)\right] \approx E\left[\hat{y}^2(n)\right]. \tag{4.47}$$

Hence,

$$E\left[\nu^2(n)\right] + E\left[q^2(n)\right] = E\left[d^2(n)\right] - E\left[\hat{y}^2(n)\right]. \tag{4.48}$$

As a result, the step-size parameter of the PVSS-NLMS algorithm given in (4.39) is obtained. Consequently, this algorithm is also suitable for the under-modeling case.

The computational complexity required, at each iteration, by the normalized step-sizes of the presented algorithms are shown in Table 4.2. In the case of the NPVSS-NLMS algorithm, the computational amount related to the estimation of the background noise power (during silences of the near-end talker) is not included, so we cannot conclude that it is the "cheapest" algo-

Table 4.2 Computational complexity of the different normalized step-size parameters.

Algorithm	$+$	$\times$	$\div$	$\sqrt{\ }$
NPVSS-NLMS	3	3	1	1
SVSS-NLMS	4	5	1	1
NEW-NPVSS-NLMS	$2L + 8$	$3L + 12$	3	1
PVSS-NLMS	6	9	1	1

rithm. Fairly, the SVSS-NLMS algorithm is the least complex one. Also, the PVSS-NLMS algorithm has a low complexity. The most expensive one is the NEW-NPVSS-NLMS algorithm, since its computational complexity depends on the filter's length [due to (4.28) and (4.30)]; it is known that in the context of echo cancellation, the value of this parameter can be very large.

4.4 VSS-APA for Echo Cancellation

The idea of the PVSS-NLMS algorithm can be extended to the APA, in order to improve the convergence rate and tracking capabilities [19]. The following expressions summarize the classical APA [8]:

$$\mathbf{e}(n) = \mathbf{d}(n) - \mathbf{X}^T(n)\hat{\mathbf{h}}(n - 1), \tag{4.49}$$

$$\hat{\mathbf{h}}(n) = \hat{\mathbf{h}}(n - 1) + \mu\mathbf{X}(n)\left[\mathbf{X}^T(n)\mathbf{X}(n)\right]^{-1}\mathbf{e}(n), \tag{4.50}$$

where

$$\mathbf{d}(n) = \left[\, d(n)\ d(n - 1)\ \cdots\ d(n - p + 1)\,\right]^T$$

is the desired signal vector of length p, with p denoting the projection order,

$$\mathbf{X}(n) = \left[\, \mathbf{x}(n)\ \mathbf{x}(n - 1)\ \cdots\ \mathbf{x}(n - p + 1)\,\right]$$

is the input signal matrix, where

$$\mathbf{x}(n - l) = \left[\, x(n - l)\ x(n - l - 1)\ \cdots\ x(n - l - L + 1)\,\right]^T$$

(with $l = 0, 1, \ldots, p - 1$) are the input signal vectors, and the constant μ is the step-size parameter of the algorithm.

Let us rewrite (4.50) in a different form:

$$\hat{\mathbf{h}}(n) = \hat{\mathbf{h}}(n - 1) + \mathbf{X}(n)\left[\mathbf{X}^T(n)\mathbf{X}(n)\right]^{-1}\mathbf{M}_\mu(n)\mathbf{e}(n), \tag{4.51}$$

where

$$\mathbf{M}_\mu(n) = \mathrm{diag}\left[\mu_1(n), \mu_2(n), \ldots, \mu_p(n)\right] \tag{4.52}$$

is a $p \times p$ diagonal matrix. It is obvious that (4.50) is obtained when $\mu_1(n) = \mu_2(n) = \cdots = \mu_p(n) = \mu$. Using the adaptive filter coefficients at time n, the a posteriori error vector can be written as

$$\boldsymbol{\varepsilon}(n) = \mathbf{d}(n) - \mathbf{X}^T(n)\hat{\mathbf{h}}(n), \tag{4.53}$$

It can be noticed that the vector $\mathbf{e}(n)$ from (4.49) plays the role of the a priori error vector. Replacing (4.51) in (4.53) and taking (4.49) into account, it results that

$$\boldsymbol{\varepsilon}(n) = \left[\mathbf{I}_p - \mathbf{M}_\mu(n)\right]\mathbf{e}(n), \tag{4.54}$$

where $\mathbf{I}_p$ denotes the $p \times p$ identity matrix. Imposing the cancellation of the p a posteriori errors, i.e., $\boldsymbol{\varepsilon}(n) = \mathbf{0}_{p\times 1}$, and assuming that $\mathbf{e}(n) \neq \mathbf{0}_{p\times 1}$, where $\mathbf{0}_{p\times 1}$ is a column vector with all its p elements equal to zero, we easily see from (4.54) that $\mathbf{M}_\mu(n) = \mathbf{I}_p$. This corresponds to the classical APA update [see (4.50)], with the step-size $\mu = 1$. In the absence of the near-end signal, i.e., $\nu(n) = 0$, the scheme from Fig. 4.1 is reduced to an ideal "system identification" configuration. In this case, the value of the step-size $\mu = 1$ makes sense, because it leads to the best performance [8].

Taking into account the basic idea of the NPVSS-NLMS algorithm (i.e., to recover the near-end signal from the error signal of the adaptive filter, which is also the main issue in echo cancellation), a more reasonable condition to impose in (4.54) is

$$\boldsymbol{\varepsilon}(n) = \boldsymbol{\nu}(n), \tag{4.55}$$

where the column vector

$$\boldsymbol{\nu}(n) = \left[\nu(n)\ \nu(n-1)\ \cdots\ \nu(n-p+1)\right]^T$$

represents the near-end signal vector of length p. Taking (4.54) into account, it results that

$$\varepsilon_{l+1}(n) = \left[1 - \mu_{l+1}(n)\right]e_{l+1}(n) = \nu(n-l), \tag{4.56}$$

where the variables $\varepsilon_{l+1}(n)$ and $e_{l+1}(n)$ denote the $(l+1)$th elements of the vectors $\boldsymbol{\varepsilon}(n)$ and $\mathbf{e}(n)$, respectively, with $l = 0, 1, \ldots, p-1$. The goal is to find an expression for the step-size parameter $\mu_{l+1}(n)$ in such a way that

$$E\left[\varepsilon_{l+1}^2(n)\right] = E\left[\nu^2(n-l)\right]. \tag{4.57}$$

Squaring (4.56) and taking the expectation of both sides, it results:

$$\left[1 - \mu_{l+1}(n)\right]^2 E\left[e_{l+1}^2(n)\right] = E\left[\nu^2(n-l)\right]. \tag{4.58}$$

By solving the quadratic equation (4.58), two solutions are obtained, i.e.,

$$\mu_{l+1}(n) = 1 \pm \sqrt{\frac{E\left[\nu^2(n-l)\right]}{E\left[e_{l+1}^2(n)\right]}}. \tag{4.59}$$

Following the analysis from [24], which states that a value of the step-size between 0 and 1 is preferable over the one between 1 and 2 (even if both solutions are stable but the former has less steady-state mean-squared error with the same convergence speed), it is reasonable to choose

$$\mu_{l+1}(n) = 1 - \sqrt{\frac{E\left[\nu^2(n-l)\right]}{E\left[e_{l+1}^2(n)\right]}}. \tag{4.60}$$

From a practical point of view, (4.60) has to be evaluated in terms of power estimates as

$$\mu_{l+1}(n) = 1 - \frac{\hat{\sigma}_\nu(n-l)}{\hat{\sigma}_{e_{l+1}}(n)}. \tag{4.61}$$

The variable in the denominator can be computed in a recursive manner as in (4.25), i.e.,

$$\hat{\sigma}_{e_{l+1}}^2(n) = \lambda \hat{\sigma}_{e_{l+1}}^2(n-1) + (1-\lambda)e_{l+1}^2(n), \tag{4.62}$$

Following the same development as in the case of the PVSS-NLMS algorithm [see (4.35)–(4.39)], the step-sizes of the proposed VSS-APA are

$$\mu_{l+1}(n) = \left| 1 - \frac{\sqrt{\left|\hat{\sigma}_d^2(n-l) - \hat{\sigma}_{\hat{y}}^2(n-l)\right|}}{\zeta + \hat{\sigma}_{e_{l+1}}(n)} \right|, \quad l = 0, 1, \ldots, p-1. \tag{4.63}$$

The absolute values that appear in (4.63) can be explained as follows. Under our assumptions, we have $E\left[d^2(n-l)\right] \geq E\left[\hat{y}^2(n-l)\right]$ and $E\left[d^2(n-l)\right] - E\left[\hat{y}^2(n-l)\right] \approx E\left[e_{l+1}^2(n)\right]$. Nevertheless, the power estimates of the involved signals could lead to some deviations from the previous theoretical conditions, so that we prevent these situations by taking the absolute values. The adaptive filter coefficients should be updated using (4.51), with the step-sizes computed according to (4.63). In practice, (4.51) has to be rewritten as

$$\hat{\mathbf{h}}(n) = \hat{\mathbf{h}}(n-1) + \mathbf{X}(n)\left[\delta\mathbf{I}_p + \mathbf{X}^T(n)\mathbf{X}(n)\right]^{-1}\mathbf{M}_\mu(n)\mathbf{e}(n), \tag{4.64}$$

where δ is the regularization parameter (as in the case of the NLMS-based algorithms). The reason behind this regularization process is to prevent the problems associated with the inverse of the matrix $\mathbf{X}^T(n)\mathbf{X}(n)$, which could become ill-conditioned especially when highly correlated inputs (e.g., speech) are involved. An insightful analysis about this parameter, in the framework of APA, can be found in [15]. Considering the context of a "system identification" configuration, the value of the regularization parameter depends on the level of the noise that corrupts the output of the system that has to be identified. A low signal-to-noise ratio requires a high value of the regularization parameter. In the AEC context, different types of noise corrupt the output of the echo path, e.g., the background noise or/and the near-end speech; in addition, the under-modeling noise (if it is present) increases the overall level of noise. Also, the value of the regularization parameter depends on the value of the projection order of the algorithm. The larger the value of the projection order of the APA, the higher is the condition number of the matrix $\mathbf{X}^T(n)\mathbf{X}(n)$; consequently, a larger value of δ is required.

Summarizing, the proposed VSS-APA is listed in Table 4.3. For a value of the projection order $p = 1$, the PVSS-NLMS algorithm presented in Section 4.3 is obtained. As compared to the classical APA, the additional computational amount of the VSS-APA consists of $3p+6$ multiplication operations, p divisions, $4p+2$ additions, and p square-root operations. Taking into account the fact that the value of the projection order in AEC applications is usually smaller than 10 and the length of the adaptive filter is large (e.g., hundreds of coefficients), it can be concluded that the computational complexity of the proposed VSS-APA is moderate and comparable with the classical APA or to any fast versions of this algorithm.

Since it is also based on the assumption that the adaptive filter coefficients have converged to a certain degree, the VSS-APA could also experience a slower initial convergence rate and a slower tracking capability as compared to the APA. Concerning the initial convergence rate, we could start the proposed algorithm using a regular APA in the first L iterations. Also, in order to deal with echo path changes, the proposed algorithm could be equipped with an echo path changes detector [25], [26]. Nevertheless, in our simulations none of the previous scenarios are considered. The experimental results prove that the performance degradation is not significant.

4.5 VFF-RLS for System Identification

The RLS algorithm [4], [5], [6] is one of the most popular adaptive filters. It belongs to the Kalman filters family [27], and many adaptive algorithms (including the NLMS) can be seen as approximations of it. As compared to the NLMS algorithm, the RLS offers a superior convergence rate especially for highly correlated input signals. The price to pay for this is an increase in the computational complexity. For this reason, it is not very often involved in echo cancellation. Nevertheless, it is interesting to show that the idea of

Table 4.3 VSS-APA.

Initialization:

$\hat{\mathbf{h}}(0) = \mathbf{0}_{L \times 1}$

$\hat{\sigma}_d^2(n) = 0, \ \hat{\sigma}_{\hat{y}}^2(n) = 0, \ \text{for } n \leq 0$

$\hat{\sigma}_{e_{l+1}}^2(0) = 0, \ l = 0, 1, \ldots, p - 1$

Parameters:

$\lambda = 1 - \dfrac{1}{KL}, \ \text{weighting factor with } K > 1$

$\delta > 0, \ \text{regularization}$

$\zeta > 0, \ \text{very small number to avoid division by zero}$

For time index $n = 1, 2, \ldots$:

$\mathbf{e}(n) = \mathbf{d}(n) - \mathbf{X}^T(n)\hat{\mathbf{h}}(n - 1)$

$\hat{y}(n) = \mathbf{x}^T(n)\hat{\mathbf{h}}(n - 1)$

$\hat{\sigma}_d^2(n) = \lambda\hat{\sigma}_d^2(n - 1) + (1 - \lambda)d^2(n)$

$\hat{\sigma}_{\hat{y}}^2(n) = \lambda\hat{\sigma}_{\hat{y}}^2(n - 1) + (1 - \lambda)\hat{y}^2(n)$

$\hat{\sigma}_{e_{l+1}}^2(n) = \lambda\hat{\sigma}_{e_{l+1}}^2(n - 1) + (1 - \lambda)e_{l+1}^2(n), \ l = 0, 1, \ldots, p - 1$

$$\mu_{l+1}(n) = \left| 1 - \frac{\sqrt{\left| \hat{\sigma}_d^2(n - l) - \hat{\sigma}_{\hat{y}}^2(n - l) \right|}}{\zeta + \hat{\sigma}_{e_{l+1}}(n)} \right|, \ l = 0, 1, \ldots, p - 1$$

$\mathbf{M}_\mu(n) = \text{diag}\left[\mu_1(n), \mu_2(n), \ldots, \mu_p(n) \right]$

$\hat{\mathbf{h}}(n) = \hat{\mathbf{h}}(n - 1) + \mathbf{X}(n)\left[\delta\mathbf{I}_p + \mathbf{X}^T(n)\mathbf{X}(n) \right]^{-1}\mathbf{M}_\mu(n)\mathbf{e}(n)$

---

the NPVSS-NLMS algorithm can be extended in the case of RLS, improving its robustness capabilities for real-world system identification problems.

Similar to the attributes of the step-size from the NLMS-based algorithms, the performance of RLS-type algorithms in terms of convergence rate, tracking, misadjustment, and stability depends on a parameter known as the forgetting factor [4], [5], [6]. The classical RLS algorithm uses a constant forgetting factor (between 0 and 1) and needs to compromise between the previous performance criteria. When the forgetting factor is very close to one, the algorithm achieves low misadjustment and good stability, but its tracking capabilities are reduced. A small value of the forgetting factor improves the tracking but increases the misadjustment, and could affect the stability of the algorithm. Motivated by these aspects, a number of variable forgetting factor RLS (VFF-RLS) algorithms have been developed, e.g., [23], [28] (and references therein). The performance and the applicability of these methods for system identification depend on several factors such as 1) the ability of detecting the changes of the system, 2) the level and the character of the noise that usually corrupts the output of the unknown system, and 3) complexity and stability issues.

It should be mentioned that in the system identification context, when the output of the unknown system is corrupted by another signal (which is usually an additive noise), the goal of the adaptive filter is not to make the error signal goes to zero, because this will introduce noise in the adaptive filter. The objective instead is to recover the "corrupting signal" from the error signal of the adaptive filter after this one converges to the true solution. This idea is consistent with the concept of the NPVSS-NLMS algorithm. Based on this approach, a new VFF-RLS algorithm [20] is presented in this section.

Let us consider the same system identification problem from Section 4.2 (see also Fig. 4.1 for notation). The RLS algorithm is immediately deduced from the normal equations which are

$$\mathbf{R}(n)\hat{\mathbf{h}}(n) = \mathbf{r}_{dx}(n), \tag{4.65}$$

where

$$\mathbf{R}(n) = \sum_{i=1}^{n} \beta^{n-i}\mathbf{x}(i)\mathbf{x}^{T}(i),$$

$$\mathbf{r}_{dx}(n) = \sum_{i=1}^{n} \beta^{n-i}\mathbf{x}(i)d(i),$$

and the parameter β is the forgetting factor. According to (4.1), the normal equations become

$$\sum_{i=1}^{n} \beta^{n-i}\mathbf{x}(i)\mathbf{x}^{T}(i)\hat{\mathbf{h}}(n) = \sum_{i=1}^{n} \beta^{n-i}\mathbf{x}(i)y(i) + \sum_{i=1}^{n} \beta^{n-i}\mathbf{x}(i)\nu(i). \tag{4.66}$$

For a value of β very close to 1 and for a large value of n, it may be assumed that

$$\frac{1}{n}\sum_{i=1}^{n} \beta^{n-i}\mathbf{x}(i)\nu(i) \approx E\left[\mathbf{x}(n)\nu(n)\right] = 0. \tag{4.67}$$

Consequently, taking (4.66) into account,

$$\sum_{i=1}^{n} \beta^{n-i}\mathbf{x}(i)\mathbf{x}^{T}(i)\hat{\mathbf{h}}(n) \approx \sum_{i=1}^{n} \beta^{n-i}\mathbf{x}(i)y(i) = \sum_{i=1}^{n} \beta^{n-i}\mathbf{x}(i)\mathbf{x}^{T}(i)\mathbf{h}, \tag{4.68}$$

thus $\hat{\mathbf{h}}(n) \approx \mathbf{h}$ and $e(n) \approx \nu(n)$. Now, for a small value of the forgetting factor, so that $\beta^k \ll 1$ for $k \geq n_0$, it can be assumed that

$$\sum_{i=1}^{n} \beta^{n-i}(\bullet) \approx \sum_{i=n-n_0+1}^{n} \beta^{n-i}(\bullet).$$

According to the orthogonality theorem [4], [5], [6], the normal equations become

$$\sum_{i=n-n_0+1}^{n} \beta^{n-i}\mathbf{x}(i)e(i) = \mathbf{0}_{L\times 1}.$$

This is a homogeneous set of L equations with n_0 unknown parameters, $e(i)$. When $n_0 < L$, this set of equations has the unique solution $e(i) = 0$, for $i = n - n_0 + 1, \ldots, n$, leading to $\hat{y}(n) = y(n) + \nu(n)$. Consequently, there is a "leakage" of $\nu(n)$ into the output of the adaptive filter. In this situation, the signal $\nu(n)$ is cancelled; even if the error signal is $e(n) = 0$, this does not lead to a correct solution from the system identification point of view. A small value of β or a high value of L intensifies this phenomenon.

Summarizing, for a low value of β the output of the adaptive system is $\hat{y}(n) \approx y(n) + \nu(n)$, while $\beta \approx 1$ leads to $\hat{y}(n) \approx y(n)$. Apparently, for a system identification application, a value of β very close to 1 is desired; but in this case, even if the initial convergence rate of the algorithm is satisfactory, the tracking capabilities suffer a lot. In order to provide fast tracking, a lower value of β is desired. On the other hand, taking into account the previous aspects, a low value of β is not good in the steady-state. Consequently, a VFF-RLS algorithm (which could provide both fast tracking and low misadjustment) can be a more appropriate solution, in order to deal with these aspects.

We start our development by writing the relations that define the classical RLS algorithm:

$$\mathbf{k}(n) = \frac{\mathbf{P}(n-1)\mathbf{x}(n)}{\beta + \mathbf{x}^T(n)\mathbf{P}(n-1)\mathbf{x}(n)}, \tag{4.69}$$

$$\hat{\mathbf{h}}(n) = \hat{\mathbf{h}}(n-1) + \mathbf{k}(n)e(n), \tag{4.70}$$

$$\mathbf{P}(n) = \frac{1}{\beta}\left[\mathbf{P}(n-1) - \mathbf{k}(n)\mathbf{x}^T(n)\mathbf{P}(n-1)\right], \tag{4.71}$$

where $\mathbf{k}(n)$ is the Kalman gain vector, $\mathbf{P}(n)$ is the inverse of the input correlation matrix $\mathbf{R}(n)$, and $e(n)$ is the a priori error signal defined in (4.2); the a posteriori error signal is defined in (4.3). Using (4.2) and (4.70) in (4.3), it results

$$\varepsilon(n) = e(n)\left[1 - \mathbf{x}^T(n)\mathbf{k}(n)\right]. \tag{4.72}$$

According to the problem statement, it is desirable to recover the system noise from the error signal. Consequently, it can be imposed the same condition from (4.8). Using (4.8) in (4.72) and taking (4.69) into account, it finally results

$$E\left\{\left[1 - \frac{\theta(n)}{\beta(n) + \theta(n)}\right]^2\right\} = \frac{\sigma_\nu^2}{\sigma_e^2(n)}, \tag{4.73}$$

where $\theta(n) = \mathbf{x}^T(n)\mathbf{P}(n-1)\mathbf{x}(n)$. In (4.73), we assumed that the input and error signals are uncorrelated, which is true when the adaptive filter has started to converge to the true solution. We also assumed that the forgetting factor is deterministic and time dependent. By solving the quadratic equation (4.73), it results a variable forgetting factor

$$\beta(n) = \frac{\sigma_\theta(n)\sigma_\nu}{\sigma_e(n) - \sigma_\nu}, \tag{4.74}$$

where $E\left[\theta^2(n)\right] = \sigma_\theta^2(n)$. In practice, the variance of the error signal is estimated based on (4.25), while the variance of $\theta(n)$ is evaluated in a similar manner, i.e.,

$$\hat{\sigma}_\theta^2(n) = \lambda\hat{\sigma}_\theta^2(n-1) + (1-\lambda)\theta^2(n). \tag{4.75}$$

The estimate of the noise power, $\hat{\sigma}_\nu^2(n)$ [which should be used in (4.74) from practical reasons], can be estimated as in (4.26).

Theoretically, $\sigma_e(n) \geq \sigma_\nu$ in (4.74). Compared to the NLMS algorithm (where there is the gradient noise, so that $\sigma_e(n) > \sigma_\nu$), the RLS algorithm with $\beta(n) \approx 1$ leads to $\sigma_e(n) \approx \sigma_\nu$. In practice (since power estimates are used), several situations have to be prevented in (4.74). Apparently, when $\hat{\sigma}_e(n) \leq \hat{\sigma}_\nu$, it could be set $\beta(n) = \beta_{\max}$, where $\beta_{\max}$ is very close or equal to 1. But this could be a limitation, because in the steady-state of the algorithm $\hat{\sigma}_e(n)$ varies around $\hat{\sigma}_\nu$. A more reasonable solution is to impose that $\beta(n) = \beta_{\max}$ when

$$\hat{\sigma}_e(n) \leq \rho\hat{\sigma}_\nu, \tag{4.76}$$

with $1 < \rho \leq 2$. Otherwise, the forgetting factor of the proposed VFF-RLS algorithm is evaluated as

$$\beta(n) = \min\left[\frac{\hat{\sigma}_\theta(n)\hat{\sigma}_\nu(n)}{\zeta + |\hat{\sigma}_e(n) - \hat{\sigma}_\nu(n)|}, \beta_{\max}\right], \tag{4.77}$$

where the small positive constant ζ prevents a division by zero. Before the algorithm converges or when there is an abrupt change of the system, $\hat{\sigma}_e(n)$ is large as compared to $\hat{\sigma}_\nu(n)$; thus, the parameter $\beta(n)$ from (4.77) takes low values, providing fast convergence and good tracking. When the algorithm

Table 4.4 VFF-RLS algorithm.

Initialization:

$\mathbf{P}(0) = \delta \mathbf{I}_L \ (\delta > 0, \text{regularization})$

$\hat{\mathbf{h}}(0) = \mathbf{0}_{L \times 1}$

$\hat{\sigma}_e^2(0) = \hat{\sigma}_\theta^2(0) = \hat{\sigma}_\nu^2(0) = 0$

Parameters:

$\lambda = 1 - \dfrac{1}{KL}$ (with $K > 1$) and $\gamma > \lambda$, weighting factors

$\beta_{\max}$, upper bound of the forgetting factor (very close or equal to 1)

$\zeta > 0$, very small number to avoid division by zero

For time index $n = 1, 2, \ldots$:

$e(n) = d(n) - \mathbf{x}^T(n)\hat{\mathbf{h}}(n-1)$

$\theta(n) = \mathbf{x}^T(n)\mathbf{P}(n-1)\mathbf{x}(n)$

$\hat{\sigma}_e^2(n) = \lambda \hat{\sigma}_e^2(n-1) + (1-\lambda)e^2(n)$

$\hat{\sigma}_\theta^2(n) = \lambda \hat{\sigma}_\theta^2(n-1) + (1-\lambda)\theta^2(n)$

$\hat{\sigma}_\nu^2(n) = \gamma \hat{\sigma}_\nu^2(n-1) + (1-\gamma)e^2(n)$

$$\beta(n) = \begin{cases} \beta_{\max}, & \text{if } \hat{\sigma}_e(n) \leq \rho\hat{\sigma}_\nu \ (\text{where } 1 < \rho \leq 2) \\[2ex] \min\left[\dfrac{\hat{\sigma}_\theta(n)\hat{\sigma}_\nu(n)}{\zeta + |\hat{\sigma}_e(n) - \hat{\sigma}_\nu(n)|}, \beta_{\max}\right], & \text{otherwise} \end{cases}$$

$\mathbf{k}(n) = \dfrac{\mathbf{P}(n-1)\mathbf{x}(n)}{\beta(n) + \theta(n)}$

$\hat{\mathbf{h}}(n) = \hat{\mathbf{h}}(n-1) + \mathbf{k}(n)e(n)$

$\mathbf{P}(n) = \dfrac{1}{\beta(n)}\left[\mathbf{P}(n-1) - \mathbf{k}(n)\mathbf{x}^T(n)\mathbf{P}(n-1)\right]$

---

converges to the steady-state solution, $\hat{\sigma}_e(n) \approx \hat{\sigma}_\nu(n)$ [so that the condition (4.76) is fulfilled] and $\beta(n)$ is equal to $\beta_{\max}$, providing low misadjustment. The resulted VFF-RLS algorithm is summarized in Table 4.4. It can be noticed that the mechanism that controls the forgetting factor is very simple and not expensive in terms of multiplications and additions.

4.6 Simulations

4.6.1 VSS-NLMS Algorithms for AEC

The simulations were performed in a typical AEC configuration, as shown in Fig. 4.1. The measured acoustic impulse response was truncated to 1000 coefficients, and the same length was used for the adaptive filter (except for one experiment performed in the under-modeling case, where the adaptive filter length was set to 512 coefficients); the sampling rate is 8 kHz. The input signal, $x(n)$, is either an AR(1) process generated by filtering a white Gaussian noise through a first-order system $1/(1 - 0.5z^{-1})$ or a speech sequence. An independent white Gaussian noise $w(n)$ is added to the echo signal $y(n)$, with 20 dB signal-to-noise ratio (SNR) for most of the experiments. The measure of performance is the normalized misalignment, defined as $20 \log_{10} \left[\left\| \mathbf{h} - \hat{\mathbf{h}}(n) \right\|_2 / \|\mathbf{h}\|_2 \right]$; for the under-modeling case, the normalized misalignment is computed by padding the vector of the adaptive filter coefficients with zeros up to the length of the acoustic impulse response. Other parameters are set as follows; the regularization is $\delta = 20\sigma_x^2$ for all the NLMS-based algorithms, the weighting factor λ is computed using $K = 6$, and the small positive constant from the denominator of the normalized step-sizes is $\zeta = 10^{-8}$.

First, we evaluate the performance of the NPVSS-NLMS algorithm as compared to the classical NLMS algorithm with two different step-sizes, i.e., (a) $\left[\delta + \mathbf{x}^T(n)\mathbf{x}(n)\right]^{-1}$ and (b) $0.05 \left[\delta + \mathbf{x}^T(n)\mathbf{x}(n)\right]^{-1}$. It is assumed that the power of the background noise is known, since it is needed in the step-size formula of the NPVSS-NLMS algorithm. A single-talk scenario is considered in Fig. 4.2, using the AR(1) process as input signal. Clearly, the NPVSS-NLMS outperforms by far the NLMS algorithm, achieving similar convergence rate with the fastest NLMS (i.e., with the largest normalized step-size) and the final misalignment of the NLMS with the smaller normalized step-size. In Fig. 4.3, an abrupt change of the echo path was introduced at time 10, by shifting the acoustic impulse response to the right by 12 samples, in order to test the tracking capabilities of the algorithms. It can be noticed that the NPVSS-NLMS algorithm tracks as fast as the NLMS with its maximum step-size. Based on these results, we will use the NPVSS-NLMS as the reference algorithm in the following simulations, instead of the classical NLMS algorithm.

Next, it is important to evaluate the influence of the parameters γ and ς over the performances of the SVSS-NLMS and NEW-NPVSS-NLMS algorithms, respectively. In order to stress the algorithms, the abrupt change of the echo path is introduced at time 10, and the SNR decreases from 20 dB to 10 dB (i.e., background noise increases) at time 20. The behavior of the SVSS-NLMS algorithm is presented in Fig. 4.4. At a first glance, a large value of γ represents a more proper choice. But in this situation the estimation of

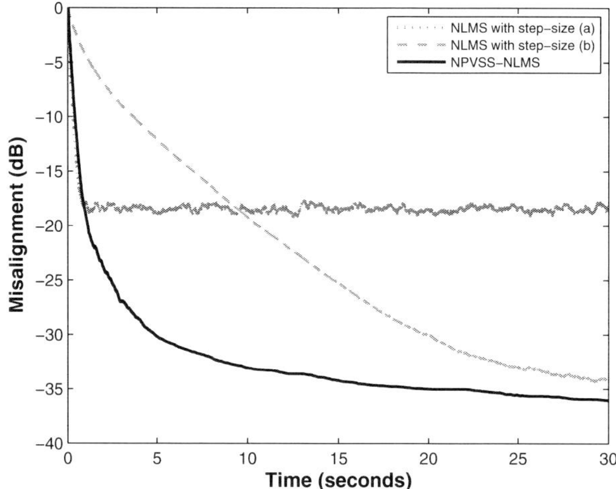

Fig. 4.2 Misalignment of the NLMS algorithm with two different step sizes (a) $\left[\delta + \mathbf{x}^T(n)\mathbf{x}(n)\right]^{-1}$ and (b) $0.05\left[\delta + \mathbf{x}^T(n)\mathbf{x}(n)\right]^{-1}$, and misalignment of the NPVSS-NLMS algorithm. The input signal is an AR(1) process, $L = 1000$, $\lambda = 1 - 1/(6L)$, and SNR = 20 dB.

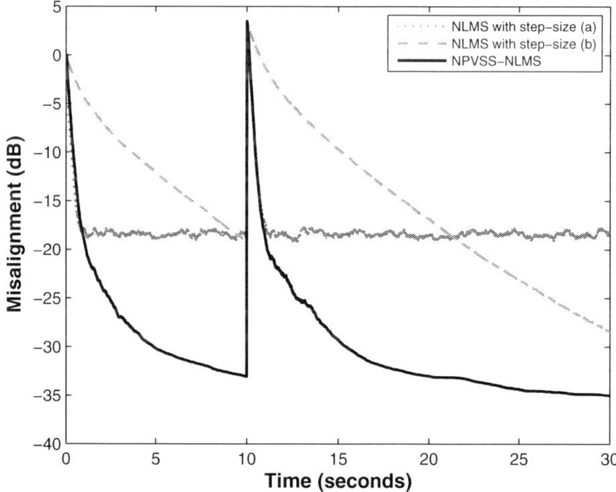

Fig. 4.3 Misalignment during impulse response change. The impulse response changes at time 10. Other conditions are the same as in Fig. 4.2.

the near-end signal power from (4.26) will also have a large latency, which is not suitable for the case when there are near-end signal variations. Consequently, a compromise should be made when choosing the value of γ. In the case of the NEW-NPVSS-NLMS algorithm (Fig. 4.5), a very small value of ς

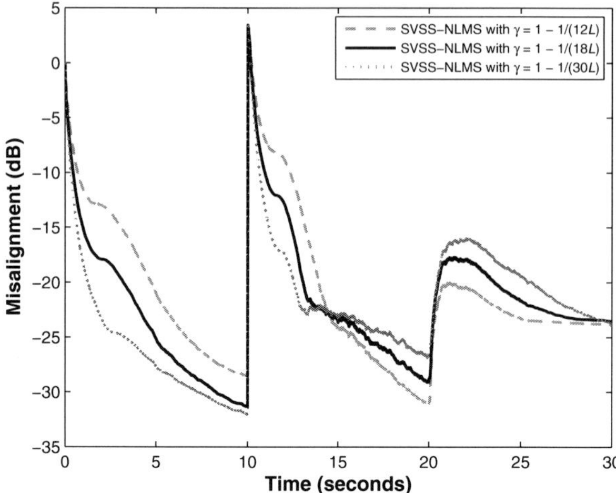

Fig. 4.4 Misalignment of the SVSS-NLMS algorithm for different values of γ. Impulse response changes at time 10, and SNR decreases from 20 dB to 10 dB at time 20. The input signal is an AR(1) process, $L = 1000$, and $\lambda = 1 - 1/(6L)$.

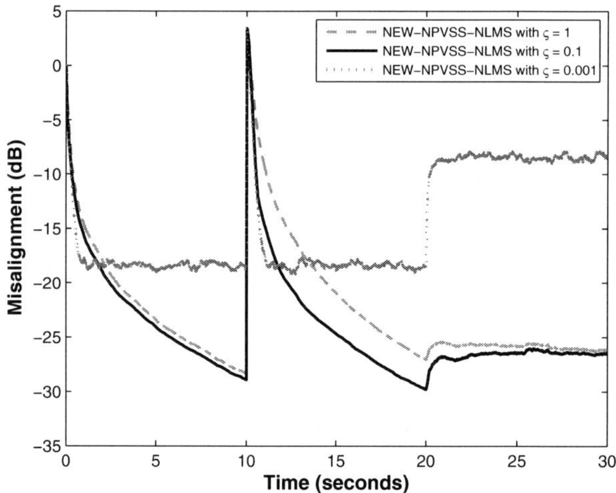

Fig. 4.5 Misalignment of the NEW-NPVSS-NLMS algorithm for different values of ς. Other conditions are the same as in Fig. 4.4.

will make the first line of (4.31) futile, since $\xi(n) > \varsigma$; thus, the algorithm acts like a regular NLMS with the normalized step-size equal to 1. In terms of robustness against near-end signal variations, the performance is improved for a larger value of ς. Nevertheless, it should be taken into account that a very

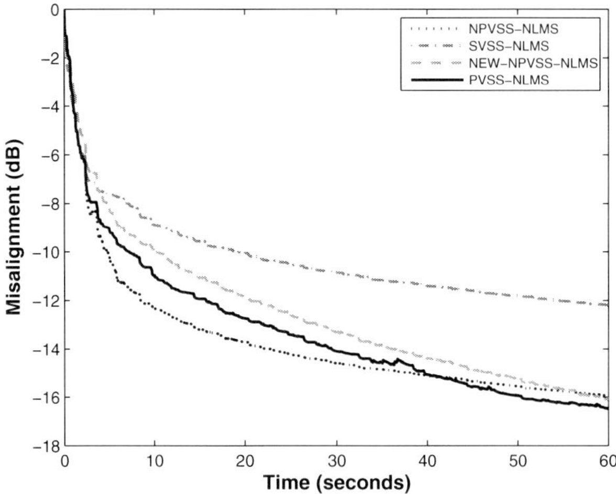

Fig. 4.6 Misalignment of the NPVSS-NLMS, SVSS-NLMS [with $\gamma = 1 - 1/(18L)$], NEW-NPVSS-NLMS (with $\varsigma = 0.1$), and PVSS-NLMS algorithms. The input signal is speech, $L = 1000$, $\lambda = 1 - 1/(6L)$, and SNR $= 20$ dB.

large value of ς could affect the tracking capabilities of the algorithm (and the initial convergence), because the condition $\xi(n) > \varsigma$ could be inhibited.

In order to approach the context of a real-world AEC scenario, the speech input is used in the following simulations. The parameters of the NEW-NPVSS-NLMS and the SVSS-NLMS algorithms were set in order to obtain similar initial convergence rates. The NPVSS-NLMS algorithm (assuming that the power of the background noise is known) and the PVSS-NLMS algorithm are also included for comparisons. First, their performances are analyzed in a single-talk case (Fig. 4.6). In terms of final misalignment, the PVSS-NLMS algorithm is close to the NPVSS-NLMS algorithm; the NEW-NPVSS-NLMS algorithm outperforms the SVSS-NLMS algorithm. In Fig. 4.7, the abrupt change of the echo path is introduced at time 20. As expected, the PVSS-NLMS algorithm has a slower reaction as compared to the other algorithms, since the assumption (4.36) is strongly biased.

Variations of the background noise are considered in Fig. 4.8. The SNR decreases from 20 dB to 10 dB between time 20 and 30, and to -5 dB between time 40 and 50. It was assumed that the new values of background noise power estimate are not available yet for the NPVSS-NLMS algorithm; consequently, it fails during these variations. Also, even if the NEW-NPVSS-NLMS algorithm is robust against the first variation, it is affected in the last situation. For such a low SNR, a higher value of ς is required, which could degrade the convergence rate and the tracking capabilities of the algorithm. For improving the robustness of the SVSS-NLMS algorithm, the value of γ

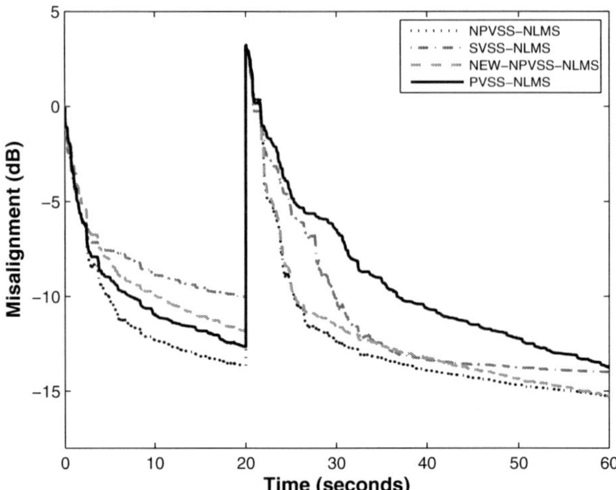

Fig. 4.7 Misalignment during impulse response change. The impulse response changes at time 20. Other conditions are the same as in Fig. 4.6.

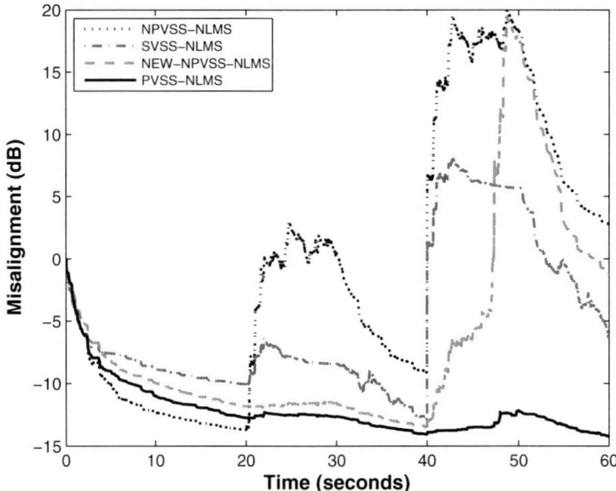

Fig. 4.8 Misalignment during background noise variations. The SNR decreases from 20 dB to 10 dB between time 20 and 30, and to −5 dB between time 40 and 50. Other conditions are the same as in Fig. 4.6.

should be increased, but with the same unwanted effects. Despite the value of SNR, the PVSS-NLMS algorithm is very robust in both situations.

The double-talk scenario is considered in Fig. 4.9, without using any double-talk detector (DTD). The NPVSS-NLMS algorithm is not included for comparison since it was not designed for such a case. Different values of

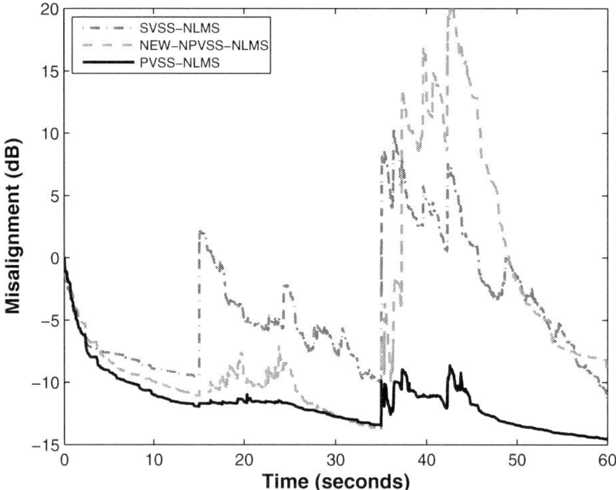

Fig. 4.9 Misalignment during double-talk, without DTD. Near-end speech appears between time 15 and 25 (with FNR = 5 dB), and between time 35 and 45 (with FNR = 3 dB). Other conditions are the same as in Fig. 4.6. (The NPVSS-NLMS algorithm is not included for comparison.)

the far-end to near-end speech ratio (FNR) were used, i.e., FNR = 5 dB between time 15 and 25, and FNR = 3 dB between time 35 and 45. Despite the small variation of the FNR, it can be noticed that the NEW-NPVSS-NLMS algorithm is very sensitive to this modification; in other words, the value of ς is not proper for the second double-talk situation. In terms of robustness, the PVSS-NLMS outperforms the other VSS-NLMS algorithms.

Finally, the under-modeling situation is considered in Fig. 4.10, in the context of a single-talk scenario; the input signal is the AR(1) process. The same echo path of 1000 coefficients is used, but setting the adaptive filter length to 512 coefficients. In terms of the final misalignment, it can be noticed that the NPVSS-NLMS is outperformed by the other algorithms, since the under-modeling noise is present (together with the background noise).

4.6.2 VSS-APA for AEC

Among the previous VSS-NLMS algorithms, the PVSS-NLMS proved to be the most robust against near-end signal variations, like background noise increase or double-talk. Furthermore, it is suitable for real-world applications, since all the parameters of the step-size formula are available. The main drawback of the PVSS-NLMS algorithm is with regards to its tracking capabilities, as a consequence of the assumption (4.36). This was the main

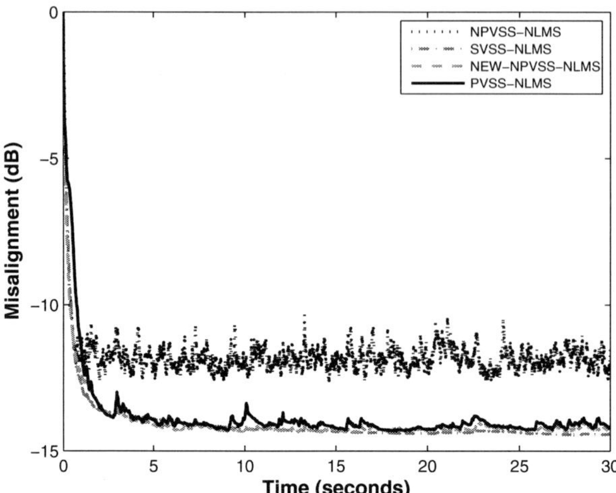

Fig. 4.10 Misalignment of the NPVSS-NLMS, SVSS-NLMS [with $\gamma = 1 - 1/(18L)$], NEW-NPVSS-NLMS (with $\varsigma = 0.1$), and PVSS-NLMS algorithms for the under-modeling case. The input signal is an AR(1) process, $L = 512$ (the echo path has 1000 coefficients), $\lambda = 1 - 1/(6L)$, and SNR = 20 dB.

motivation behind the development of the VSS-APA in Section 4.4. In the following simulations, the conditions of the AEC application are the same as described in the beginning of the previous subsection. The input signal is speech in all the experiments.

The first set of simulations is performed in a single-talk scenario. In Fig. 4.11, the VSS-APA is compared to the classical APA with different values of the step-size. The value of the projection order is $p = 2$ and the regularization parameter is $\delta = 50\sigma_x^2$ for both algorithms. Since the requirements are for both high convergence rate and low misalignment, a compromise choice has to be made in the case of the APA. Even if the value $\mu = 1$ leads to the fastest convergence mode, the value $\mu = 0.25$ seems to offer a more proper solution, taking into account the previous performance criteria. In this case, the convergence rate is slightly reduced as compared to the situation when $\mu = 1$, but the final misalignment is significantly lower. The VSS-APA has an initial convergence rate similar to the APA with $\mu = 0.25$ [it should be noted that the assumption (4.36) is not yet fulfilled in the first part of the adaptive process], but it achieves a significant lower misalignment, which is close to the one obtained by the APA with $\mu = 0.025$.

The effects of different projection orders for the VSS-APA are evaluated in Fig. 4.12. Following the discussion about the regularization parameter from the end of Section 4.4, the value of this parameter increases with the projection order; we set $\delta = 20\sigma_x^2$ for $p = 1$, $\delta = 50\sigma_x^2$ for $p = 2$, and $\delta = 100\sigma_x^2$ for $p = 4$. It can be noticed that the most significant performance

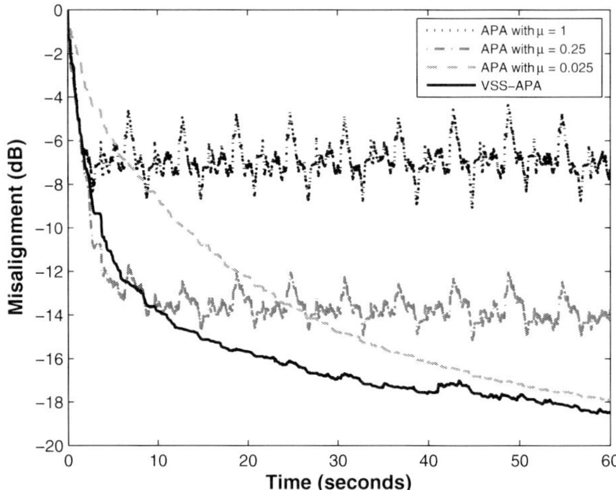

Fig. 4.11 Misalignment of the APA with three different step-sizes ($\mu = 1$, $\mu = 0.25$, and $\mu = 0.025$), and VSS-APA. The input signal is speech, $p = 2$, $L = 1000$, $\lambda = 1 - 1/(6L)$, and SNR $= 20$ dB.

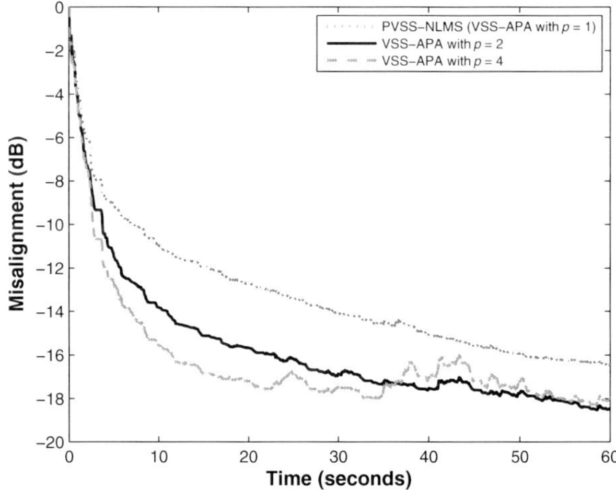

Fig. 4.12 Misalignment of the VSS-APA with different projection orders, i.e., $p = 1$ (PVSS-NLMS algorithm), $p = 2$, and $p = 4$. Other conditions are the same as in Fig. 4.11.

improvement is for the VSS-APA with $p = 2$ as compared to the case when $p = 1$ (i.e., the PVSS-NLMS algorithm). Consequently, from practical reasons related to the computational complexity, a projection order $p = 2$ could be good enough for real-world AEC applications; this value will be used in all the following simulations.

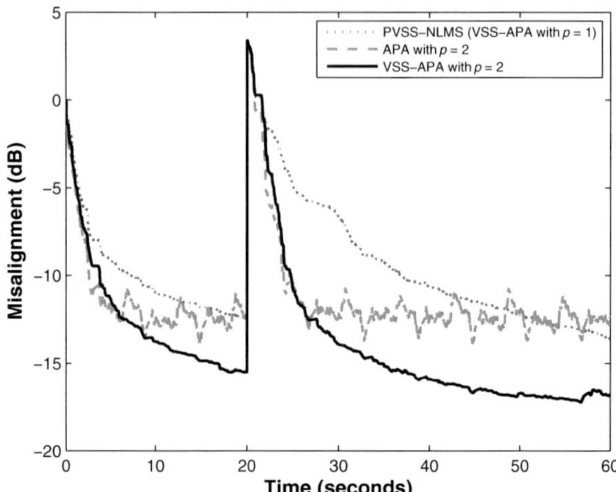

Fig. 4.13 Misalignment during impulse response change. The impulse response changes at time 20. Algorithms: PVSS-NLMS algorithm, APA (with $\mu = 0.25$), and VSS-APA. Other conditions are the same as in Fig. 4.11.

The tracking capabilities of the VSS-APA are shown in Fig. 4.13, as compared to the classical APA with a step-size $\mu = 0.25$ (this value of the step-size will be used in all the following experiments). The abrupt change of the echo path is introduced at time 20. As a reference, the PVSS-NLMS algorithm is also included. It can be noticed that the VSS-APA has a tracking reaction similar to the APA. Also, it tracks significantly faster than the PVSS-NLMS algorithm (i.e., VSS-APA with $p = 1$).

The robustness of the VSS-APA is evaluated in the following simulations. In the experiment presented in Fig. 4.14, the SNR decreases from 20 dB to 10 dB after 20 seconds from the debut of the adaptive process, for a period of 20 seconds. It can be noticed that the VSS-APA is very robust against the background noise variation, while the APA is affected by this change in the acoustic environment.

Finally, the double-talk scenario is considered. The near-end speech appears between time 25 and 35, with FNR = 4 dB. In Fig. 4.15, the algorithms are not equipped with a DTD. It can be noticed that the VSS-APA outperforms by far the APA. In practice, a simple DTD can be involved in order to enhance the performance. In Fig. 4.16, the previous experiment is repeated using the Geigel DTD; its settings are chosen assuming a 6 dB attenuation, i.e., the threshold is equal to 0.5 and the hangover time is set to 240 samples. As expected, the algorithms perform better; but it is interesting to notice that the VSS-APA without a DTD is more robust as compared to the APA with a DTD.

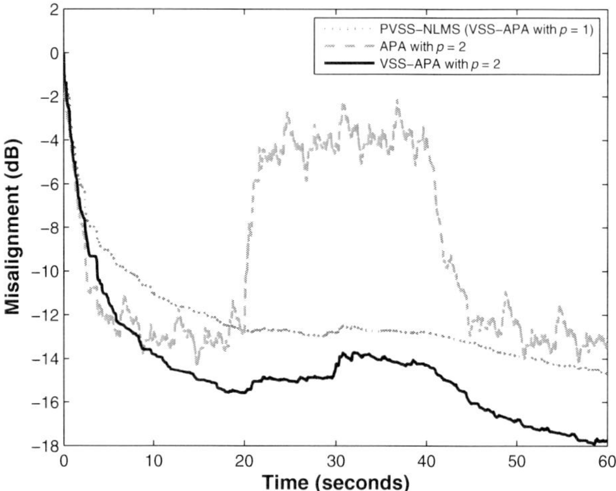

Fig. 4.14 Misalignment during background noise variations. The SNR decreases from 20 dB to 10 dB between time 20 and 40. Other conditions are the same as in Fig. 4.11.

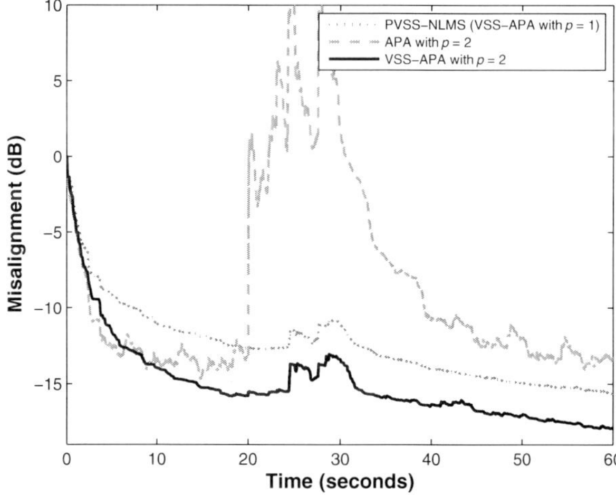

Fig. 4.15 Misalignment during double-talk, without a DTD. Near-end speech appears between time 20 and 30 (with FNR = 4 dB). Other conditions are the same as in Fig. 4.11.

4.6.3 VFF-RLS for System Identification

Due to their computational complexity, the RLS-based algorithms are not common choices for AEC. Nevertheless, they are attractive for different system identification problems, especially when the length of the unknown sys-

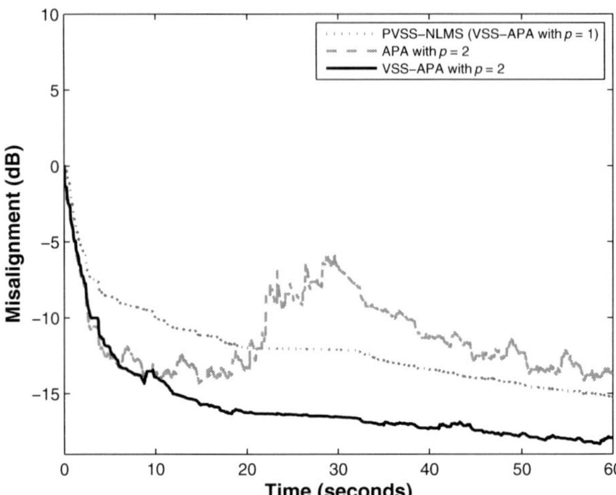

Fig. 4.16 Misalignment during double-talk with the Geigel DTD. Other conditions are the same as in Fig. 4.15.

tem is not too large. We should note that the RLS-based algorithms are less sensitive to highly correlated input data, as compared to the gradient-based algorithms, like NLMS and APA. In order to be more realistic in results, in the next experiments we aim to identify an impulse response of length $L = 100$ (the most significant part of the acoustic impulse response used in the previous subsections); the same length is used for the adaptive filter.

In Fig. 4.17, the VFF-RLS algorithm presented in Section 4.5 is compared with the classical RLS algorithm with two different forgetting factors, i.e., $\beta = 1 - 1/(3L)$ and $\beta = 1$. The input signal is an AR(1) process generated by filtering a white Gaussian noise through a first-order system $1/\left(1 - 0.98z^{-1}\right)$. The output of the unknown system is corrupted by a white Gaussian noise, with an SNR = 20 dB. The parameters of the VFF-RLS algorithm are set as follows (see Table 4.4): $\delta = 100$, $\lambda = 1 - 1/(6L)$, $\gamma = 1 - 1/(18L)$, $\beta_{\max} = 1$, $\zeta = 10^{-8}$, and $\rho = 1.5$. An abrupt change of the system is introduced at time 10 (by shifting the impulse response to the right by 5 samples), and the SNR decreases from 20 dB to 10 dB at time 20. It can be noticed that the VFF-RLS algorithm achieves the same initial misalignment as the RLS with its maximum forgetting factor, but it tracks as fast as the RLS with the smaller forgetting factor. Also, it is very robust against the SNR variation.

The same scenario is considered in Fig. 4.18, comparing the VFF-RLS algorithm with the PVSS-NLMS algorithm and with VSS-APA (using $p = 2$). It can be noticed that the VFF-RLS outperforms by far the other algorithms in terms of both convergence rate and misalignment. As expected, the highly correlated input data influences the performance of the gradient-based algorithms.

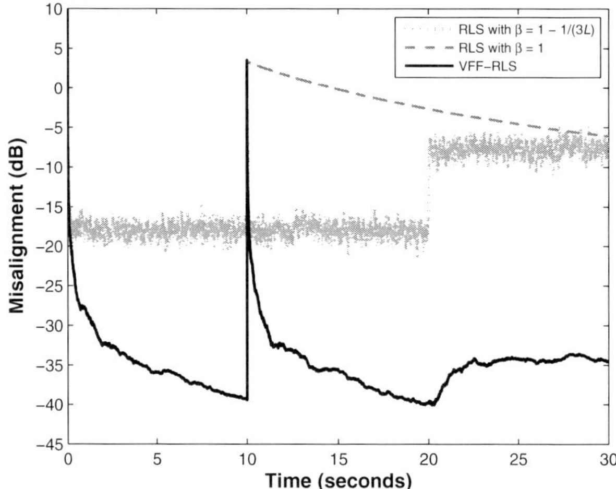

Fig. 4.17 Misalignment of the RLS algorithm with two different forgetting factors (a) $\beta = 1$ and (b) $\beta = 1 - 1/(3L)$, and misalignment of the VFF-RLS algorithm. Impulse response changes at time 10, and SNR decreases from 20 dB to 10 dB at time 20. The input signal is an AR(1) process, $L = 100$, $\lambda = 1 - 1/(6L)$, $\gamma = 1 - 1/(18L)$, $\beta_{\max} = 1$, and $\rho = 1.5$.

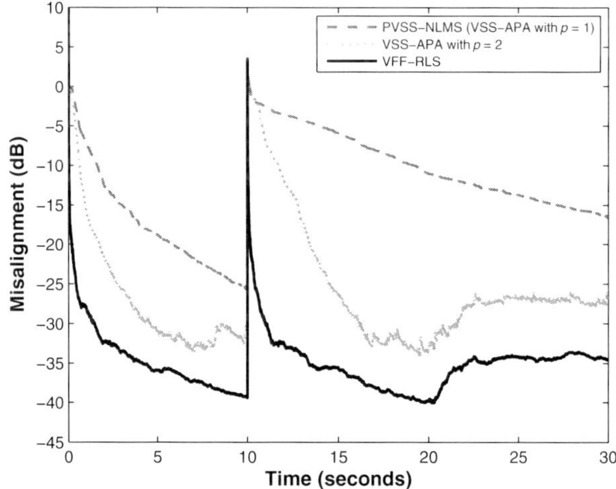

Fig. 4.18 Misalignment of the PVSS-NLMS, VSS-APA (with $p = 2$), and VFF-RLS algorithms. Other conditions are the same as in Fig. 4.17.

4.7 Conclusions

The main objective of this chapter has been to present several VSS algorithms suitable for AEC. The root of this family of algorithms is the NPVSS-NLMS derived in [14]. Its basic principle is to recover the system noise from the error signal of the adaptive filter in the context of a system identification problem. The only parameter that is needed is the power estimate of the system noise. Translating this problem in terms of AEC configuration, there is a strong need in the power estimation of the near-end signal. Following this issue, three algorithms were presented in this chapter, i.e., SVSS-NLMS, NEW-NPVSS-NLMS, and PVSS-NLMS. Among these, the most suitable for real-world AEC applications seems to be the PVSS-NLMS algorithm, since it is entirely non-parametric. In order to improve its tracking capabilities, the idea of the PVSS-NLMS algorithm was generalized to a VSS-APA. It was shown that this algorithm is also robust against near-end signal variations, like the increase of the background noise or double-talk. Concerning the last scenario, the VSS-APA can be combined with a simple Geigel DTD in order to further enhance its robustness. This is also a low complexity practical solution, taking into account that in AEC applications more complex DTDs or techniques based on robust statistics are involved. We also extended the basic idea of the NPVSS-NLMS to the RLS algorithm (which is controlled by a forgetting factor), resulting a VFF-RLS algorithm. Even if the computational complexity limits its use in AEC systems, this algorithm could be suitable in subbands or for other practical system identification problems.

Finally, we should mention that the PVSS-NLMS, VSS-APA, and VFF-RLS algorithms have simple and efficient mechanisms to control their variable parameters; as a result, we recommend them for real-world applications.

References

1. S. L. Gay and J. Benesty, Eds., *Acoustic Signal Processing for Telecommunication.* Boston, MA: Kluwer Academic Publisher, 2000.
2. J. Benesty, T. Gaensler, D. R. Morgan, M. M. Sondhi, and S. L. Gay, *Advances in Network and Acoustic Echo Cancellation.* Berlin, Germany: Springer-Verlag, 2001.
3. J. Benesty and Y. Huang, Eds., *Adaptive Signal Processing–Applications to Real-World Problems.* Berlin, Germany: Springer-Verlag, 2003.
4. B. Widrow and S. D. Stearns, *Adaptive Signal Processing.* Englewood Cliffs, NJ: Prentice-Hall, 1985.
5. S. Haykin, *Adaptive Filter Theory*, Fourth ed. Upper Saddle River, NJ: Prentice-Hall, 2002.
6. A. H. Sayed, *Adaptive Filters.* New York: Wiley, 2008.
7. C. Breining, P. Dreiseitel, E. Haensler, A. Mader, B. Nitsch, H. Puder, T. Schertler, G. Schmidt, and J. Tilp, "Acoustic echo control–an application of very-high-order adaptive filters," *IEEE Signal Process. Magazine*, vol. 16, pp. 42–69, July 1999.
8. K. Ozeki and T. Umeda, "An adaptive filtering algorithm using an orthogonal projection to an affine subspace and its properties," *Electron. Commun. Jpn.*, vol. 67-A, pp. 19–27, May 1984.
9. S. L. Gay and S. Tavathia, "The fast affine projection algorithm," in *Proc. IEEE ICASSP*, 1995, vol. 5, pp. 3023–3026.
10. M. Tanaka, Y. Kaneda, S. Makino, and J. Kojima, "A fast projection algorithm for adaptive filtering," *IEICE Trans. Fundamentals*, vol. E78-A, pp. 1355–1361, Oct. 1995.

11. F. Albu and C. Kotropoulos, "Modified Gauss-Seidel affine projection algorithm for acoustic echo cancellation," in *Proc. IEEE ICASSP*, 2005, vol. 3, pp. 121–124.
12. A. Mader, H. Puder, and G. U. Schmidt, "Step-size control for acoustic echo cancellation filters–an overview," *Signal Process.*, vol. 80, pp. 1697–1719, Sept. 2000.
13. H.-C. Shin, A. H. Sayed, and W.-J. Song, "Variable step-size NLMS and affine projection algorithms," *IEEE Signal Process. Lett.*, vol. 11, pp. 132–135, Feb. 2004.
14. J. Benesty, H. Rey, L. Rey Vega, and S. Tressens, "A nonparametric VSS NLMS algorithm," *IEEE Signal Process. Lett.*, vol. 13, pp. 581–584, Oct. 2006.
15. H. Rey, L. Rey Vega, S. Tressens, and J. Benesty, "Variable explicit regularization in affine projection algorithm: Robustness issues and optimal choice," *IEEE Trans. Signal Process.*, vol. 55, pp. 2096–2108, May 2007.
16. C. Paleologu, S. Ciochina, and J. Benesty, "Variable step-size NLMS algorithm for under-modeling acoustic echo cancellation," *IEEE Signal Process. Lett.*, vol. 15, pp. 5–8, 2008.
17. M. A. Iqbal and S. L. Grant, "Novel variable step size NLMS algorithms for echo cancellation," in *Proc. IEEE ICASSP*, 2008, pp. 241–244.
18. C. Paleologu, S. Ciochina, and J. Benesty, "Double-talk robust VSS-NLMS algorithm for under-modeling acoustic echo cancellation," in *Proc. IEEE ICASSP*, 2008, pp. 245–248.
19. C. Paleologu, J. Benesty, and S. Ciochina, "A variable step-size affine projection algorithm designed for acoustic echo cancellation," *IEEE Trans. Audio, Speech, Language Process.*, vol. 16, pp. 1466–1478, Nov. 2008.
20. C. Paleologu, J. Benesty, and S. Ciochina, "A robust variable forgetting factor recursive least-squares algorithm for system identification," *IEEE Signal Process. Lett.*, vol. 15, pp. 597–600, 2008.
21. D. R. Morgan and S. G. Kratzer, "On a class of computationally efficient, rapidly converging, generalized NLMS algorithms," *IEEE Signal Process. Lett.*, vol. 3, pp. 245–247, Aug. 1996.
22. S. Gollamudi, S. Nagaraj, S. Kapoor, and Y.-F. Huang, "Set-membership filtering and a set-membership normalized LMS algorithm with an adaptive step size," *IEEE Signal Process. Lett.*, vol. 5, pp. 111–114, May 1998.
23. S.-H. Leung and C. F. So, "Gradient-based variable forgetting factor RLS algorithm in time-varying environments," *IEEE Trans. Signal Process.*, vol. 53, pp. 3141–3150, Aug. 2005.
24. S. G. Sankaran and A. A. L. Beex, "Convergence behavior of affine projection algorithms," *IEEE Trans. Signal Process.*, vol. 48, pp. 1086–1096, Apr. 2000.
25. J. C. Jenq and S. F. Hsieh, "Decision of double-talk and time-variant echo path for acoustic echo cancellation," *IEEE Signal Process. Lett.*, vol. 10, pp. 317–319, Nov. 2003.
26. J. Huo, S. Nordholm, and Z. Zang, "A method for detecting echo path variation," in *Proc. IWAENC*, 2003, pp. 71–74.
27. A. H. Sayed and T. Kailath, "A state-space approach to adaptive RLS filtering," *IEEE Signal Process. Magazine*, vol. 11, pp. 18–60, July 1994.
28. S. Song, J. S. Lim, S. J. Baek, and K. M. Sung, "Gauss Newton variable forgetting factor recursive least squares for time varying parameter tracking," *Electronics Lett.*, vol. 36, pp. 988–990, May 2000.

Chapter 5
Simultaneous Detection and Estimation Approach for Speech Enhancement and Interference Suppression

Ari Abramson and Israel Cohen

Abstract [1]In this chapter, we present a simultaneous detection and estimation approach for speech enhancement. A detector for speech presence in the short-time Fourier transform domain is combined with an estimator, which jointly minimizes a cost function that takes into account both detection and estimation errors. Cost parameters control the trade-off between speech distortion, caused by missed detection of speech components, and residual musical noise resulting from false-detection. Furthermore, a modified decision-directed *a priori* signal-to-noise ratio (SNR) estimation is proposed for transient-noise environments. Experimental results demonstrate the advantage of using the proposed simultaneous detection and estimation approach with the proposed *a priori* SNR estimator, which facilitate suppression of transient noise with a controlled level of speech distortion.

5.1 Introduction

In many signal processing applications as well as communication applications, the signal to be estimated is not surely present in the available noisy observation. Therefore, algorithms often try to estimate the signal under uncertainty by using some *a priori* probability for the existence of the signal, e.g., [1, 2, 3, 4], or alternatively, apply an independent detector for signal presence, e.g., [5, 6, 7, 8, 9, 10]. The detector may be designed based on the noisy observation, or, on the estimated signal. Considering speech sig-

Ari Abramson

Technion–Israel Institute of Technology, Israel, e-mail: `aari@tx.technion.ac.il`

Israel Cohen

Technion–Israel Institute of Technology, Israel e-mail: `icohen@ee.technion.ac.il`

[1] This work was supported by the Israel Science Foundation under Grant 1085/05 and by the European Commission under project Memories FP6-IST-035300.

I. Cohen et al. (Eds.): Speech Processing in Modern Communication, STSP 3, pp. 127–150.
springerlink.com   © Springer-Verlag Berlin Heidelberg 2010

nals, the spectral coefficients are generally sparse in the short-time Fourier transform (STFT) domain in the sense that speech is present only in some of the frames, and in each frame only some of the frequency-bins contain the significant part of the signal energy. Therefore, both signal estimation and detection are generally required while processing noisy speech signals. The well-known spectral-subtraction algorithm [11, 12] contains an elementary detector for speech activity in the time-frequency domain, but it generates musical noise caused by falsely detecting noise peaks as bins that contain speech, which are randomly scattered in the STFT domain. Subspace approaches for speech enhancement [13, 14, 15, 16] decompose the vector of the noisy signal into a signal-plus-noise subspace and a noise subspace, and the speech spectral coefficients are estimated after removing the noise subspace. Accordingly, these algorithms are aimed at detecting the speech coefficients and subsequently estimating their values. McAulay and Malpass [2] were the first to propose a speech spectral estimator under a two-state model. They derived a maximum likelihood (ML) estimator for the speech spectral amplitude under speech-presence uncertainty. Ephraim and Malah followed this approach of signal estimation under speech presence uncertainty and derived an estimator which minimizes the mean-squared error (MSE) of the short-term spectral amplitude (STSA) [3]. In [17], speech presence probability is evaluated to improve the minimum MSE (MMSE) of the LSA estimator, and in [4] a further improvement of the MMSE-LSA estimator is achieved based on a two-state model.

Middleton *et al.* [18, 19] were the first to propose simultaneous signal detection and estimation within the framework of statistical decision theory. This approach was recently generalized to speech enhancement, as well as single sensor audio source separation [20, 21]. The speech enhancement problem is formulated by incorporating simultaneous operations of detection and estimation. A detector for the speech coefficients is combined with an estimator, which jointly minimizes a cost function that takes into account both estimation and detection errors. Under speech-presence, the cost is proportional to some distortion between the desired and estimated signals, while under speech-absence, the distortion depends on a certain attenuation factor [12, 4, 22]. A combined detector and estimator enables to control the trade-off between speech distortion, caused by missed detection of speech components, and residual musical noise resulting from false-detection. The combined solutions generalize standard algorithms, which involve merely estimation under signal presence uncertainty.

In some speech processing applications, an indicator for the transient noise activity may be available, *e.g.*, a siren noise in an emergency car, lens-motor noise of a digital video camera or a keyboard typing noise in a computer-based communication system. The transient spectral variances can be estimated in such cases from training signals. However, applying a standard estimator to the spectral coefficients may result in removal of critical speech components in case of falsely detecting the speech components, or under-suppression of transient noise in case of miss detecting the noise transients. For cases where some indicator (or detector) for the presence of noise transients in the STFT domain is available, the speech enhancement problem is reformulated using two hypotheses. Cost parameters control the trade-off between speech distortion and residual transient noise. The optimal signal estimator is derived which employs the available detector. The resulting estimator generalizes the optimally-modified log-spectral amplitude (OM-LSA) estimator [4].

This chapter is organized as follows. In Section 5.2, we briefly review classical speech enhancement under signal presence uncertainty. In Section 5.3, the speech enhancement problem is reformulated as a simultaneous detection and estimation problem in the STFT domain. A detector for the speech coefficients is combined with an estimator, which jointly minimizes a cost function that takes into account both estimation and detection errors. The combined solution is derived for the quadratic distortion measure as well as the quadratic spectral amplitude distortion measure. In Section 5.4, we consider the integration of a spectral estimator with a *given* detector for noise transients and derive an optimal estimator which minimizes the mean-square error of the log-spectral amplitude. In Section 5.5, a modification of the decision-directed *a priori* signal-to-noise ratio (SNR) estimator is presented which better suites transient-noise environments. Experimental results are given in Section 5.6. It shows that the proposed approaches facilitate improved noise reduction with a controlled level of speech distortion.

5.2 Classical Speech Enhancement in Nonstationary Noise Environments

Let us start with a short presentation of classical approach for spectral speech enhancement while considering nonstationary noise environments. Specifically, we may assume that some indicator for transient noise activity is available.

Let $x(n)$ and $d(n)$ denote speech and uncorrelated additive noise signals, and let $y(n) = x(n) + d(n)$ be the observed signal. Applying the STFT to the observed signal, we have

$$Y_{\ell k} = X_{\ell k} + D_{\ell k} \,, \tag{5.1}$$

where $\ell = 0, 1, \ldots$ is the time frame index and $k = 0, 1, \ldots, K - 1$ is the frequency-bin index. Let $H_1^{\ell k}$ and $H_0^{\ell k}$ denote, respectively, speech presence and absence hypotheses in the time-frequency bin (ℓ, k), *i.e.*,

$$H_1^{\ell k} : Y_{\ell k} = X_{\ell k} + D_{\ell k} \,,$$
$$H_0^{\ell k} : Y_{\ell k} = D_{\ell k} \,. \tag{5.2}$$

The noise expansion coefficients can be represented as the sum of two uncorrelated noise components $D_{\ell k} = D_{\ell k}^s + D_{\ell k}^t$, where $D_{\ell k}^s$ denotes a quasi-stationary noise component and $D_{\ell k}^t$ denotes a highly nonstationary transient component. The transient components are generally rare, but they may be of high energy and thus cause significant degradation to speech quality and intelligibility. However, in many applications, a reliable indicator for the transient noise activity may be available in the system. For example, in an emergency car (*e.g.*, police or ambulance) the engine noise may be considered as quasi-stationary, but activating a siren results in a highly nonstationary noise which is perceptually very annoying. Since the sound generation in the siren is nonlinear, linear echo cancelers may be inappropriate. In a computer-based communication system, a transient noise such as a keyboard typing noise may be present in addition to quasi-stationary background office noise. Another example is a digital camera, where activating the lens-motor (zooming in/out) may result in high-energy transient noise components, which degrade the recorded audio. In the above examples, an indicator for the transient noise activity may be available, *i.e.*, siren source signal, keyboard output signal and the lens-motor controller output. Furthermore, given that a transient noise source is active, a detector for the transient noise in the STFT domain may be designed and its spectrum can be estimated based on training data.

The objective of a speech enhancement system is to reconstruct the spectral coefficients of the speech signal such that under speech-presence a certain distortion measure between the spectral coefficient and its estimate, $d\left(X_{\ell k}, \hat{X}_{\ell k}\right)$, is minimized. Under speech-absence a constant attenuation of the noisy coefficient would be desired to maintain a natural background noise [22, 4]. Although the speech expansion coefficients are not necessarily present, most classical speech enhancement algorithms try to estimate the spectral coefficients rather than detecting their existence, or try to independently design detectors and estimators. The well-known spectral subtraction algorithm estimates the speech spectrum by subtracting the estimated noise spectrum from the noisy squared absolute coefficients [11, 12], and thresholding the result by some desired residual noise level. Thresholding the spectral coefficients is in fact a detection operation in the time-frequency domain, in the sense that speech coefficients are assumed to be absent in the low-energy time-frequency bins and present in noisy coefficients whose energy is above the threshold.

McAulay and Malpass were the first to propose a two-state model for the speech signal in the time-frequency domain [2]. Accordingly, the MMSE estimator follows

$$
\begin{aligned}
\hat{X}_{\ell k} &= E\left\{X_{\ell k} \mid Y_{\ell k}\right\} \\
&= E\left\{X_{\ell k} \mid Y_{\ell k}, H_1^{\ell k}\right\} p\left(H_1^{\ell k} \mid Y_{\ell k}\right) .
\end{aligned}
\tag{5.3}
$$

The resulting estimator does not detect speech components, but rather, a soft-decision is performed to further attenuate the signal estimate by the *a posteriori* speech presence probability. Ephraim and Malah followed the same approach and derived an estimator which minimizes the MSE of the STSA under signal presence uncertainty [3]. Accordingly,

$$\left|\hat{X}_{\ell k}\right| = E\left\{|X_{\ell k}| \mid Y_{\ell k}, H_1^{\ell k}\right\} p\left(H_1^{\ell k} \mid Y_{\ell k}\right). \tag{5.4}$$

Both in [2] and [3], under $H_0^{\ell k}$ the speech components are assumed zero and the *a priori* probability of speech presence is both time and frequency invariant, *i.e.*, $p\left(H_1^{\ell k}\right) = p\left(H_1\right)$. In [17, 4], the speech presence probability is evaluated for each frequency-bin and time-frame to improve the performance of the MMSE-LSA estimator [23]. Further improvement of the MMSE-LSA suppression rule can be achieved by considering under $H_0^{\ell k}$ a constant attenuation factor $G_f << 1$, which is determined by subjective criteria for residual noise naturalness, see also [22]. The OM-LSA estimator [4] is given by

$$\left|\hat{X}_{\ell k}\right| = \left(\exp\left[E\left\{\log|X_{\ell k}| \mid Y_{\ell k}, H_1^{\ell k}\right\}\right]\right)^{p\left(H_1^{\ell k} \mid Y_{\ell k}\right)} \left(G_f \left|Y_{\ell k}\right|\right)^{1-p\left(H_1^{\ell k} \mid Y_{\ell k}\right)}. \tag{5.5}$$

Suppose that an indicator for the presence of transient noise components is available in a highly nonstationary noise environment, then high-energy transients may be attenuated by using one of the above-mentioned estimators (5.3)–(5.5) and heuristically setting the *a priori* speech presence probability $p\left(H_1^{\ell k}\right)$ to a sufficiently small value. Unfortunately, this also results in suppression of desired speech components and intolerable degradation of speech quality. In general, an estimation-only approach under signal presence uncertainty produces larger speech degradation for small $p\left(H_1^{\ell k}\right)$, since the optimal estimate is attenuated by the *a posteriori* speech presence probability. On the other hand, increasing $p\left(H_1^{\ell k}\right)$ prevents the estimator from sufficiently attenuating noise components. Integrating a jointly optimal detector and estimator into the speech enhancement system may significantly improve the speech enhancement performance under highly non-stationary noise conditions and may allow further reduction of transient components without much degradation of the desired signal.

5.3 Simultaneous Detection and Estimation for Speech Enhancement

Middleton and Esposito [18] were the first to propose simultaneous signal detection and estimation within the framework of statistical decision theory. This approach was generalized to speech enhancement formalism in [20]. A decision space, $\left\{\eta_0^{\ell k}, \eta_1^{\ell k}\right\}$, is assumed for the detection operation where under the decision $\eta_j^{\ell k}$, signal hypothesis $H_j^{\ell k}$ is accepted and a corresponding estimate $\hat{X}_{\ell k} = \hat{X}_{\ell k,j}$ is considered. The detection and estimation are strongly coupled so that the detector is optimized with the knowledge of the specific structure of the estimator, and the estimator is optimized in the sense of minimizing a Bayesian risk associated with the combined operations. For notation simplification, we omit the time-frequency indices (ℓ, k). Let

$$C_j\left(X,\hat{X}\right) \geq 0 \tag{5.6}$$

denote the cost of making a decision η_j (and choosing an estimator $\hat{X}_j$) where X is the desired signal. Then, the Bayes risk of the two operations associated with simultaneous detection and estimation is defined by [18, 19, 20]

$$R = \sum_{j=0}^{1} \int_{\Omega_y} \int_{\Omega_x} C_j\left(X,\hat{X}\right) p\left(\eta_j \mid Y\right) p\left(Y \mid X\right) p\left(X\right) dX dY , \tag{5.7}$$

where Ω_x and Ω_y are the spaces of the speech and noisy signals, respectively. The simultaneous detection and estimation approach is aimed at jointly minimizing the Bayes risk over both the decision rule and the corresponding signal estimate. Let $q \triangleq p\left(H_1\right)$ denote the *a priori* speech presence probability. Then, the *a priori* distribution of the speech expansion coefficient follows

$$p\left(X\right) = q\,p\left(X \mid H_1\right) + (1-q)\,p\left(X \mid H_0\right) , \tag{5.8}$$

where $p\left(X \mid H_0\right) = \delta\left(X\right)$ denotes the Dirac-delta function. The cost function $C_j\left(X,\hat{X}\right)$ may be defined differently whether H_1 or H_0 is true. Therefore, we let

$$C_{ij}\left(X,\hat{X}\right) \triangleq C_j\left(X,\hat{X} \mid H_i\right) \tag{5.9}$$

denote the cost which is conditioned on the true hypothesis[2]. The cost function $C_{ij}\left(X,\hat{X}\right)$ depends on both the true signal value and its estimate under the decision η_j and therefore couples the operations of detection and estimation. By substituting (5.8) and (5.9) into (5.7) we obtain [20]

$$R = \int_{\Omega_y} p\left(\eta_0 \mid Y\right) \left[q\,r_{10}\left(Y\right) + (1-q)\,r_{00}\left(Y\right)\right] dY$$
$$+ \int_{\Omega_y} p\left(\eta_1 \mid Y\right) \left[q\,r_{11}\left(Y\right) + (1-q)\,r_{01}\left(Y\right)\right] dY , \tag{5.10}$$

where

$$r_{ij}\left(Y\right) = \int_{\Omega_x} C_{ij}\left(X,\hat{X}\right) p\left(X \mid H_i\right) p\left(Y \mid X\right) dX \tag{5.11}$$

---

[2] Note that $X = 0$ implies that H_0 is true and $X \neq 0$ implies H_1 so the sub-index i may seem to be redundant. However, this notation simplifies the subsequent formulations.

denotes the risk associated with the pair $\{H_i, \eta_j\}$ and the observation Y.

Since the detector's decision under a given observation is binary, *i.e.*, $p(\eta_j \,|\, Y) \in \{0, 1\}$, for minimizing the combined risk we first evaluate the optimal estimator under each of the decisions, then the optimal decision rule is derived based on the optimal estimators $\hat{X}_0$, $\hat{X}_1$ to further minimize the combined risk. The two-stage minimization guaranties minimum combined risk [19]. The optimal *nonrandom* decision rule which minimizes the combined risk (5.10) is given by

Decide η_1 (*i.e.*, $p(\eta_1 \,|\, Y) = 1$) if

$$q\left[r_{10}(Y) - r_{11}(Y)\right] \geq (1-q)\left[r_{01}(Y) - r_{00}(Y)\right] , \qquad (5.12)$$

otherwise, decide η_0.

The optimal estimator under a decision η_j is obtained from (5.10) by

$$\arg\min_{\hat{X}_j} \left\{ q\, r_{1j}(Y) + (1-q)\, r_{0j}(Y) \right\} . \qquad (5.13)$$

Note that $r_{ij}(Y)$ depends on the estimate $\hat{X}_j$ through the cost function.

The cost function associated with the pair $\{H_i, \eta_j\}$ is generally defined by

$$C_{ij}\left(X, \hat{X}\right) = b_{ij}\, d_{ij}\left(X, \hat{X}\right) , \qquad (5.14)$$

where $d_{ij}\left(X, \hat{X}\right)$ is an appropriate distortion measure and the cost parameters b_{ij} control the trade-off between the costs associated with the pairs $\{H_i, \eta_j\}$. That is, a high valued b_{01} raises the cost of a false alarm (*i.e.*, decision of speech presence when speech is actually absent), which may result in residual musical noise. Similarly, b_{10} is associated with the cost of missed detection of a signal component, which may cause perceptual signal distortion. Under a correct classification, normalized cost parameters are generally used, $b_{00} = b_{11} = 1$. However, $d_{ii}(\cdot, \cdot)$ is not necessarily zero since estimation errors are still possible even when there is no detection error.

Contrary to the approach in [18, 19, 24], in speech enhancement application the signal is not rejected when a decision η_0 is made. Instead, the estimator $\hat{X}_0 \neq 0$ compensates for any detection errors to reduce potential musical noise and audible distortions. Furthermore, when speech is indeed absent the distortion function is defined to allow some natural background noise level such that under H_0 the attenuation factor will be lower bounded by a constant gain floor $G_f << 1$ as proposed in [12, 4, 25, 22].

In the following subsections we specify two distortions measures which are of interest for spectral speech enhancement and derive their combined solution of detection and estimation.

5.3.1 Quadratic Distortion Measure

The distortion measure associated with the MMSE estimation is the quadratic distortion measure which follows:

$$
d_{ij}\left(X,\hat{X}\right) = \begin{cases} \left|X - \hat{X}_j\right|^2, & i = 1 \\ \left|G_f Y - \hat{X}_j\right|^2, & i = 0 \end{cases}.
\tag{5.15}
$$

We assume that both X and D are statistically independent, zero-mean, complex-valued Gaussian random variables with variances λ_x and λ_d, respectively. Let $\xi \triangleq \lambda_x/\lambda_d$ denote the *a priori* SNR under hypothesis H_1, let $\gamma \triangleq |Y|^2/\lambda_d$ denote the *a posteriori* SNR and let $v \triangleq \gamma \xi/(1+\xi)$. Then, the optimal estimation under the quadratic cost function is obtained by substituting (5.15) and (5.14) into (5.13) and set its derivation to zero [19, 21]

$$
\begin{aligned}
\hat{X}_j &= \left[b_{1j}\,\Lambda\left(\xi,\gamma\right) G_{MSE}\left(\xi\right) + b_{0j}\,G_f\right]\phi_j\left(\xi,\gamma\right)^{-1} Y \\
&\triangleq G_j\,Y, \qquad j = 0,1,
\end{aligned}
\tag{5.16}
$$

where $G_{MSE}\left(\xi\right) = \left(1 + \xi^{-1}\right)^{-1}$ is the MSE gain function under signal presence, $\Lambda\left(\xi,\gamma\right)$ is the generalized likelihood ratio

$$
\begin{aligned}
\Lambda\left(\xi,\gamma\right) &= \frac{q}{(1-q)}\frac{p\left(Y\,|\,H_1\right)}{p\left(Y\,|\,H_0\right)} \\
&= \frac{q}{(1-q)}\frac{e^v}{1+\xi},
\end{aligned}
\tag{5.17}
$$

and the normalization factor $\phi_j\left(\xi,\gamma\right)$ is given by

$$
\phi_j\left(\xi,\gamma\right) = b_{0j} + b_{1j}\,\Lambda\left(\xi,\gamma\right).
\tag{5.18}
$$

The risk $r_{ij}\left(Y\right)$, required for the detector decision is obtained by substituting the quadratic cost function (5.15) into (5.11). Let X_R and X_I denote the real and imaginary parts of the expansion coefficient X. Then, for the case where H_1 is true we have

$$
r_{1j}\left(Y\right) =
$$

$$
b_{1j} \int\int_{-\infty}^{\infty} \left[X_R^2 + X_I^2 + G_j^2\left(\xi,\gamma\right)|Y|^2 - 2G_j\left(\xi,\gamma\right)\left(X_R Y_R + X_I Y_I\right)\right]
$$

$$
\times\, p\left(X_R, X_I\,|\,H_1\right) p\left(Y\,|\,X_R, X_I\right) dX_R\,dX_I,
\tag{5.19}
$$

where

$$p(X_R, X_I \mid H_1) \;=\; \frac{1}{\pi\lambda_x} \exp\left(-\frac{|X|^2}{\lambda_x}\right), \tag{5.20}$$

$$p(Y \mid X_R, X_I) \;=\; \frac{1}{\pi\lambda_d} \exp\left(-\frac{|Y-X|^2}{\lambda_d}\right). \tag{5.21}$$

By using [26, eq. 3.323.2] we have

$$\iint_{-\infty}^{\infty} p(X_R, X_I \mid H_1)\, p(Y \mid X_R, X_I)\, dX_R\, dX_I = \frac{\exp\left(-\frac{\gamma}{1+\xi}\right)}{\pi\lambda_d(1+\xi)}, \tag{5.22}$$

and using [26, eq. 3.462.2] we obtain

$$\iint_{-\infty}^{\infty} (X_R Y_R + X_I Y_I)\, p(X_R, X_I \mid H_1)\, p(Y \mid X_R, X_I)\, dX_R\, dX_I =$$
$$\frac{\upsilon}{\pi(1+\xi)} \exp\left(-\frac{\gamma}{1+\xi}\right), \tag{5.23}$$

and

$$\iint_{-\infty}^{\infty} (X_R^2 + X_I^2)\, p(X_R, X_I \mid H_1)\, p(Y \mid X_R, X_I)\, dX_R\, dX_I =$$
$$\frac{\xi(1+\upsilon)}{\pi(1+\xi)^2} \exp\left(-\frac{\gamma}{1+\xi}\right). \tag{5.24}$$

Substituting (5.22)–(5.24) into (5.19) we obtain

$$r_{1j}(Y) = \frac{b_{1j}}{\pi} \frac{\exp\left(-\frac{\gamma}{1+\xi}\right)}{1+\xi} \left[G_{MSE}(\xi) + (G_j(\xi,\gamma) - G_{MSE}(\xi))^2 \gamma\right], \tag{5.25}$$

where $G_j(\xi,\gamma)$ is defined by (5.16).

Under hypothesis H_0 speech is absent and $p(X_R, X_I \mid H_0) = \delta(X_R, X_I)$. Consequently,

$$\begin{aligned}
r_{0j}(Y) &= b_{0j} \iint_{-\infty}^{\infty} \left[(G_j(\xi,\gamma) - G_f)^2 |Y|^2\right] \\
&\quad \times p(X_R, X_I \mid H_i)\, p(Y \mid X_R, X_I)\, dX_R\, dX_I \\
&= \frac{b_{0j}}{\pi} [G_j(\xi,\gamma) - G_f]^2\, \gamma\, e^{-\gamma}.
\end{aligned} \tag{5.26}$$

By substituting (5.25) and (5.26) into (5.12) we obtain the optimal decision rule for the detection operation under the quadratic distortion measure:

$$\frac{(1-q)(1+\xi)}{q\, e^{\upsilon}} \left[(G_0 - G_f)^2 - b_{01}(G_1 - G_f)^2\right] \underset{\eta_0}{\overset{\eta_1}{\gtrless}}$$
$$\frac{1-b_{10}}{\gamma} G_{MSE} + (G_1 - G_{MSE})^2 - b_{10}(G_0 - G_{MSE})^2. \tag{5.27}$$

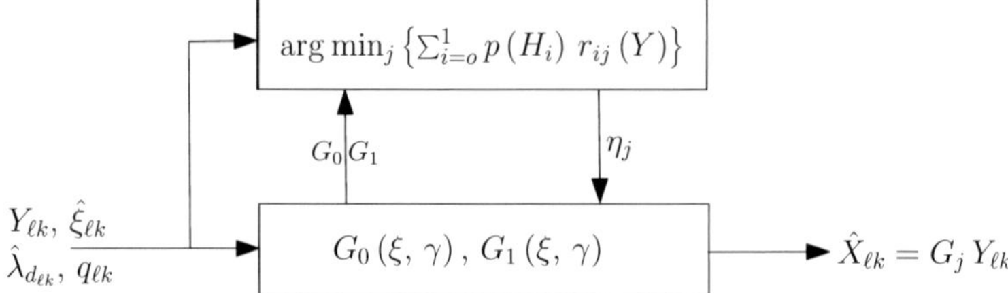

Fig. 5.1 Block diagram of the simultaneous detection and estimation.

Here, G_j, $j = 0, 1$ and G_{MSE} hold for $G_j\left(\xi, \gamma\right)$ and $G_{MSE}\left(\xi\right)$, respectively.

To conclude, a block diagram of the signal simultaneous detection and estimation from a noisy observation is shown in Fig. 5.1. The gain factor G_j is calculated under any of the decisions using (5.16). Then, the optimal decision η_j is calculated using (5.27) and accordingly, the estimated signal is given by applying selected gain to the noisy signal.

It is of interest to examine the asymptotic behavior of the estimator (5.16) under each of the decisions made by the detector. When the cost parameter associated with false alarm is much smaller than the generalized likelihood ratio, i.e., $b_{01} << \Lambda\left(\xi, \gamma\right)$, the spectral gain under the decision η_1 implies $G_1\left(\xi, \gamma\right) \cong G_{MSE}\left(\xi\right)$ which is the optimal estimation under no uncertainty. However, if $b_{01} >> \Lambda\left(\xi, \gamma\right)$, the spectral gain under η_1 needs to compensate the possibility of a high-cost false-decision made by the detector and thus $G_1\left(\xi, \gamma\right) \cong G_f$. On the other hand, if the cost parameter associated with missed detection is small and we have $b_{10} << \Lambda\left(\xi, \gamma\right)^{-1}$, than $G_0\left(\xi, \gamma\right) \cong G_f$ (i.e., estimation where speech is surely absence) but under $b_{10} >> \Lambda\left(\xi, \gamma\right)^{-1}$, in order to overcome the high cost related to missed detection, we have $G_0\left(\xi, \gamma\right) \cong G_{MSE}\left(\xi\right)$.

Recall that

$$\frac{\Lambda\left(\xi, \gamma\right)}{1 + \Lambda\left(\xi, \gamma\right)} = p\left(H_1 \,|\, Y\right) \tag{5.28}$$

is the *a posteriori* probability of speech presence, it can be seen that the proposed estimator (5.16) generalizes existing estimators. For the case of equal parameters $b_{ij} = 1 \; \forall i, j$ and $G_f = 0$ we get the estimation under signal presence uncertainty (5.3). In that case the detection operation is not needed since the estimation is independent of the detection rule.

Figure 5.2 shows attenuation curves under quadratic cost function as a function of the *a priori* SNR, ξ, with *a posteriori* SNR of $\gamma = 5$ dB, $q = 0.8$, $G_f = -15$ dB and cost parameters $b_{01} = 4$, $b_{10} = 2$. In (a), the gains G_1 (dash line), G_0 (dotted line) and the total detection and estimation system gain G (solid line) are shown and compared with (b) the MSE gain function under no uncertainty G_{MSE} (dashed line) and the MMSE estimation under signal presence uncertainty which is defined by (5.3) (dashed line). It can be seen that for *a priori* SNRs higher than about -10 dB the detector decision is η_1

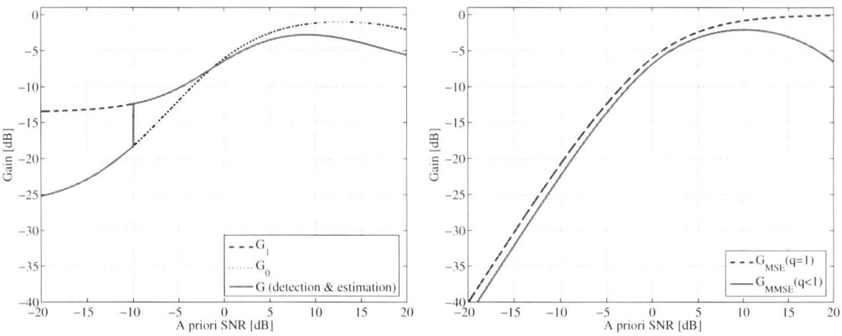

Fig. 5.2 Gain curves under quadratic cost function with $\gamma = 5$ [dB], $q = 0.8$, and $G_f = -15$, [dB]; (a) G_1, G_0 and the detection and estimation gain G with $b_{01} = 4$, $b_{10} = 2$, and (b) G_{MSE} gain curve with $q = 1$ and the MMSE gain curve under uncertainty ($q = 0.8$).

and therefore the total gain is $G = G_1$. For lower *a priori* SNRs, η_0 is decided and consequently the total gain is G_0. Note that if an ideal detector for the speech coefficients would be available, a more significantly non-continuous gain would be desired to block the noise-only coefficients. However, in the simultaneous detection and estimation approach the detector is not ideal but optimized to minimize the combined risk and the non-continuity of the system gain depends on the chosen cost parameters as well as on the gain floor.

5.3.2 Quadratic Spectral Amplitude Distortion Measure

The distortion measure of the quadratic spectral amplitude (QSA) is defined by

$$
d_{ij}\left(X, \hat{X}\right) = \begin{cases} \left(|X| - \left|\hat{X}_j\right|\right)^2, & i = 1 \\ \left(G_f\,|Y| - \left|\hat{X}_j\right|\right)^2, & i = 0 \end{cases}, \tag{5.29}
$$

and is related to the STSA suppression rule of Ephraim and Malah [3]. For evaluating the optimal detector and estimator under the QSA distortion measure we denote by $X \triangleq A\,e^{j\alpha}$ and $Y \triangleq R\,e^{j\theta}$ the clean and noisy spectral coefficients, respectively, where $A = |X|$ and $R = |Y|$. Accordingly, the pdf of the speech expansion coefficient under H_1 satisfies

$$
p\left(a, \alpha \mid H_1\right) = \frac{a}{\pi \lambda_x} \exp\left(-\frac{a^2}{\lambda_x}\right). \tag{5.30}
$$

Since the combined risk under the QSA distortion measure is independent of the signal phase nor the estimation phase, the estimated amplitude under η_j

is given by

$$
\hat{A}_j \;\; = \arg\min_{\hat{a}} \;\; \left\{ q\, b_{1j} \int_0^\infty \int_0^{2\pi} (a - \hat{a})^2\, p(a, \alpha \mid H_1)\, p(Y \mid a, \alpha)\, d\alpha\, da \right.
$$

$$
\left. + (1 - q)\, b_{0j}\, (G_f\, R - \hat{a})^2\, p(Y \mid H_0) \right\} . \tag{5.31}
$$

By using the phase of the noisy signal, the optimal estimation under the decision η_j, $j \in \{0, 1\}$ is given by [20]

$$
\begin{aligned}
\hat{X}_j \;\; &= \;\; [b_{1j}\, \Lambda(\xi, \gamma)\, G_{STSA}(\xi, \gamma) + b_{0j}\, G_f]\, \phi_j(\xi, \gamma)^{-1}\, Y \\
&\triangleq \;\; G_j(\xi, \gamma)\, Y ,
\end{aligned} \tag{5.32}
$$

where

$$
G_{STSA}(\xi, \gamma) = \left[(1 + v)\, I_0\left(\frac{v}{2}\right) + v\, I_1\left(\frac{v}{2}\right) \right] \frac{\sqrt{\pi\, v}}{2\gamma} \exp\left(-\frac{v}{2}\right) \tag{5.33}
$$

denotes the STSA gain function of Ephraim and Malah [3], and $I_\nu(\cdot)$ denotes the modified Bessel function of order ν. For evaluating the optimal decision rule under the QSA distortion measure we need to compute the risk $r_{ij}(Y)$. Under H_1 we have [20]

$$
r_{1j}(Y) = \frac{b_{1j}}{\pi} \frac{\exp\left(-\frac{\gamma}{1+\xi}\right)}{1 + \xi} \left\{ G_j^2\, \gamma + \frac{\xi}{1 + \xi}(1 + v) - 2\gamma\, G_j\, G_{STSA} \right\} , \tag{5.34}
$$

and under H_0 we have

$$
r_{0j}(Y) = \frac{b_{0j}}{\pi} [G_j(\xi, \gamma) - G_f]^2\, \gamma\, e^{-\gamma} . \tag{5.35}
$$

Substituting (5.34) and (5.35) into (5.12), we obtain the optimal decision rule under the QSA distortion measure:

$$
\frac{b_{01}(G_1 - G_f)^2 - (G_0 - G_f)^2}{\Lambda(\xi, \gamma)} \underset{\eta_0}{\overset{\eta_1}{\gtrless}}
$$

$$
b_{10}\, G_0^2 - G_1^2 + \frac{\xi(1 + v)(b_{10} - 1)}{(1 + \xi)\gamma} + 2(G_1 - b_{10}\, G_0)\, G_{STSA} . \tag{5.36}
$$

Figure 5.3 demonstrates attenuation curves under QSA cost function as a function of the *instantaneous* SNR defined by $\gamma - 1$, for several *a priori* SNRs, using the parameters $q = 0.8$, $G_f = -25$ dB and cost parameters $b_{01} = 5$ and $b_{10} = 1.1$. The gains G_1 (dashed line), G_0 (dotted line) and the total detection and estimation system gain (solid line) are compared to the STSA gain under signal presence uncertainty of Ephraim and Malah [3] (dashed-dotted line). The *a priori* SNRs range from -15 dB to 15 dB. Not only

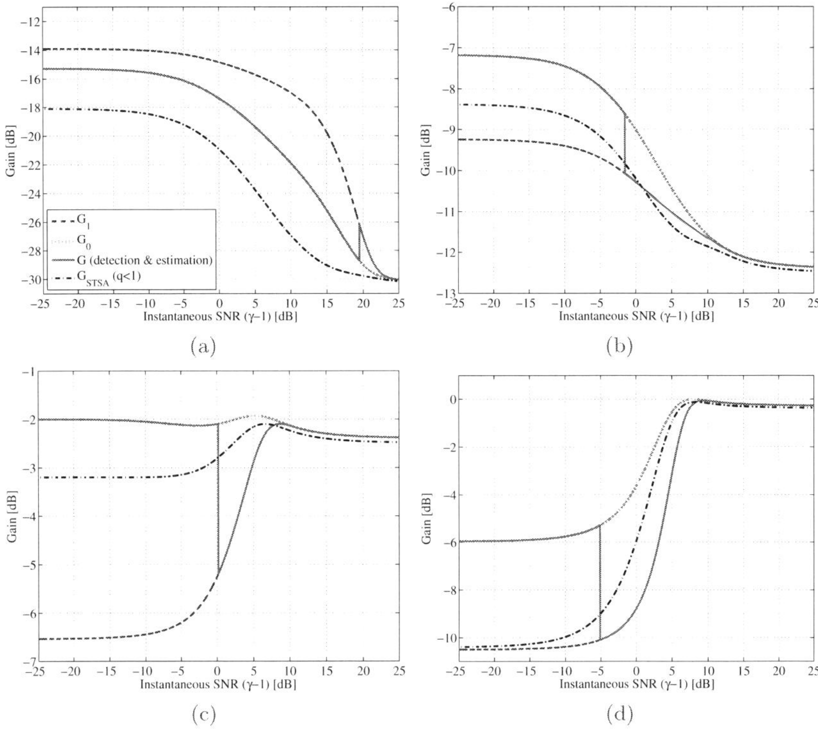

Fig. 5.3 Gain curves of G_1 (dashed line), G_0 (dotted line) and the total detection and estimation system gain curve (solid line), compared with the STSA gain under signal presence uncertainty (dashed-dotted line). The *a priori* SNRs are (a) $\xi = -15$ dB, (b) $\xi = -5$ dB, (c) $\xi = 5$ dB, and (d) $\xi = 15$ dB.

that the cost parameters shape the STSA gain curve, when combined with the detector the simultaneous detection and estimation provides a significant non-continuous modification of the standard STSA estimator. For example, for *a priori* SNRs of $\xi = -5$ and $\xi = 15$ dB, as shown in Fig. 5.3(b) and (d) respectively, as long as the instantaneous SNR is higher than about -2 dB (for $\xi = -5$ dB) or -5 dB (for $\xi = 15$ dB), the detector decision is η_1, while for lower instantaneous SNRs, the detector decision is η_0. The resulting non-continues gain function may yield greater noise reduction with slightly higher level of musicality, while not degrading speech quality.

Similarly to the case of a quadratic distortion measure, when the false alarm parameter is much smaller than the generalized likelihood ratio, $b_{01} << \Lambda(\xi, \gamma)$, the spectral gain of the estimator under the decision η_1 is $G_1(\xi, \gamma) \cong G_{STSA}(\xi, \gamma)$, which is optimal when the signal is surely present. When $b_{01} >> \Lambda(\xi, \gamma)$, the spectral gain under η_1 is $G_1(\xi, \gamma) \cong G_f$ to compensate for false decision made by the detector. If the cost parameter associated with missed detection is small and we have $b_{10} << \Lambda(\xi, \gamma)^{-1}$, then

$G_0(\xi,\gamma) \cong G_f$, and under $b_{10} >> \Lambda(\xi,\gamma)^{-1}$ we have $G_0(\xi,\gamma) \cong G_{STSA}(\xi)$ in order to overcome the high cost related to missed detection.

If one chooses constant cost parameters $b_{ij} = 1\ \forall i,j$, than the detection operation is not required (the estimation is independent of the decision rule), and we have

$$\begin{aligned} \hat{X}_0 &= [p(H_1\,|\,Y)\,G_{STSA}(\xi,\gamma) + (1 - p(H_1\,|\,Y))\,G_f]\,Y \\ &= \hat{X}_1. \end{aligned} \tag{5.37}$$

If we also set G_f to zero, the estimation reduces to the STSA suppression rule under signal presence uncertainty [3].

5.4 Spectral Estimation Under a Transient Noise Indication

In the previous section we introduced a method for optimal integration of a detector and an estimator for speech spectral components. In this section, we consider the integration of a spectral estimator with a *given* detector for noise transients. In many speech enhancement applications, an indicator for the transient source may be available, *e.g.*, siren noise in an emergency car, keyboard typing in computer-based communication system and a lens-motor noise in a digital video camera. In such cases, *a priori* information based on a training phase may yield a reliable detector for the transient noise. However, false detection of transient noise components when signal components are present may significantly degrade the speech quality and intelligibility. Furthermore, missed detection of transient noise components may result in a residual transient noise, which is perceptually annoying. The transient spectral variances can be estimated in such cases from training signals. However, applying a standard estimator to the spectral coefficients may result in removal of critical speech components in case of falsely detecting the speech components, or under-suppression of transient noise in case of miss detecting the noise transients. Consider a reliable detector for transient noise, we can define a cost (5.14) by using the quadratic log-spectral amplitude (QLSA) distortion measure is given by

$$d_{ij}\left(X,\hat{X}\right) = \begin{cases} \left(\log A - \log \hat{A}_j\right)^2, & i = 1 \\ \left(\log (G_f R) - \log \hat{A}_j\right)^2, & i = 0 \end{cases}, \tag{5.38}$$

and is related with the LSA estimation [23].

Similarly to the case of a QSA distortion measure, the average risk is independent of the signal phase nor on the estimation phase. Thus, by substituting the cost function into (5.13) we have

$$\hat{A}_j \;=\; \arg\min_{\hat{a}_j} \left\{ q\, b_{1j} \int_0^\infty \int_0^{2\pi} (\log a - \log \hat{a}_j)^2 \, p(a, \alpha \mid H_1)\, p(Y \mid a, \alpha)\, d\alpha\, da \right.$$

$$\left. + (1-q)\, b_{0j} \left(\log(G_f R) - \log \hat{a}_j\right)^2 p(Y \mid H_0) \right\}. \tag{5.39}$$

By setting the derivative of (5.39) according to $\hat{a}_j$ equal to zero, we obtain[3]

$$q\, b_{1j} \int_0^\infty \int_0^{2\pi} \left(\log a - \log \hat{A}_j\right) p(a, \alpha \mid H_1)\, p(Y \mid a, \alpha)\, d\alpha\, da$$

$$+ (1-q)\, b_{0j} \left(\log(G_f R) - \log \hat{A}_j\right) p(Y \mid H_0) = 0. \tag{5.40}$$

The integration over $\log a$ yields

$$\int_0^\infty \int_0^{2\pi} \log a \, p(a, \alpha \mid H_1)\, p(Y \mid a, \alpha)\, d\alpha\, da$$

$$= \int_0^\infty \int_0^{2\pi} \log a \, p(a, \alpha \mid Y, H_1)\, p(Y \mid H_1)\, d\alpha\, da$$

$$= E\{\log a \mid Y, H_1\}\, p(Y \mid H_1)\,, \tag{5.41}$$

and

$$\int_0^\infty \int_0^{2\pi} p(a, \alpha \mid H_1)\, p(Y \mid a, \alpha)\, d\alpha\, da$$

$$= \int_0^\infty \int_0^{2\pi} p(a, \alpha, Y \mid H_1)\, d\alpha\, da = p(Y \mid H_1)\,. \tag{5.42}$$

Substituting (5.41) and (5.42) into (5.40) we obtain

$$\hat{A}_j = \exp\left\{ b_{1j} \Lambda(Y)\, E\{\log a \mid Y\}\, \phi_j(Y)^{-1} + b_{0j} \log(G_f R)\, \phi_j(Y)^{-1} \right\}, \tag{5.43}$$

where

$$\exp[E\{\log a \mid Y\}] \;=\; \frac{\xi}{1+\xi} \exp\left(\frac{1}{2} \int_v^\infty \frac{e^{-t}}{t}\, dt\right) R$$

$$\triangleq\; G_{LSA}(\xi, \gamma)\, R\,, \tag{5.44}$$

is the LSA suppression rule [23].

Substituting (5.44) into (5.43) and applying the noisy phase, we obtain the optimal estimation under a decision η_j, $j \in \{0, 1\}$:

$$\hat{X}_j = \left[G_f^{b_{0j}} G_{LSA}(\xi, \gamma)^{b_{1j}\Lambda(\xi,\gamma)} \right]^{\phi_j(\xi,\gamma)^{-1}} Y \triangleq G_j(\xi, \gamma)\, Y\,. \tag{5.45}$$

---

[3] Note that this solution is not dependent on the basis of the log and in addition, the optimal solution is strictly positive.

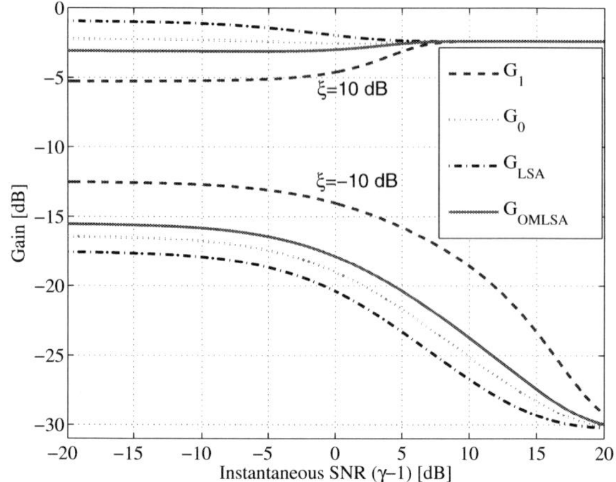

Fig. 5.4 Gain curves under QLSA distortion measure with $q = 0.8$, $b_{01} = 4$, $b_{10} = 2$, and $G_f = -15$ dB.

If we consider the simultaneous detection and estimation which was formulated in Section 5.3, the derivation of the optimal decision rule for the QLSA distortion measure is mathematically intractable. However, the estimation (5.45) under *any given detector* is still optimal in the sense of minimizing the average cost under the given decision. Therefore, the estimation (5.45) can be incorporated with any reliable detector to yield a sub-optimal detection and estimation system. Even where a non optimal detector is considered, the use of the cost parameters enables better control on the spectral gain under any decision made by the detector. In [27], a further generalization is considered by incorporating the probabilities for the detector decisions.

Figure 5.4 shows attenuation curves under QLSA distortion measure as a function of the instantaneous SNR with different *a priori* SNRs and with $q = 0.8$, $G_f = -15$ dB and cost parameters $b_{01} = 4$ and $b_{10} = 2$.

The signal estimate (5.45) generalizes existing suppression rules. For equal cost parameters (5.45) reduces to the OM-LSA estimator [4] which is given by (5.5). If we also let $G_f = 0$ and $q = 1$ we get the LSA suppression rule [23]. Both these estimators are shown in Fig. 5.4 with comparison to the spectral gains under any potential decision made by the detector.

5.5 A Priori SNR Estimation

In spectral speech enhancement applications, the *a priori* SNR is often estimated by using the decision-directed approach [3]. Accordingly, in each time-frequency bin we compute

$$\hat{\xi}_{\ell k} = \max \left\{ \alpha\, G^2 \left(\hat{\xi}_{\ell-1,k}, \gamma_{\ell-1,k} \right) \gamma_{\ell-1,k} \left(1 - \alpha \right) \left(\gamma_{\ell k} - 1 \right), \xi_{\min} \right\}, \quad (5.46)$$

where α $(0 \leq \alpha \leq 1)$ is a weighting factor that controls the trade-off between noise reduction and transient distortion introduced into the signal, and $\xi_{\min}$ is a lower bound for the *a priori* SNR which is necessary for reducing the residual musical noise in the enhanced signal [3, 22]. Since the *a priori* SNR is defined under the assumption that $H_1^{\ell k}$ is true, it is proposed in [4] to replace the gain G in (5.46) by G_{H_1} which represents the spectral gain when the signal is surely present (*i.e.*, $q = 1$). Increasing the value of α results in a greater reduction of the musical noise phenomena, at the expense of further attenuation of transient speech components (*e.g.*, speech onsets) [22]. By using the proposed approach with high cost for false speech detection, the musical noise can be reduced without increasing the value of α, which enables rapid changes in the *a priori* SNR estimate. The lower bound for the *a priori* SNR is related to the spectral gain floor G_f since both imply a lower bound on the spectral gain. The latter parameter is used to evaluate both the optimal detector and estimator while taking into account the desired residual noise level.

The decision-directed estimator is widely used, but is not suitable for transient noise environments, since a high-energy noise burst may yield an instantaneous increase in the *a posteriori* SNR and a corresponding increase in $\hat{\xi}_{\ell k}$ as can be seen from (5.46). The spectral gain would then be higher than the desired value, and the transient noise component would not be sufficiently attenuated. Let $\hat{\lambda}_{d_{\ell k}}^{s}$ denote the estimated spectral variance of the stationary noise component and let $\hat{\lambda}_{d_{\ell k}}^{t}$ denote the estimated spectral variance of the transient noise component. The former may be practically estimated by using the improved minima-controlled recursive averaging (IMCRA) algorithm [4, 28] or by using the minimum-statistics approach [29], while $\lambda_{d_{\ell k}}^{t}$ may be evaluated based on a training signals as assumed in [27]. The total variance of the noise component is $\hat{\lambda}_{d_{\ell k}} = \hat{\lambda}_{d_{\ell k}}^{s} + \hat{\lambda}_{d_{\ell k}}^{t}$. Note that $\lambda_{d_{\ell k}}^{t} = 0$ in time-frequency bins where the transient noise source is inactive. Since the *a priori* SNR is highly dependent on the noise variance, we first estimate the speech spectral variance by

$$\hat{\lambda}_{x_{\ell k}} = \max \left\{ \alpha\, G_{H_1}^2 \left(\hat{\xi}_{\ell-1,k}, \gamma_{\ell-1,k} \right) |Y_{\ell-1,k}|^2 \left(1 - \alpha \right) \left(|Y_{\ell k}|^2 - \hat{\lambda}_{d_{\ell k}} \right), \lambda_{\min} \right\},$$
$$(5.47)$$

where $\lambda_{\min} = \xi_{\min}\, \hat{\lambda}_{d_{\ell k}}^{s}$. Then, the *a priori* SNR is evaluated by $\hat{\xi}_{\ell k} = \hat{\lambda}_{x_{\ell k}}/\hat{\lambda}_{d_{\ell k}}$. In a stationary noise environment this estimator reduces to the decision-directed estimator (5.46), with G_{H_1} substituting G. However, under the presence of a transient noise component, this method yields a lower *a priori* SNR estimate, which enables higher attenuation of the high-energy transient noise component. Furthermore, to allow further reduction of the transient noise component to the level of the residual stationary noise, the gain floor is modified by $\tilde{G}_f = G_f\, \hat{\lambda}_{d_{\ell k}}^{s}/\hat{\lambda}_{d_{\ell k}}$ as proposed in [30].

The different behaviors under transient noise conditions of this modified decision-directed *a priori* SNR estimator and the decision-directed estimator as proposed in [4] are illustrated in Figs 5.5 and 5.6. Figure 5.5 shows the signals in the time domain: the analyzed signal contains a sinusoidal wave which is active in only two specific segments. The noisy signal contains both additive white Gaussian noise with 5 dB SNR and high-energy transient noise components. The signal enhanced by using the decision-directed estimator and the STSA suppression rule is shown in Fig. 5.5(c). The signal enhanced by using the modified *a priori* SNR estimator and the STSA suppression rule is shown in Fig. 5.5(d), and the result obtained by using the proposed modified *a priori* SNR estimation with the detection and estimation approach is shown in Fig. 5.5(d) (using the same parameters as in the previous section). Both the decision-directed estimator and the modified *a priori* SNR estimator are applied with $\alpha = 0.98$ and $\xi_{\min} = -20$ dB. Clearly, in stationary noise intervals, and where the SNR is high, similar results are obtained by both *a priori* SNR estimators. However, the proposed modified *a priori* SNR estimator obtain higher attenuation of the transient noise, whether it is incorporated with the STSA or the simultaneous detection and estimation approach. Figure 5.6 shows the amplitudes of the STFT coefficients of the noisy and enhanced signals at the frequency band which contains the desired sinusoidal component. Accordingly, the modified *a priori* SNR estimator enables a greater reduction of the background noise, particularly transient noise components. Moreover, it can be seen that using the simultaneous detection and estimation yields better attenuation of both the stationary and background noise compared to the STSA estimator, even while using the same *a priori* SNR estimator.

5.6 Experimental Results

For the experimental study, speech signals from the TIMIT database [31] were sampled at 16 kHz and degraded by additive noise. The noisy signals are transformed into the STFT domain using half-overlapping Hamming windows of 32 msec length, and the background-noise spectrum is estimated by using the IMCRA algorithm (for all the considered enhancement algorithms) [28, 4]. The performance evaluation includes objective quality measures, a subjective study of spectrograms and informal listening tests. The first quality measure is the segmental SNR defined by [32]

$$SegSNR = \frac{1}{|\mathcal{L}|} \sum_{\ell \in \mathcal{L}} \mathcal{T} \left\{ 10 \log_{10} \frac{\sum_{n=0}^{K-1} x^2 \left(n + \ell K/2\right)}{\sum_{n=0}^{K-1} \left[x \left(n + \ell K/2\right) - \hat{x} \left(n + \ell K/2\right)\right]^2} \right\},$$

$$(5.48)$$

where $\mathcal{L}$ represents the set of frames which contain speech, $|\mathcal{L}|$ denotes the number of elements in $\mathcal{L}$, $K = 512$ is the number of samples per frame and

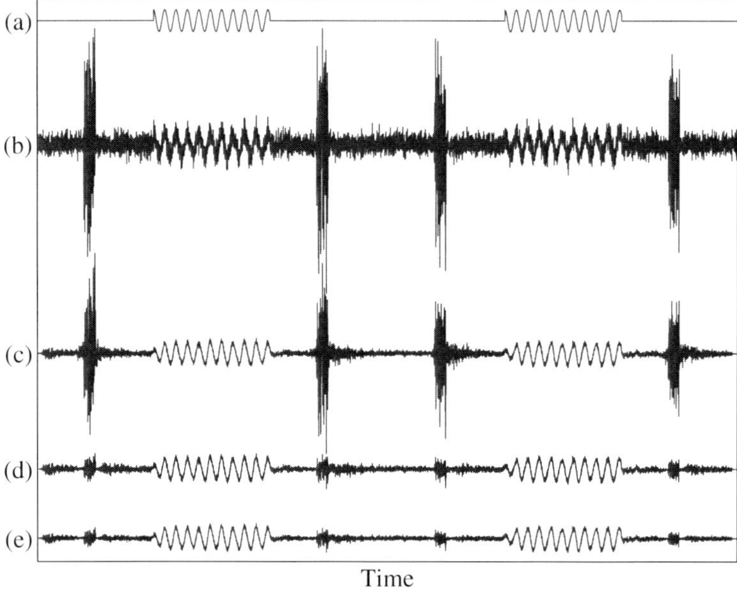

Fig. 5.5 Signals in the time domain. (a) Clean sinusoidal signal; (b) noisy signal with both stationary and transient components; (c) enhanced signal obtained by using the STSA and the decision-directed estimators; (d) enhanced signal obtained by using the STSA and the modified *a priori* SNR estimators; (e) enhanced signal obtained by using the detection and estimation approach and the modified *a priori* SNR estimator.

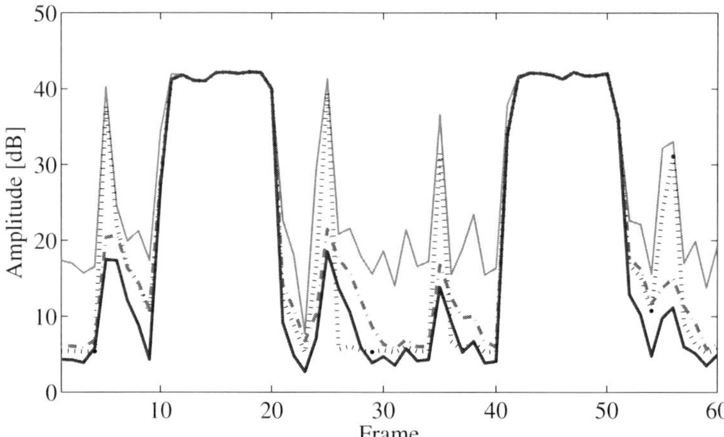

Fig. 5.6 Amplitudes of the STFT coefficients along time-trajectory corresponding to the frequency of the sinusoidal signal: noisy signal (light solid line), STSA with decision-directed estimation (dotted line), STSA with the modified *a priori* SNR estimator (dashed-dotted line) and simultaneous detection and estimation with the modified *a priori* SNR estimator (dark solid line).

Table 5.1 Segmental SNR and log spectral distortion obtained by using either the simultaneous detection and estimation approach or the STSA estimator in stationary noise environment.

Input SNR dB	Input Signal		Detection & Estimation		STSA ($\alpha = 0.98$)		STSA ($\alpha = 0.92$)	
	SegSNR	LSD	SegSNR	LSD	SegSNR	LSD	SegSNR	LSD
-5	-6.801	20.897	1.255	7.462	0.085	9.556	-0.684	10.875
0	-3.797	16.405	4.136	5.242	3.169	6.386	2.692	7.391
5	0.013	12.130	5.98	3.887	5.266	4.238	5.110	4.747
10	4.380	8.194	6.27	3.143	5.93	3.167	6.014	3.157

the operator $\mathcal{T}$ confines the SNR at each frame to a perceptually meaningful range between -10 dB and 35 dB. The second quality measure is log-spectral distortion (LSD) which is defined by

$$LSD = \frac{1}{L} \sum_{\ell=0}^{L-1} \left\{ \frac{1}{K/2+1} \sum_{k=0}^{K/2} \left[10 \log_{10} \mathcal{C}X_{\ell k} - 10 \log_{10} \mathcal{C}\hat{X}_{\ell k} \right]^2 \right\}^{\frac{1}{2}} , \quad (5.49)$$

where $\mathcal{C}X \triangleq \max\left\{|X|^2, \epsilon\right\}$ is a spectral power clipped such that the log-spectrum dynamic range is confined to about 50 dB, that is, $\epsilon = 10^{-50/10} \cdot \max_{\ell,k}\left\{|X_{\ell k}|^2\right\}$. The third quality measure (used in Section 5.6-B) is the perceptual evaluation of speech quality (PESQ) score [33].

5.6.1 Simultaneous Detection and Estimation

The suppression rule results from the proposed simultaneous detection and estimation approach with the QSA distortion measure is compared to the STSA estimation [3] for stationary white Gaussian noise with SNRs in the range $[-5, 10]$ dB. For both algorithms the *a priori* SNR is estimated by the decision-directed approach (5.46) with $\xi_{\min} = -15$ dB, and the *a priori* speech presence probability is $q = 0.8$. For the STSA estimator a decision-directed estimation [4] with $\alpha = 0.98$ reduces the residual musical noise but generally implies transient distortion of the speech signal [3, 22]. However, the inherent detector obtained by the simultaneous detection and estimation approach may improve the residual noise reduction and therefore a lower weighting factor α may be used to allow lower speech distortion. Indeed, for the simultaneous detection and estimation approach $\alpha = 0.92$ implies better results, while for the STSA algorithm, better results are achieved with $\alpha = 0.98$. The cost parameters for the simultaneous detection and estimation should be chosen according to the system specification, *i.e.*, whether the quality of the speech signal or the amount of noise reduction is of higher importance. Table 5.1 summarizes the average segmental SNR and LSD for these two enhancement algorithms, with cost parameters $b_{01} = 10$ and $b_{10} = 2$,

Table 5.2 Objective quality measures.

Method	SegSNR [dB]	LSD [dB]	PESQ
Noisy speech	−2.23	7.69	1.07
OM-LSA	−1.31	6.77	0.97
Proposed Alg.	5.41	1.67	2.87

and $G_f = -15$ dB for the simultaneous detection and estimation algorithm. The results for the STSA algorithm are presented for $\alpha = 0.98$ as well as for $\alpha = 0.92$ (note that for the STSA estimator $G_f = 0$ is considered as originally proposed). It shows that the simultaneous detection and estimation yields improved segmental SNR and LSD, while a greater improvement is achieved for lower input SNR. Informal subjective listening tests and inspection of spectrograms demonstrate improved speech quality with higher attenuation of the background noise. However, since the weighting factor used for the *a priori* SNR estimate is lower, and the gain function is discontinuous, the residual noise resulting from the simultaneous detection and estimation algorithm is slightly more musical than that resulting from the STSA algorithm.

5.6.2 Spectral Estimation Under a Transient Noise Indication

The application of the spectral estimation under an indicator for the transient noise presented in Section 5.4, with the a priori SNR estimation for nonstationary environment of Section 5.5, is demonstrated in a computer-based communication system. The background office noise is slowly-varying while possible keyboard typing interference may exist. Since the keyboard signal is available to the computer, a reliable detector for the transient-like keyboard noise is assumed to be available based on a training phase but still, erroneous detections are reasonable. The speech signals degraded by a stationary background noise with 15 dB SNR and a keyboard typing noise such that the total SNR is 0.8 dB. The transient noise detector is assumed to have an error probability of 10% and the missed detection and false detection costs are set to 1.2.

Figure 5.7 demonstrates the spectrograms and waveforms of a signal enhanced by using the proposed algorithm, compared to using the OM-LSA algorithm. It can be seen that using the proposed approach, the transient noise is significantly attenuated, while the OM-LSA is unable to eliminate the keyboard transients.

The results of the objective measures are summarized in Table 5.2. It can be seen that the proposed detection and estimation approach significantly

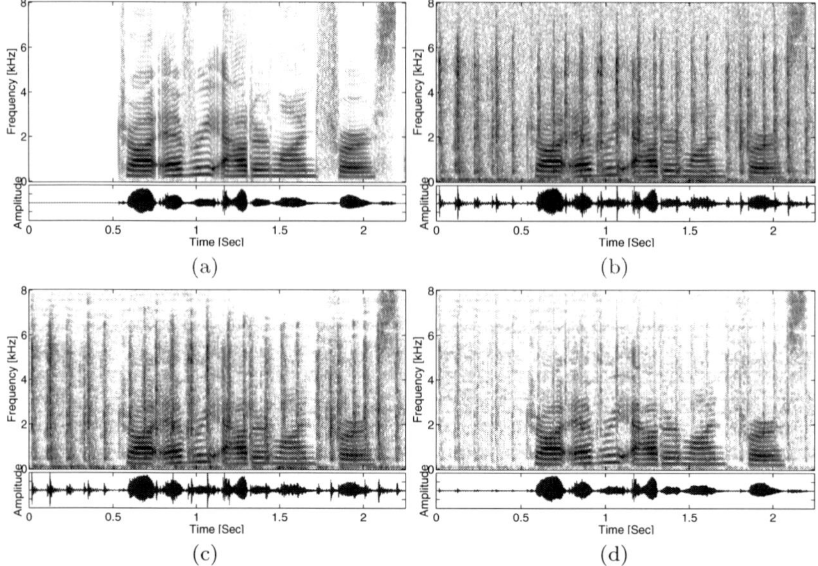

Fig. 5.7 Speech spectrograms and waveforms. (a) Clean signal ("Draw any outer line first"); (b) noisy signal (office noise including keyboard typing noise, SNR=0.8 dB); (c) speech enhanced by using the OM-LSA estimator; (d) speech enhanced by using the proposed algorithm.

improves speech quality compared to using the OM-LSA algorithm. Informal listening tests confirm that the annoying keyboard typing noise is dramatically reduced and the speech quality is significantly improved.

5.7 Conclusions

We have presented a novel formulation of the single-channel speech enhancement problem in the time-frequency domain. The formulation relies on coupled operations of detection and estimation in the STFT domain, and a cost function that combines both the estimation and detection errors. A detector for the speech coefficients and a corresponding estimator for their values are jointly designed to minimize a combined Bayes risk. In addition, cost parameters enable to control the trade-off between speech quality, noise reduction and residual musical noise. The proposed method generalizes the traditional spectral enhancement approach which considers estimation-only under signal presence uncertainty. In addition we propose a modified decision-directed *a priori* SNR estimator which is adapted to transient noise environment. Experimental results show greater noise reduction with improved speech quality when compared with the STSA suppression rules under stationary noise. Furthermore, it is demonstrated that under transient noise environment, greater reduction of transient noise components may be achieved by exploiting a reliable detector for interfering transients.

References

1. I. Cohen and S. Gannot, "Spectral enhancement methods," in *Springer Handbook of Speech Processing*, J. Benesty, M. M. Sondhi, and Y. Huang, Eds. Springer, 2007, ch. 45.
2. R. J. McAulay and M. L. Malpass, "Speech enhancement using a soft-decision noise suppression filter," *IEEE Trans. Acoust., Speech, Signal Processing*, vol. ASSP-28, no. 2, pp. 137–145, Apr. 1980.
3. Y. Ephraim and D. Malah, "Speech enhancement using a minimum mean-square error short-time spectral amplitude estimator," *IEEE Trans. Acoust., Speech, Signal Processing*, vol. ASSP-32, no. 6, pp. 1109–1121, Dec. 1984.
4. I. Cohen and B. Berdugo, "Speech enhancement for non-stationary environments," *Signal Processing*, vol. 81, pp. 2403–2418, Nov. 2001.
5. J. Sohn and W. Sung, "A voice activity detector employing soft decision based noise spectrum adaptation," in *Proc. 23rd IEEE Int. Conf. Acoust., Speech Signal Process., ICASSP-98*, vol. 1, Seattle, Washington, May 1998, pp. 365–368.
6. J. Sohn, N. S. Kim, and W. Sung, "A statistical model-based voice activity detection," *IEEE Signal Processing Lett.*, vol. 6, no. 1, pp. 1–3, Jan. 1999.
7. Y. D. Cho and A. Kondoz, "Analysis and improvement of a statistical model-based voice activity detector," *IEEE Signal Processing Lett.*, vol. 8, no. 10, pp. 276–278, Oct. 2001.
8. S. Gazor and W. Zhang, "A soft voice activity detector based on a Laplacian-Gaussian model," *IEEE Trans. Speech Audio Processing*, vol. 11, no. 5, pp. 498–505, Sep. 2003.
9. A. Davis, S. Nordholm, and R. Tongneri, "Statistical voice activity detection using low-variance spectrum estimation and an adaptive threshold," *IEEE Trans. Audio, Speech, and Language Processing*, vol. 14, no. 2, pp. 412–423, Mar. 2006.
10. J.-H. Chang, N. S. Kim, and S. K. Mitra, "Voice activity detection based on multiple statistical models," *IEEE Trans. Signal Processing*, vol. 54, no. 6, pp. 1965–1976, Jun. 2006.
11. S. F. Boll, "Suppression of acousting noise in speech using spectral subtraction," *IEEE Trans. Acoust., Speech, Signal Processing*, vol. ASSP-27, no. 2, pp. 113–120, Apr. 1979.
12. M. Berouti, R. Schwartz, and J. Makhoul, "Enhancement of speech corrupted by acoustic noise," in *Proc. IEEE Int. Conf. Acoust., Speech Signal Process., ICASSP-79*, vol. 4, Apr. 1979, pp. 208–211.
13. Y. Ephraim and H. L. Van Trees, "A signal subspace approach for speech enhancement," *IEEE Trans. Speech Audio Processing*, vol. 3, no. 4, pp. 251–266, Jul. 1995.
14. H. Lev-Ari and Y. Ephraim, "Extension of the signal subspace speech enhancement approach to colored noise," *IEEE Signal Processing Lett.*, vol. 10, no. 4, pp. 104–106, Apr. 2003.
15. Y. Hu and P. C. Loizou, "A generalized subspace approach for enhancing speech corrupted by colored noise," *IEEE Trans. Speech Audio Processing*, vol. 11, no. 4, pp. 334–341, Jul. 2003.
16. F. Jabloun and B. Champagne, "A perceptual signal subspace approach for speech enhancement in colored noise," in *Proc. 27th IEEE Int. Conf. Acoust., Speech Signal Process., ICASSP-02*, Orlando, Florida, May 2002, pp. 569–572.
17. D. Malah, R. V. Cox, and A. J. Accardi, "Tracking speech-presence uncertainty to improve speech enhancement in non-stationary noise environments," in *Proc. 24th*

IEEE Int. Conf. Acoust., Speech Signal Process., ICASSP-99, Phoenix, Arizona, Mar. 1999, pp. 789–792.

18. D. Middleton and F. Esposito, "Simultaneous optimum detection and estimation of signals in noise," *IEEE Trans. Inform. Theory*, vol. IT-14, no. 3, pp. 434–444, May 1968.

19. A. Fredriksen, D. Middleton, and D. Vandelinde, "Simultaneous signal detection and estimation under multiple hypotheses," *IEEE Trans. Inform. Theory*, vol. IT-18, no. 5, pp. 607–614, 1972.

20. A. Abramson and I. Cohen, "Simultaneous detection and estimation approach for speech enhancement," *IEEE Trans. Audio, Speech, and Language Processing*, vol. 15, no. 8, pp. 2348–2359, Nov. 2007.

21. ——, "Single-sensor blind source separation using classification and estimation approach and GARCH modeling," *IEEE Trans. Audio, Speech, and Language Processing*, vol. 16, no. 8, Nov. 2008.

22. O. Cappé, "Elimination of the musical noise phenomenon with the Ephraim and Malah noise suppressor," *IEEE Trans. Speech Audio Processing*, vol. 2, no. 2, pp. 345–349, Apr. 1994.

23. Y. Ephraim and D. Malah, "Speech enhancement using a minimum mean-square error log-spectral amplitude estimator," *IEEE Trans. Acoust., Speech, Signal Processing*, vol. 33, no. 2, pp. 443–445, Apr. 1985.

24. A. G. Jaffer and S. C. Gupta, "Coupled detection-estimation of gaussian processes in gaussian noise," *IEEE Trans. Inform. Theory*, vol. IT-18, no. 1, pp. 106–110, Jan. 1972.

25. I. Cohen, "Speech spectral modeling and enhancement based on autoregressive conditional heteroscedasticity models," *Signal Processing*, vol. 86, no. 4, pp. 698–709, Apr. 2006.

26. I. S. Gradshteyn and I. M. Ryzhik, *Table of Integrals, Series, and Products*, 6th ed., A. Jefferey and D. Zwillinger, Eds.   Academic Press, 2000.

27. A. Abramson and I. Cohen, "Enhancement of speech signals under multiple hypotheses using an indicator for transient noise presence," in *Proc. 32nd IEEE Int. Conf. Acoust., Speech Signal Process., ICASSP-07*, Honolulu, Hawaii, Apr. 2007, pp. 553–556.

28. I. Cohen, "Noise spectrum estimation in adverse environments: Improved minima controlled recursive averaging," *IEEE Trans. Speech Audio Processing*, vol. 11, no. 5, pp. 466–475, Sept. 2003.

29. R. Martin, "Noise power spectral density estimation based on optimal smoothing and minimum statistics," *IEEE Trans. Speech Audio Processing*, vol. 9, pp. 504–512, Jul. 2001.

30. E. Habets, I. Cohen, and S. Gannot, "MMSE log-spectral amplitude estimator for multiple interferences," in *Proc. Int. Workshop on Acoust. Echo and Noise Control., IWAENC-06*, Paris, France, Sept. 2006.

31. J. S. Garofolo, "Getting started with the DARPA TIMIT CD-ROM: an acoustic phonetic continuous speech database," *Technical report, National Institute of Standards and Technology (NIST)*, Gaithersburg, Maryland (prototype as of December 1988).

32. S. R. Quackenbush, T. P. Barnwell, and M. A. Clements, *Objective Meaasures of Speech Quality*.   Englewood Cliffs, NJ: Prentice-Hall, 1988.

33. ITU-T Rec. P.862, "Perceptual evaluation of speech quality (PESQ), an objective method for end-to-end speech quality assessment of narrowband telephone networks and speech codecs," *International Telecommunication Union, Geneva, Switzerland*, Feb. 2001.

Chapter 6
Speech Dereverberation and Denoising Based on Time Varying Speech Model and Autoregressive Reverberation Model

Takuya Yoshioka, Tomohiro Nakatani, Keisuke Kinoshita, and Masato Miyoshi

Abstract Speech dereverberation and denoising have been important problems for decades in the speech processing field. As regards to denoising, a model-based approach has been intensively studied and many practical methods have been developed. In contrast, research on dereverberation has been relatively limited. It is in very recent years that studies on a model-based approach to dereverberation have made rapid progress. This chapter reviews a model-based dereverberation method developed by the authors. This dereverberation method is effectively combined with a traditional denoising technique, specifically a multichannel Wiener filter. This combined method is derived by solving a dereverberation and denoising problem with a model-based approach. The combined dereverberation and denoising method as well as the original dereverberation method are developed by using a multichannel autoregressive model of room acoustics and a time-varying power spectrum model of clean speech signals.

6.1 Introduction

The great advances in speech processing technologies made over the past few decades, in combination with the growth in computing and networking capabilities, have led to the recent dramatic spread of mobile telephony and videoconferencing. Along with this evolution of speech processing products, there is a growing demand to be able to use these products in a hands-free manner. The disadvantage of hands-free use is that room reverberation, ambient noise, and acoustic echo degrade the quality of speech picked up by microphones as illustrated in Fig. 6.1. Such degraded speech quality limits the

Takuya Yoshioka, Tomohiro Nakatani, Keisuke Kinoshita, and Masato Miyoshi
NTT Communication Science Laboratories, Japan, e-mail: `{takuya,nak,kinoshita,
miyo}@cslab.kecl.ntt.co.jp`

I. Cohen et al. (Eds.): Speech Processing in Modern Communication, STSP 3, pp. 151–182.
springerlink.com

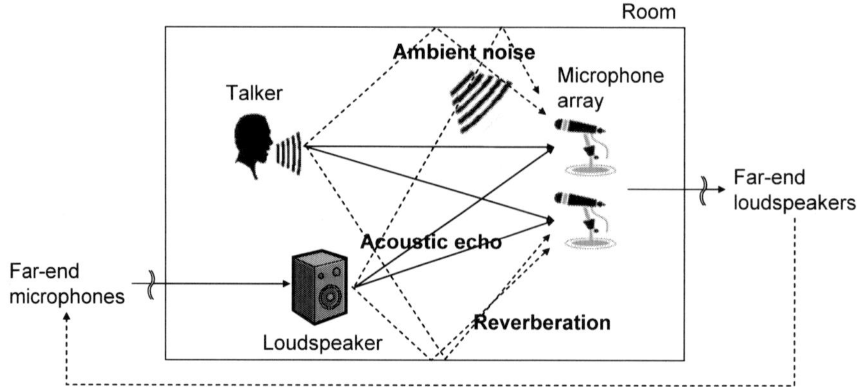

Fig. 6.1 Acoustic disturbances in a room.

applications of speech processing products. Thus, techniques for filtering out such unfavorable acoustic disturbances are vital for the further expansion of existing speech processing products as well as the development of new ones.

6.1.1 Goal

The task of interest is as follows. We assume that the voice of one talker is recorded in a room with an M-element microphone array. Let $s_{n,l}$ be the speech signal from the talker represented in the short-time Fourier transform (STFT) domain, where n and l are time frame and frequency bin indices, respectively. This speech signal is contaminated by reverberation, ambient noise, and acoustic echo while propagating through the room from the talker to the microphone array (see Fig. 6.1). Hence, the audio signals observed by the microphone array are distorted versions of the original speech signal. We represent the observed signals in vector form as

$$\mathbf{y}_{n,l} = [y_{n,l}^1, \cdots, y_{n,l}^M]^T, \tag{6.1}$$

where $y_{n,l}^m$ is the signal observed by the mth microphone and superscript T is a non-conjugate transpose. Assume that $\mathbf{y}_{n,l}$ is observed over N consecutive time frames. Let us represent the set of observed samples and the set of corresponding clean speech samples as

$$y = \{\mathbf{y}_{n,l}\}_{0 \leq n \leq N-1, 0 \leq l \leq L-1} \quad \text{and} \quad s = \{s_{n,l}\}_{0 \leq n \leq N-1, 0 \leq l \leq L-1}, \tag{6.2}$$

respectively, where L is the number of frequency bins. Now, our goal is to estimate s from y, or in other words, to cancel all acoustic disturbances in the room without precise knowledge of the acoustical properties of the room

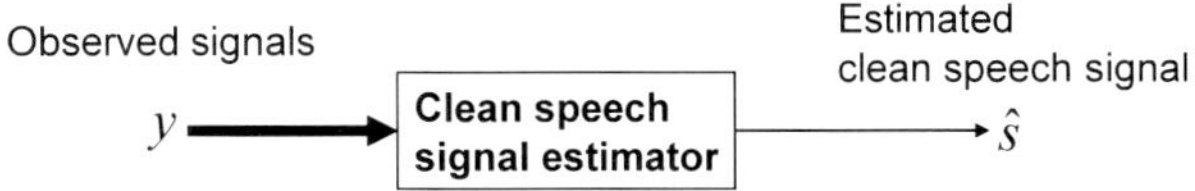

Fig. 6.2 Clean speech signal estimator. A multichannel signal (vector) is indicated by a thick arrow while a single-channel signal (scalar) is represented by a thin arrow.

as illustrated in Fig. 6.2. We denote the estimates of s and $s_{n,l}$ as $\hat{s}$ and $\hat{s}_{n,l}$, respectively. Note that, as indicated by this task definition, we consider the clean speech signal estimation based on batch processing throughout this chapter.

The task defined above consists of three topics: dereverberation, denoising, and acoustic echo cancellation. In this chapter, we focus on dereverberation and its combination with denoising. Specifically, our aim is to achieve dereverberation and denoising simultaneously. Acoustic echo cancellation is beyond the scope of this chapter[1].

6.1.2 Technological Background

There are many microphone array control techniques for dereverberation and denoising. The effectiveness of each technique depends on the size and element number of the microphone array.

When using a large-scale microphone array, a delay-and-sum beamformer may be the best selection [3, 4]. A delay-and-sum beamformer steers an acoustic beam in the direction of the talker by superimposing the delayed versions of the individual microphone signals. To achieve a high dereverberation and denoising performance, this beamforming technique requires a very large size and the use of many microphones. Thus, a delay-and-sum beamformer has been used with microphone arrays deployed in such large rooms as auditoriums; for example, a 1-m square two-dimensional microphone array with 63 electret microphones is reported in [4].

If we wish to make microphone arrays available in a wide variety of situations, the arrays must be sufficiently small. However, because small microphone arrays based on a delay-and-sum beamformer do not provide sufficient gains, we require an alternative methodology for denoising and dereverberation.

Many denoising techniques have been developed including the generalized sidelobe canceller and the multichannel Wiener filter[2]. Several techniques for dereverberation have also been proposed; they include cepstrum filtering [6],

---

[1] We refer the reader to [1] and [2] for details on acoustic echo cancellation.

[2] References [2] and [5] provide excellent reviews of microphone array techniques.

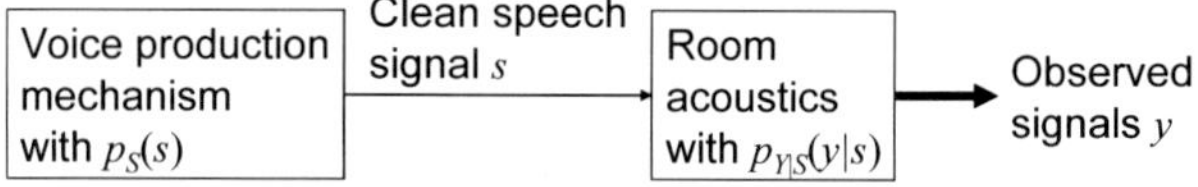

Fig. 6.3 Generation process of observed signals.

blind deconvolution [7], and spectral subtraction [8][3]. A few methods that combine dereverberation and denoising techniques [10, 11, 12] have recently been proposed. This chapter reviews one such method proposed by the authors [11], which was designed using a model-based approach.

6.1.3 Minimum Mean-Squared Error Signal Estimation and Model-Based Approach

One of the most commonly used approaches for general signal estimation tasks is the minimum mean-squared error (MMSE) estimation. With the MMSE estimation method, we regard signals $s_{n,l}$ and $\mathbf{y}_{n,l}$ as realizations of random variables $S_{n,l}$ and $\mathbf{Y}_{n,l}$, respectively. In response to this, data sets s and y are also regarded as realizations of random variable sets $S = \{S_{n,l}\}_{n,l}$ and $Y = \{\mathbf{Y}_{n,l}\}_{n,l}$, respectively[4]. The MMSE estimate is defined as the mean of the posterior distribution of the clean speech signal as follows [13]:

$$\hat{s} = \int s \cdot p_{S|Y}(s|y)ds. \tag{6.3}$$

By using the Bayes theorem, we may rewrite the posterior pdf, $p_{S|Y}(s|y)$, as

$$p_{S|Y}(s|y) = \frac{p_{Y|S}(y|s)p_S(s)}{\int p_{Y|S}(y|s)p_S(s)ds}. \tag{6.4}$$

Therefore, the MMSE estimation requires explicit knowledge of $p_{Y|S}(y|s)$ and $p_S(s)$.

The first pdf, $p_{Y|S}(y|s)$, represents the characteristics of the distortion caused by the reverberation and ambient noise. The second pdf, $p_S(s)$, defines how likely a given signal is to be of clean speech. As illustrated in Fig. 6.3, these two pdfs define the generation process of the observed signals. Hereafter, $p_{Y|S}(y|s)$ and $p_S(s)$ are referred to as the room acoustics pdf and the clean speech pdf, respectively.

---

[3] Reference [9] may be helpful in overviewing existing dereverberation techniques. In [9], some dereverberation techniques are reviewed and compared experimentally.

[4] Hereafter, when representing a set of indexed variables, we sometimes omit the range of index values.

Unfortunately, neither pdf is available in practice. Thus, as a suboptimal solution, we define parametric models of these pdfs and determine the values of the model parameters using the observed signals to approximate the true pdfs. The parameter values are determined so as to minimize a prescribed cost function. This is the basic concept behind the model-based approach, which has been widely used to solve signal estimation problems. Below, we denote the room acoustics pdf and the clean speech pdf as $p_{Y|S}(y|s; \Psi)$ and $p_S(s; \Phi)$, respectively, to make it clear that these pdfs are modeled ones with parameter sets Ψ and Φ.

Thus, the procedure for deriving a clean speech signal estimator can be summarized as follows:

1. Define the room acoustics pdf, $p_{Y|S}(y|s; \Psi)$, with parameter set Ψ.
2. Define the clean speech pdf, $p_S(s; \Phi)$, with parameter set Φ.
3. Derive the MMSE clean speech signal estimator according to (6.3) and (6.4).
4. Define a cost function and an algorithm for optimizing the values of the parameters in Ψ and Φ.

The dereverberation and denoising performance is greatly dependent on how well the above two models with the optimal parameter values simulate the characteristics of the true pdfs and how accurately the parameters are optimized. Hence, the main issues to be investigated are the selection of the models, cost function, and optimization algorithm.

The aim of this chapter is to present recently-proposed effective clean speech and room acoustics models. Specifically, a multichannel auto-regressive (MCAR) model is used to describe the room acoustics pdf, $p(y|s; \Psi)$. On the other hand, the clean speech pdf, $p(s; \Phi)$, is represented by a time-varying power spectrum (TVPS) model. Based on these models, we derive the corresponding MMSE clean speech signal estimator as well as a cost function and a parameter optimization algorithm for minimizing this cost function. We define the cost function based on the maximum likelihood estimation (MLE) method [14] as described later.

The remainder of this chapter consists mainly of two parts. The first part, Section 6.2, considers the dereverberation issue. The main objective of Section 6.2 is to show the MCAR model of the room acoustics pdf and the TVPS model of the clean speech pdf. In the second part, Section 6.3, we describe a combined dereverberation and denoising method by extending the dereverberation method described in Section 6.2. Both the original dereverberation method and the combined dereverberation and denoising method are derived according to the four steps mentioned above.

6.2 Dereverberation Method

In this section, we consider the dereverberation issue. In the context of dereverberation, we describe the MCAR model of the room acoustics pdf, the TVPS model of the clean speech pdf, and the dereverberation method based on these models. This dereverberation method was proposed in [15], and is referred to as the weighted prediction error (WPE) method.

In practice, the WPE method can be derived from several different standpoints. We begin by deriving the WPE method in a heuristic manner to help the reader to understand this dereverberation method intuitively. Then, we describe the MCAR and TVPS models and derive the WPE method based on the model-based approach.

6.2.1 Heuristic Derivation of Weighted Prediction Error Method

Apart from the model-based approach, we here describe the WPE method from a heuristic viewpoint. The WPE method consists of two components as illustrated in the block diagram in Fig. 6.4:

- a dereverberation filter, which is composed of M delay operators, M reverberation prediction filters, and two adders to compute the output signal of the dereverberation process;
- a filter controller, which updates the reverberation prediction filters in response to the output of the dereverberation filter.

These two components work iteratively and alternately. After convergence, an estimate of the clean speech signal, $\hat{s}_{t,l}$, is obtained at the output of the dereverberation filter.

Let $g_{D_l+1,l}^m, \ldots, g_{D_l+K_l,l}^m$ denote the tap weights of the reverberation prediction filter for the mth microphone and lth frequency bin. These tap weights are called room regression coefficients[5].

For each l, the reverberation prediction filters for all the microphones in combination estimate the reverberation component of the first observed signal from the signals observed by all the microphones during the preceding $D_l + K_l$ time frames as

$$\hat{y}_{n,l}^1 = \sum_{m=1}^{M} \sum_{k=D_l+1}^{D_l+K_l} (g_{k,l}^m)^* y_{n-k,l}^m, \tag{6.5}$$

---

[5] Although it may be more natural to call $g_{k,l}^m$, for example, a reverberation prediction coefficient at this point, we call $g_{k,l}^m$ a room regression coefficient since it can be regarded as a coefficient of a regression system as described later.

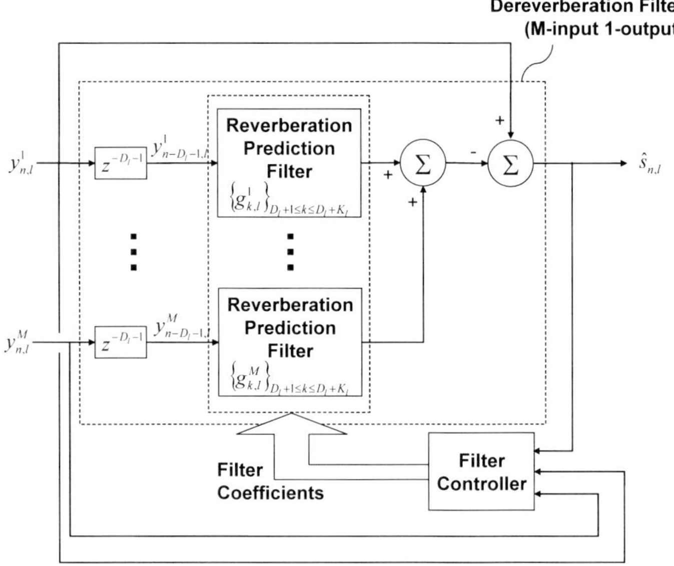

Fig. 6.4 Block diagram representation of the weighted prediction error method.

where superscript $*$ is a complex conjugate. A typical value of D_l is 1 or 2, and K_l is usually set at a value greater than the reverberation time at the lth frequency bin. Then, the estimated reverberation component, $\hat{y}_{n,l}^1$, is subtracted from $y_{n,l}^1$ to compute the output signal:

$$\hat{s}_{n,l} = y_{n,l}^1 - \hat{y}_{n,l}^1. \tag{6.6}$$

For the sake of readability, we rewrite (6.5) and (6.6) as

$$\hat{s}_{n,l} = y_{n,l}^1 - \sum_{k=D_l+1}^{D_l+K_l} \mathbf{g}_{k,l}^H \mathbf{y}_{n-k,l} \tag{6.7}$$

$$= y_{n,l}^1 - \bar{\mathbf{g}}_l^H \bar{\mathbf{y}}_{n-D_l-1,l}, \tag{6.8}$$

where superscript H stands for a conjugate transpose, and $\mathbf{g}_{k,l}$, $\bar{\mathbf{g}}_l$, and $\bar{\mathbf{y}}_{n,l}$ are defined as

$$\mathbf{g}_{k,l} = [g_{k,l}^1, \cdots, g_{k,l}^M]^T, \tag{6.9}$$

$$\bar{\mathbf{g}}_l = [\mathbf{g}_{D_l+1,l}^T, \cdots, \mathbf{g}_{D_l+K_l,l}^T]^T, \tag{6.10}$$

$$\bar{\mathbf{y}}_{n,l} = [\mathbf{y}_{n,l}^T, \cdots, \mathbf{y}_{n-K_l+1,l}^T]^T. \tag{6.11}$$

$\bar{\mathbf{g}}_l$ is referred to as a room regression vector.

Table 6.1 Summary of the algorithm for the WPE method.

Initialize the values of the room regression vectors at zero as

$$\bar{\mathbf{g}}_l[0] = [0, \cdots, 0] \text{ for } 0 \leq l \leq L - 1$$

For $i = 0, 1, 2, \cdots$, compute

$$\hat{s}_{n,l}[i+1] = y_{n,l}^1 - \bar{\mathbf{g}}_l[i]^H \bar{\mathbf{y}}_{n-D_l-1,l}$$

$$\bar{\mathbf{g}}_l[i+1] = \left(\frac{\sum_{n=0}^{N-1} \bar{\mathbf{y}}_{n-D_l-1,l} \bar{\mathbf{y}}_{n-D_l-1,l}^H}{|\hat{s}_{n,l}[i+1]|^2} \right)^{-1} \left(\frac{\sum_{n=0}^{N-1} \bar{\mathbf{y}}_{n-D_l-1,l} y_{n,l}^*}{|\hat{s}_{n,l}[i+1]|^2} \right) \text{ for } 0 \leq l \leq L - 1$$

$\hat{s}_{n,l}[i+1]$ after convergence is the output of the dereverberation process.

The value of room regression vector $\bar{\mathbf{g}}_l$ is determined by the iterative process composed of the dereverberation filter and the filter controller. Let i denote the iteration index and $\bar{\mathbf{g}}_l[i]$ denote the value of $\bar{\mathbf{g}}_l$ after the ith iteration has been completed[6].

Suppose that we have completed the ith iteration. At the $(i+1)$th iteration, the dereverberation filter computes the output signal, denoted by $\hat{s}_{n,l}[i+1]$, by using the current room regression vector value, $\bar{\mathbf{g}}_l[i]$, according to (6.8). Then, the filter controller updates the room regression vector value from $\bar{\mathbf{g}}_l[i]$ to $\bar{\mathbf{g}}_l[i+1]$. The updated room regression vector, $\bar{\mathbf{g}}_l[i+1]$, is defined as the room regression vector value that minimizes

$$\mathcal{F}^{[i+1]}(\bar{\mathbf{g}}_l) = \sum_{n=0}^{N-1} \frac{|y_{n,l}^1 - \bar{\mathbf{g}}_l^H \bar{\mathbf{y}}_{n-D_l-1,l}|^2}{|\hat{s}_{n,l}[i+1]|^2}. \tag{6.12}$$

Such a room regression vector value is computed by

$$\bar{\mathbf{g}}_l[i+1] = \left(\frac{\sum_{n=0}^{N-1} \bar{\mathbf{y}}_{n-D_l-1,l} \bar{\mathbf{y}}_{n-D_l-1,l}^H}{|\hat{s}_{n,l}[i+1]|^2} \right)^{-1} \left(\frac{\sum_{n=0}^{N-1} \bar{\mathbf{y}}_{n-D_l-1,l} y_{n,l}^*}{|\hat{s}_{n,l}[i+1]|^2} \right). \tag{6.13}$$

Thus, the algorithm for the WPE method may be summarized as in Table 6.1.

The key to the success of the WPE method is the appropriateness of the cost function, $\mathcal{F}^{[i+1]}(\bar{\mathbf{g}}_l)$, given by (6.12). The appropriateness may be intuitively explained from the perspective of speech sparseness.

We first explain the concept of speech sparseness. Fig. 6.5 shows example spectrograms of clean (left panel) and reverberant (right panel) speech signals. We see that the sample powers of a time-frequency-domain clean speech signal are sparsely distributed. This is because a clean speech signal may contain short pauses and has a harmonic spectral structure. In contrast, the sample power distribution for a reverberant speech signal appears

---

[6] The same notation is used to represent a value of a variable whose value changes every iteration throughout this paper.

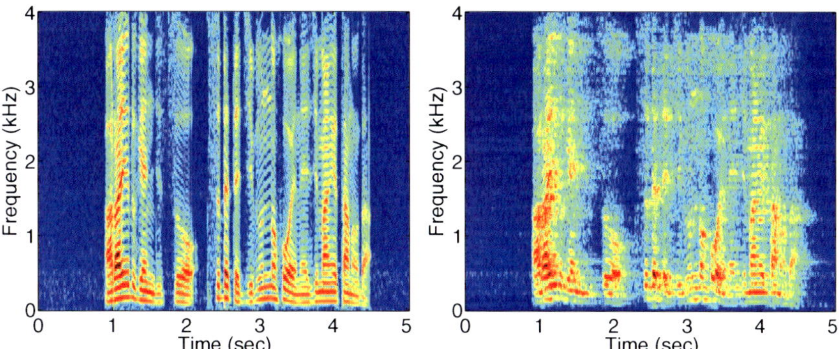

Fig. 6.5 Example spectrograms of clean (left panel) and reverberant (right panel) speech signals.

less sparse than that for a clean speech signal since reverberation spreads the sample powers at each time frame into the subsequent time frames and smears the harmonicity. Hence, the dereverberation process is expected to increase the degree of sparseness.

Now, with this in mind, let us look at (6.12). We find that $\mathcal{F}^{[i+1]}(\bar{\mathbf{g}}_l)$ is the weighted sum of the sample powers of the output signal, where the weight for time frequency point (n, l) is given by $1/|\hat{s}_{n,l}[i+1]|^2$. Therefore, minimizing $\mathcal{F}^{[i+1]}(\bar{\mathbf{g}}_l)$ is expected to decrease the sample powers of the output signal at time frequency points with small $|\hat{s}_{n,l}[i+1]|^2$ values. Thus, repeatedly updating the room regression vectors may increase the number of time-frequency points with small sample powers. This indicates that the degree of sparseness of the sample power distribution for the output signal is increased, and dereverberation is thereby achieved.

The above describes the WPE method intuitively. The above description clarifies the reason for calling this dereverberation technique as the "weighted prediction error" method. With the above description, however, we cannot answer the following important questions:

- Does the room regression vector value always converge to a stationary point?
- How can we effectively combine the WPE method with a denoising technique?

To obtain further insight and answer these questions, we need to understand the theoretical background to the WPE method. Thus, in the remainder of Section 6.2, we look into the derivation, which is based on the four steps presented in Section 6.1.3.

6.2.2 Reverberation Model

The first step is to define room acoustics pdf $p_{Y|S}(y|s;\Psi)$, which characterizes the reverberation effect (because we ignore ambient noise in this section). Here, we present three models of the room acoustics pdf:

- a multichannel moving average (MCMA) model;
- a multichannel autoregressive (MCAR) model;
- a simplified MCAR model.

The WPE method is based on the simplified MCAR model, and we use this model in this section. The remaining two models are by-products of the derivation of the simplified MCAR model. Although the MCMA model is the most basic model and accurately simulates a physical reverberation process, the dereverberation method based on this model provides a limited dereverberation performance. In contrast, the MCAR model and its simplified version, which are derived on the basis of the MCMA model, provide a high dereverberation performance. The complete MCAR model is used to derive a combined dereverberation and denoising method in Section 6.3.

6.2.2.1 Multichannel Moving Average Model

We begin by considering the process for reverberating a clean speech signal in the continuous-time domain. Let $s(t)$ denote the continuous-time signal of clean speech and $\mathbf{y}(t)$ denote the continuous-time signal vector observed by the microphone array. The process for generating observed signal vector $\mathbf{y}(t)$ may be described as the linear convolution of clean speech signal $s(t)$ and the vector of the impulse responses from the talker to the microphone array as [16]

$$\mathbf{y}(t) = \int_0^{\infty} \mathbf{h}(\tau)s(t - \tau)d\tau, \tag{6.14}$$

where $\mathbf{h}(t)$ denotes the impulse response vector. Such impulse responses are called room impulse responses.

By analogy with the continuous-time-domain reverberation model (6.14), we may define a time-frequency-domain reverberation model as

$$\mathbf{y}_{n,l} = \sum_{k=0}^{J_l} \mathbf{h}_{k,l}s_{n-k,l} + \mathbf{e}_{n,l}. \tag{6.15}$$

$\{\mathbf{h}_{k,l}\}_k$ is a room impulse response representation at the lth frequency bin, and J_l is the order of this room impulse response. $\mathbf{e}_{n,l}$ is a signal vector representing the modelling error. In general, the magnitude of $\mathbf{e}_{n,l}$ is very small, and $\mathbf{e}_{n,l}$ is assumed to be normally distributed with mean $\mathbf{0}_M$ and covariance matrix $\sigma^2 I_M$, where $\mathbf{0}_M$ is the M-dimensional zero vector, I_M

is the M-dimensional identity matrix, and σ^2 is a prescribed constant satisfying $0 < \sigma^2 \ll 1$. Equation (6.15) means that the reverberation effect may be modeled as a multichannel moving average (MCMA) system at each frequency bin.

Based on (6.15), we can define room acoustics pdf $p_{Y|S}(y|s;\Psi)$[7]. Although we can develop a dereverberation (and denoising) method based on this room acoustics pdf [17], this reverberation model did not yield a high dereverberation performance in the authors' tests. To overcome this limitation, we would like to introduce an MCAR model, which we obtain by further modifying the MCMA model (6.15).

6.2.2.2 Multichannel Autoregressive Model

Assume that $\sigma^2 \to 0$, or equivalently, that the error signal vector, $\mathbf{e}_{n,l}$, in (6.15) can be negligible as

$$\mathbf{y}_{n,l} = \sum_{k=0}^{J_l} \mathbf{h}_{k,l} s_{n-k,l}. \tag{6.16}$$

Let $\mathbf{h}_l(z)$ denote the transfer function vector corresponding to the room impulse response vector given by $\{\mathbf{h}_{k,l}\}_k$. It can be proven that if $M \geq 2$ and the elements of $\mathbf{h}_l(z)$ are coprime, for any integer K_l with $K_l \geq J_l/(M-1)$, there exists an M-by-M matrix set $\{G_{k,l}\}_{1 \leq k \leq K_l}$ such that [18]

$$\mathbf{y}_{n,l} = \sum_{k=1}^{K_l} G_{k,l}^H \mathbf{y}_{n-k,l} + \mathbf{h}_{0,l} s_{n,l}. \tag{6.17}$$

Equation (6.17) states that the reverberation effect can be expressed as a multichannel autoregressive (MCAR) system. To be more precise, the reverberant speech signal vector is the output of an MCAR system with a regression matrix set $\{G_{k,l}\}_k$ driven by a clean speech signal multiplied by

---

[7] Since the MCMA model is of little interest in this chapter, we do not go into this model in detail. Nonetheless, for the sake of completeness, we present the derived MCMA-model-based room acoustics pdf. The pdf is represented as

$$p_{Y|S}(y|s;\Psi) = \frac{1}{\pi^{LN}(\sigma^2)^{LNM}} \times$$

$$\exp\left\{-\frac{1}{\sigma^2} \sum_{l=0}^{L-1} \sum_{n=0}^{N-1} \left(\mathbf{y}_{n,l} - \sum_{k=0}^{J_l} \mathbf{h}_{k,l} s_{n-k,l}\right)^H \left(\mathbf{y}_{n,l} - \sum_{k=0}^{J_l} \mathbf{h}_{k,l} s_{n-k,l}\right)\right\}$$

$$\Psi = \{\{\mathbf{h}_{k,l}\}_{0 \leq l \leq J_l}\}_{0 \leq l \leq L-1}.$$

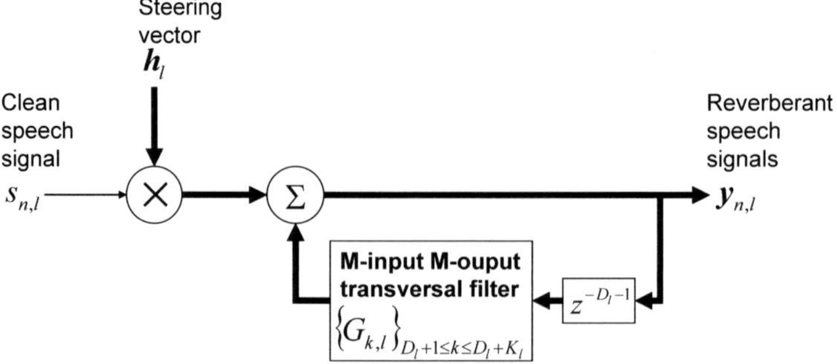

Fig. 6.6 MCAR model of reverberation.

$\mathbf{h}_{0,l}$. $G_{k,l}$ is called a room regression matrix while $\mathbf{h}_{0,l}$ is called a steering vector[8]. We rewrite the steering vector, $\mathbf{h}_{0,l}$, as $\mathbf{h}_l$.

Let us generalize (6.17) to introduce an additional D_l-frame delay into the MCAR system as

$$\mathbf{y}_{n,l} = \sum_{k=D_l+1}^{D_l+K_l} G_{k,l}^H \mathbf{y}_{n-k,l} + \mathbf{h}_l s_{n,l}. \tag{6.18}$$

We empirically found that such a delay was beneficial in preventing the excessive whitening problem, which may be caused by a dereverberation process [19]. Fig. 6.6 shows the diagram for generating $\mathbf{y}_{n,l}$ in accordance with (6.18).

Room acoustics pdf $p_{Y|S}(y|s;\Psi)$ can be defined by taking the error signals into consideration again and by using (6.18). We skip the details of this reverberation model[9]. This reverberation model is revisited in Section 6.3.

---

[8] In (6.17), $G_{k,l}$ and $\mathbf{h}_l$ express the effects of inter-frame and intra-frame distortion, respectively. If we assume that $G_{k,l}$ is zero for any k values, or in other words, if we ignore the inter-frame distortion, (6.17) reduces to

$$\mathbf{y}_{n,l} = \mathbf{h}_l s_{n,l}.$$

This is the conventional room acoustics model used in the field of adaptive beamformers. In that context, $\mathbf{h}_l$ is called a steering vector because it is used to steer an acoustic beam. Hence, we also call $\mathbf{h}_l$ a steering vector in (6.17).

[9] We describe only the resultant room acoustics pdf here. The MCAR-model-based room acoustics pdf is represented as

$$p_{\mathbf{Y}|S}(\mathbf{y}|s;\Psi) = \frac{1}{\pi^{LN}(\sigma^2)^{LNM}} \exp\Bigg\{-\frac{1}{\sigma^2}\sum_{l=0}^{L-1}\sum_{n=0}^{N-1}\Bigg(\mathbf{y}_{n,l} - \sum_{k=D_l+1}^{D_l+K_l} G_{k,l}^H \mathbf{y}_{n-k,l} - \mathbf{h}_l s_{n,l}\Bigg)^H$$
$$\times \Bigg(\mathbf{y}_{n,l} - \sum_{k=D_l+1}^{D_l+K_l} G_{k,l}^H \mathbf{y}_{n-k,l} - \mathbf{h}_l s_{n,l}\Bigg)\Bigg\},$$

6.2.2.3 Simplified Multichannel Autoregressive Model

Although the MCAR model is effective as regards to dereverberation, it requires $M^2(K_l + 1)$ parameters, namely many more than the number of parameters of the MCMA model. For the purpose of dereverberation, we can reduce the number of parameters to MK_l by simplifying the MCAR model.

Now, let $\mathbf{g}_{k,l}^m$ denote the mth column of $G_{k,l}$. On the basis of (6.18), we obtain a clean speech signal recovery formula as

$$s_{n,l} = \frac{1}{h_l^1}\left(y_{n,l}^1 - \sum_{k=D_l+1}^{D_l+K_l} (\mathbf{g}_{k,l}^1)^H \mathbf{y}_{n-k,l}\right). \tag{6.19}$$

Moreover, by using (6.18) and (6.19), we obtain the following generation model for the microphone signals:

$$y_{n,l}^1 = \sum_{k=D_l+1}^{D_l+K_l} (\mathbf{g}_{k,l}^1)^H \mathbf{y}_{n-k,l} + h_l^1 s_{n,l}, \tag{6.20}$$

$$y_{n,l}^m = \sum_{k=D_l+1}^{D_l+K_l} (\mathbf{g}_{k,l}^m)^H \mathbf{y}_{n-k,l} + \frac{h_l^m}{h_l^1}\left(y_{n,l}^1 - \sum_{k=D_l+1}^{D_l+K_l} (\mathbf{g}_{k,l}^1)^H \mathbf{y}_{n-k,l}\right) \text{ for } m \geq 2.$$
$$\tag{6.21}$$

The set of (6.20) and (6.21) indicates that only the first microphone signal is generated based on the clean speech signal, and that the microphone signals where $m \geq 2$ can be determined independently of the clean speech signal. Hence, from the viewpoint of dereverberation, we only require the $\{h_l^1\}_l$ and $\{\mathbf{g}_{k,l}^1\}_{k,l}$ values. Indeed, these parameter values suffice to recover the clean speech signal as indicated by (6.19). Furthermore, since the main information represented by h_l^1 is the signal transmission delay from the talker to the first microphone, which does not require compensation, we assume that $h_l^1 = 1$ for all l values.

At this point, we observe that (6.19) with $h_l^1 = 1$ is equivalent to (6.7). Therefore, we rewrite $\mathbf{g}_{k,l}^1$ as $\mathbf{g}_{k,l}$. The reason for calling each element of $\mathbf{g}_{k,l}$ a room regression coefficient is now clear.

From the above discussion, we can simplify the MCAR model as the block diagram shown in Fig. 6.7. With this model, the microphone signals where $m \geq 2$, i.e., $y_{n,l}^2, \cdots, y_{n,l}^M$, are regarded not as stochastic signals but as deterministic ones. Let us take into account of the error signal in (6.20) again, and assume that the error signal is normally distributed with mean 0 and

---

where σ^2 is the variance of the error signals, and parameter set Ψ consists of the room regression matrices and steering vectors, i.e.,

$$\Psi = \{\{G_{k,l}\}_{D_l+1 \leq k \leq D_l+K_l}, \mathbf{h}_l\}_{0 \leq l \leq L-1}.$$

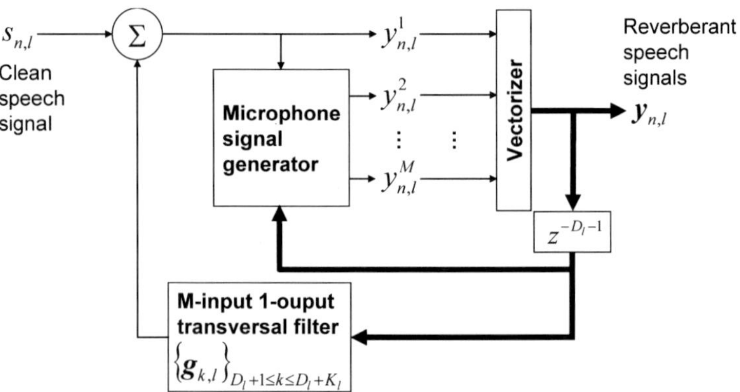

Fig. 6.7 Simplified MCAR model.

variance σ^2. Then, it is obvious that the simplified MCAR model gives the following conditional pdf:

$$p_{Y^1_{n,l}|\mathbf{Y}_{n-1,l},\cdots,\mathbf{Y}_{n-D_l-K_l,l},S_{n,l}}(y^1_{n,l}|\mathbf{y}_{n-1,l},\cdots,\mathbf{y}_{n-D_l-K_l,l},s_{n,l};\Psi)$$

$$= \mathcal{N}_{\mathbb{C}}\Big\{y^1_{n,l};\sum_{k=D_l+1}^{D_l+K_l}\mathbf{g}^H_{k,l}\mathbf{y}_{t-k,l}+s_{n,l},\sigma^2\Big\}, \quad (6.22)$$

where the parameter set, Ψ, is given by

$$\Psi = \{\{\mathbf{g}_{k,l}\}_{D_l+1\leq k\leq D_l+K_l}\}_{0\leq l\leq L-1}. \quad (6.23)$$

$\mathcal{N}_{\mathbb{C}}\{\mathbf{x};\mu,\Sigma\}$ is the pdf of a complex-valued normal distribution with mean μ and covariance matrix Σ, which is defined as [20]

$$\mathcal{N}_{\mathbb{C}}\{\mathbf{x};\mu,\Sigma\} = \frac{1}{|\Sigma|}\exp\{-(\mathbf{x}-\mu)^H\Sigma^{-1}(\mathbf{x}-\mu)\}, \quad (6.24)$$

where $|\cdot|$ is a determinant. Therefore, we have the room acoustics pdf based on the simplified MCAR model as

$$p_{\mathbf{Y}|S}(\mathbf{y}|s;\Psi) =$$

$$\prod_{l=0}^{L-1}\prod_{n=0}^{N-1}p_{Y^1_{n,l}|\mathbf{Y}_{n-1,l},\cdots,\mathbf{Y}_{n-D_l-K_l,l},S_{n,l}}(y^1_{n,l}|\mathbf{y}_{n-1,l},\cdots,\mathbf{y}_{n-D_l-K_l,l},s_{n,l};\Psi)$$

$$= \frac{1}{\pi^{LN}(\sigma^2)^{LN}}\exp\Big\{-\frac{1}{\sigma^2}\sum_{l=0}^{L-1}\sum_{n=0}^{N-1}\Big|y^1_{n,l}-\sum_{k=D_l+1}^{D_l+K_l}\mathbf{g}^H_{k,l}\mathbf{y}_{n-k,l}-s_{n,l}\Big|^2\Big\}.$$

$$(6.25)$$

Thus, we have derived the simplified MCAR room acoustics model, and we use this model to derive the WPE method.

6.2.3 Clean Speech Model

The next step in deriving the dereverberation algorithm is to define clean speech pdf $p_S(s; \Phi)$ (see Section 6.1.3). To define the clean speech pdf, we use a simple time-varying power spectrum (TVPS) model, which is described below.

With the TVPS model, we assume the following two conditions:

1. $s_{n,l}$ is normally distributed with mean 0 and variance $\lambda_{n,l}$:

$$p_{S_{n,l}}(s_{n,l}; \lambda_{n,l}) = \mathcal{N}_{\mathbb{C}}\{s_{n,l}; 0, \lambda_{n,l}\}. \tag{6.26}$$

Note that $\{\lambda_{n,l}\}_l$ corresponds to the short-time power spectral density (psd) of the time-domain clean speech signal at time frame n.

2. If $(n_1, l_1) \neq (n_2, l_2)$, s_{n_1,l_1} and s_{n_2,l_2} are statistically independent.

Based on these assumptions, we then have the following clean speech pdf:

$$p_S(s; \Phi) = \prod_{l=0}^{L-1} \prod_{n=0}^{N-1} p_{S_{n,l}}(s_{n,l}; \lambda_{n,l})$$

$$= \frac{1}{\pi^{LN} \prod_{l=0}^{L-1} \prod_{n=0}^{N-1} \lambda_{n,l}} \exp\left\{-\sum_{l=0}^{L-1} \sum_{n=0}^{N-1} \frac{|s_{n,l}|^2}{\lambda_{n,l}}\right\}, \tag{6.27}$$

where model parameter set Φ consists of the short-time psd components as

$$\Phi = \{\lambda_{n,l}\}_{0 \leq n \leq N-1, 0 \leq l \leq L-1}. \tag{6.28}$$

6.2.4 Clean Speech Signal Estimator and Parameter Optimization

The third step is to define the MMSE clean speech signal estimator. When using the simplified MCAR model, the clean speech signal is recovered simply by computing (6.19) with the condition $h_l^1 = 1$.

In the final step, we specify a cost function and an optimization algorithm for the model parameters, i.e., $\Phi = \{\lambda_{n,l}\}_{n,l}$ and $\Psi = \{\mathbf{g}_{k,l}\}_{k,l}$. We define the cost function based on the maximum likelihood estimation (MLE) method. With the MLE method, the cost function is defined as the negative logarithm of the marginal pdf of the observed signals as

$$\mathcal{L}(\Phi, \Psi) = -\log p_Y(y; \Phi, \Psi). \tag{6.29}$$

The optimum parameter values are obtained as those that minimize the negative log likelihood function (cost function) $\mathcal{L}(\Phi, \Psi)$[10].

Let us begin by deriving the negative log likelihood function. Looking at Fig. 6.7 and (6.27), it is obvious that the marginal pdf is expressed as[11]

$$p_Y(y; \Phi, \Psi) = \prod_{l=0}^{L-1} \prod_{n=0}^{N-1} \mathcal{N}_{\mathbb{C}} \left\{ y_{n,l}^1; \sum_{k=D_l+1}^{D_l+K_l} \mathbf{g}_{k,l}^H \mathbf{y}_{n-k,l}, \lambda_{n,l} \right\}. \tag{6.31}$$

The negative log likelihood function is hence written as

$$\mathcal{L}(\Phi, \Psi) = \sum_{l=0}^{L-1} \sum_{n=0}^{N-1} \left(\log \lambda_{n,l} + \frac{|y_{n,l}^1 - \bar{\mathbf{g}}_l^H \bar{\mathbf{y}}_{n-D_l-1,l}|^2}{\lambda_{n,l}} \right), \tag{6.32}$$

where $\bar{\mathbf{g}}_l$ is the room regression vector defined by (6.10) and $\bar{\mathbf{y}}_{n,l}$ is defined by (6.11).

Unfortunately, we cannot analytically minimize the negative log likelihood function specified by (6.32) with respect to Φ and Ψ. Thus, we optimize the Φ and Ψ values alternately. This optimization algorithm starts with $\Psi[0] = \{\bar{\mathbf{g}}_l[0]\}_l$. The value of the psd component set, $\Phi = \{\lambda_{n,l}\}_{n,l}$, is updated by solving

$$\Phi[i+1] = \underset{\Phi}{\operatorname{argmin}} \, \mathcal{L}(\Phi, \Psi[i]). \tag{6.33}$$

We may find that this is easily accomplished by computing

$$s_{n,l}[i+1] = y_{n,l} - \bar{\mathbf{g}}_l[i]^H \bar{\mathbf{y}}_{n-D_l-1,l}, \tag{6.34}$$

$$\lambda_{n,l}[i+1] = |s_{n,l}[i+1]|^2, \tag{6.35}$$

for all n and l values. The value of the room regression vector set, $\Psi = \{\bar{\mathbf{g}}_l\}_l$, is updated by solving

$$\Psi[i+1] = \underset{\Psi}{\operatorname{argmin}} \, \mathcal{L}(\Phi[i+1], \Psi). \tag{6.36}$$

We find that $\mathcal{L}(\Phi[i+1], \Psi)$ is equivalent to $\mathcal{F}^{[i+1]}(\bar{\mathbf{g}}_l)$ of (6.12). Hence, (6.36) leads to the update formula defined by (6.13). The above pair of update processes is repeated until convergence. We observe that this parameter op-

---

[10] The MLE method is usually defined as the maximization of a log likelihood function. We define the MLE method in terms of minimization because this definition leads to a least-squares-type formulation, which is more familiar in the microphone array field.

[11] Marginal pdf (6.31) is, of course, derived analytically by substituting (6.25) and (6.27) into

$$p_Y(y; \Phi, \Psi) = \int p_{Y|S}(y|s; \Psi) p_S(s; \Phi) ds. \tag{6.30}$$

timization algorithm followed by a clean speech signal estimator (6.19) is exactly the same as the WPE method summarized in Table 6.1.

Thus, we have derived the WPE method based on the model-based approach. It is now obvious that the negative log likelihood function value increases monotonically, and that the room regression vector values always converge.

6.3 Combined Dereverberation and Denoising Method

Thus far, we have concentrated on dereverberation, assuming that a clean speech signal is contaminated only by room reverberation. In this section, we assume that both reverberation and ambient noise are present. That is to say, we consider the acoustical system depicted in Fig. 6.8.

In Fig. 6.8, each microphone signal is the sum of the reverberant and noise signals. A simple approach to clean speech signal estimation is thus the tandem connection of denoising and dereverberation processes as illustrated in Fig. 6.9. The dereverberation process may be implemented by the WPE method. On the other hand, the denoising process is realized by using power spectrum modification techniques such as Wiener filtering [21] and spectral subtraction [22].

Unfortunately, however, the tandem approach does not provide a good estimate of a clean speech signal [11]. This is mainly because the denoised (yet reverberant) speech signals at the output of the denoising process no longer have any consistent linear relationship with the clean speech signal. This results in a significant degradation in dereverberation performance since the successive dereverberation process relies on such linearity.

In view of this, this section describes an effectively combined dereverberation and denoising method that is illustrated in Fig. 6.10. This method consists of four components, i.e., a dereverberation filter, a multichannel Wiener filter, a Wiener filter controller, and a dereverberation filter controller. These components work iteratively. The important features of this method are as follows:

- The dereverberation filter precedes the multichannel Wiener filter for denoising.
- The controllers for the dereverberation and denoising filters are coupled with each other; the dereverberation filter controller uses the Wiener filter parameters while the Wiener filter controller uses the dereverberated speech signals.

We derive this combined dereverberation and denoising method in accordance with the four steps described in Section 6.1.3.

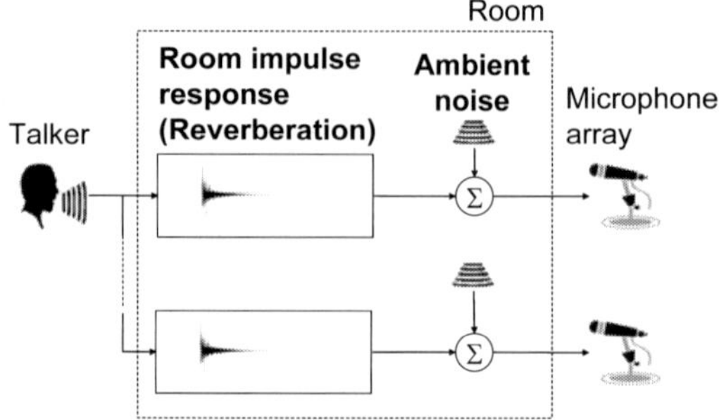

Fig. 6.8 Acoustical system of a room where reverberation and ambient noise are present.

Fig. 6.9 Tandem connection of denoising and dereverberation processes.

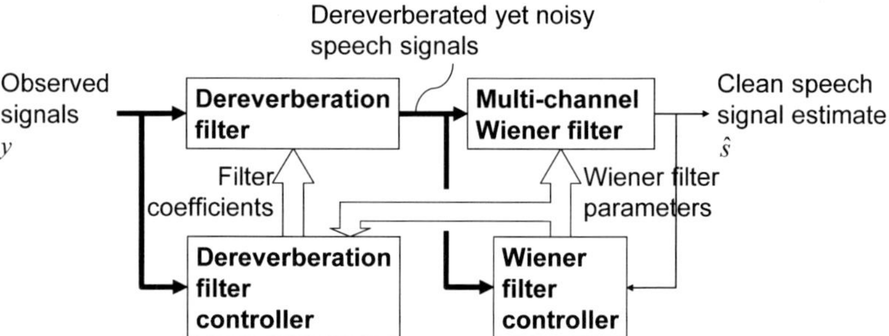

Fig. 6.10 Block diagram of combined dereverberation and denoising.

6.3.1 Room Acoustics Model

The first step is to define the room acoustics pdf $p_{Y|S}(y|s; \Psi)$. For this purpose, we extend the complete MCAR reverberation model (6.18) to take ambient noise into account. We use the complete MCAR model rather than the simplified version because the complete version explicitly expresses relations among different microphone signals. Such information is useful especially for improving the denoising performance.

Looking at Fig. 6.8, we see that observed signal vector $\mathbf{y}_{n,l}$ is given by

$$\mathbf{y}_{n,l} = \mathbf{u}_{n,l} + \mathbf{d}_{n,l}, \tag{6.37}$$

where $\mathbf{u}_{n,l} = [u_{n,l}^1, \cdots, u_{n,l}^M]^T$ and $\mathbf{d}_{n,l} = [d_{n,l}^1, \cdots, d_{n,l}^M]$ denote the noise-free reverberant speech signal vector and the noise signal vector, respectively. On the basis of the MCAR reverberation model, defined by (6.18), the noise-free reverberant speech signal vector, $\mathbf{u}_{n,l}$, is given by

$$\mathbf{u}_{n,l} = \sum_{k=D_l+1}^{D_l+K_l} G_{k,l}^H \mathbf{u}_{n-k,l} + \mathbf{h}_l s_{n,l}. \tag{6.38}$$

Substituting (6.38) into (6.37), we derive

$$\mathbf{y}_{n,l} = \sum_{k=D_l+1}^{D_l+K_l} G_{k,l}^H \mathbf{u}_{n-k,l} + \mathbf{h}_l s_{n,l} + \mathbf{d}_{n,l}. \tag{6.39}$$

By using (6.37), (6.39) is further transformed as

$$\mathbf{y}_{n,l} = \sum_{k=D_l+1}^{D_l+K_l} G_{k,l}^H \mathbf{y}_{n-k,l} + \mathbf{x}_{n,l}, \tag{6.40}$$

where $\mathbf{x}_{n,l}$ is defined as

$$\mathbf{x}_{n,l} = \mathbf{h}_l s_{n,l} + \mathbf{v}_{n,l}, \tag{6.41}$$

$$\mathbf{v}_{n,l} = \mathbf{d}_{n,l} - \sum_{k=D_l+1}^{D_l+K_l} G_{k,l}^H \mathbf{d}_{n-k,l}. \tag{6.42}$$

We find that $\mathbf{v}_{n,l}$, given by (6.42), is a filtered version of the noise signal vector $\mathbf{d}_{n,l}$. With this in mind, (6.40) along with (6.41) and (6.42) may be interpreted as follows. In (6.41), the clean speech signal, $s_{n,l}$, is first scaled by the steering vector $\mathbf{h}_l$. This scaled clean speech signal vector is then contaminated by the filtered noise, $\mathbf{v}_{n,l}$, to yield noisy anechoic speech signal vector $\mathbf{x}_{n,l}$. Finally, this noisy anechoic speech signal vector is reverberated via the MCAR system to produce the noisy reverberant speech signal vector, $\mathbf{y}_{n,l}$, observed by the microphone array. Fig. 6.11 illustrates this process for generating $\mathbf{y}_{n,l}$ from $s_{n,l}$. Hereafter, the filtered noise signal vector, $\mathbf{v}_{n,l}$, is referred to as a noise signal vector.

The important point of this model is that a clean speech signal is first mixed by noise signals and then reverberated. In fact, this generative model naturally leads to a signal estimation process which first removes the reverberation effect and then suppresses the noise signals as the upper branch of the block diagram shown in Fig. 6.10.

We can define room acoustics pdf $p_{Y|S}(y|s; \Psi)$ based on (6.40). Assume that $\mathbf{v}_{n_1,l_1}$ and $\mathbf{v}_{n_2,l_2}$ are statistically independent unless $(n_1, l_1) = (n_2, l_2)$, and that $\mathbf{v}_{n,l}$ is normally distributed with mean $\mathbf{0}_M$ and covariance matrix $\Gamma_{n,l}$ as

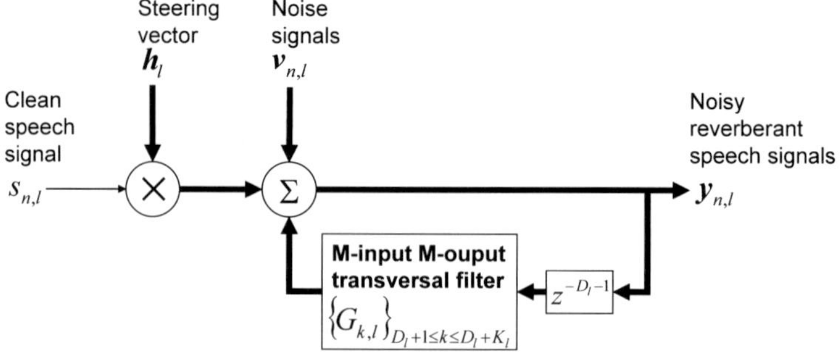

Fig. 6.11 Block diagram of generation process of noisy reverberant speech signals.

$$p_{V_{n,l}}(\mathbf{v}_{n,l}; \Gamma_{n,l}) = \mathcal{N}_{\mathbb{C}}\{\mathbf{v}_{n,l}; \mathbf{0}_M, \Gamma_{n,l}\}. \tag{6.43}$$

Let us denote the (m_1, m_2)th element of $\Gamma_{n,l}$ as $\gamma_{n,l}^{m_1,m_2}$. Then, $\{\gamma_{n,l}^{m,m}\}_l$ represents the short-time psd at the nth time frame of the mth noise signal. On the other hand, if $m_1 \neq m_2$, $\{\gamma_{n,l}^{m_1,m_2}\}_l$ represents the short-time cross spectral density (csd) at the nth time frame between the m_1th and m_2th noise signals.

By combining (6.40), (6.41), and (6.43), we obtain the following conditional pdf:

$$p_{\mathbf{Y}_{n,l}|\mathbf{Y}_{n-1,l},\cdots,\mathbf{Y}_{n-D_l-K_l,l},S_{n,l}}(\mathbf{y}_{n,l}|\mathbf{y}_{n-1,l},\cdots,\mathbf{y}_{n-D_l-K_l,l},s_{n,l};\Psi)$$
$$= \mathcal{N}_{\mathbb{C}}\left\{\mathbf{y}_{n,l}; \sum_{k=D_l+1}^{D_l+K_l} G_{k,l}\mathbf{y}_{n-k,l} + \mathbf{h}_l s_{n,l}, \Gamma_{n,l}\right\}, \tag{6.44}$$

where parameter set Ψ is defined as

$$\Psi = \{\{G_{k,l}\}_{D_l+1\leq k\leq D_l+K_l}, \mathbf{h}_l, \{\Gamma_{n,l}\}_{0\leq n\leq N-1}\}_{0\leq l\leq L-1}. \tag{6.45}$$

Therefore, the room acoustics pdf that takes both reverberation and ambient noise into account may be given by the following equation:

$$p_{\mathbf{Y}|S}(\mathbf{y}|s;\Psi) =$$

$$\prod_{l=0}^{L-1} \prod_{n=0}^{N-1} p_{\mathbf{Y}_{n,l}|\mathbf{Y}_{n-1,l},\cdots,\mathbf{Y}_{n-D_l-K_l,l},S_{n,l}}\left(\mathbf{y}_{n,l}|\mathbf{y}_{n-1,l},\cdots,\mathbf{y}_{n-D_l-K_l,l},s_{n,l};\Psi\right)$$

$$= \frac{1}{\pi^{LN}\prod_{l=0}^{L-1}\prod_{n=0}^{N-1}|\Gamma_{n,l}|}$$

$$\times \exp\left\{-\sum_{l=0}^{L-1}\sum_{n=0}^{N-1}\left(\mathbf{y}_{n,l} - \sum_{k=D_l+1}^{D_l+K_l} G_{k,l}^H \mathbf{y}_{n-k,l} - \mathbf{h}_l s_{n,l}\right)^H\right.$$

$$\left.\times \Gamma_{n,l}^{-1}\left(\mathbf{y}_{n,l} - \sum_{k=D_l+1}^{D_l+K_l} G_{k,l}^H \mathbf{y}_{n-k,l} - \mathbf{h}_l s_{n,l}\right)\right\}.$$

$$(6.46)$$

Thus, the room acoustics pdf has been defined.

6.3.2 Clean Speech Model

In the second step, we define clean speech pdf $p_S(s;\Phi)$. For the WPE method described in Section 6.2, we made no assumption as regards to the short-time psd of the clean speech signal. Although this unconstrained psd model sufficed for dereverberation, we need a more restrictive model to achieve both dereverberation and denoising. An all-pole model is one of the most widely used models that can accurately represent the voice production process with a few parameters.

The all-pole model assumes that short-time psd component $\lambda_{n,l}$ is expressed as

$$\lambda_{n,l} = \frac{\nu_n}{|1 - \alpha_{n,1}e^{-j\omega_l} - \alpha_{n,P}e^{-jP\omega_l}|^2}, \qquad (6.47)$$

where P is the prescribed order of this model. P is typically set at 12 when the sampling frequency is 8 kHz. ω_l is the angular frequency corresponding to the lth frequency bin, which is given by $\omega_l = 2\pi l/L$. $\alpha_{n,k}$ and ν_n are the kth linear prediction coefficient (LPC) and the prediction residual power (PRP) at the nth time frame. Therefore, the model parameter set, Φ, now consists of the LPCs and PRPs as

$$\Phi = \{\alpha_{n,1},\cdots,\alpha_{n,P},\nu_n\}_{0\leq n\leq N-1}, \qquad (6.48)$$

while the clean speech pdf, $p_S(s;\Phi)$, is still represented as (6.27). It should be noted that when we compare (6.48) with (6.28), the number of parameters per time frame is reduced from L to $P+1$.

6.3.3 Clean Speech Signal Estimator

In the third step, we derive the posterior pdf of the clean speech signal, $p_{S|Y}(s|y; \Phi, \Psi)$, to define the MMSE clean speech signal estimator. This posterior pdf is derived by substituting room acoustics pdf (6.46) and clean speech pdf (6.27) into (6.4). Here, we describe only the resultant posterior pdf.

At the computation of $p_{S|Y}(s|y; \Phi, \Psi)$, $\mathbf{x}_{n,l}$, originally defined by (6.41), is first computed from observed signal vector $\mathbf{y}_{n,l}$ according to

$$\mathbf{x}_{n,l} = \mathbf{y}_{n,l} - \sum_{k=D_l+1}^{D_l+K_l} G_{k,l}^H \mathbf{y}_{n-k,l}. \tag{6.49}$$

Recalling that $\mathbf{x}_{n,l}$ is the noisy anechoic speech signal vector, we can interpret (6.49) as a dereverberation process. Now, let us define $\tilde{\mathbf{w}}_{n,l}$ and $\mathbf{w}_{n,l}$ as

$$\tilde{\mathbf{w}}_{n,l} = \frac{\Gamma_{n,l}^{-1} \mathbf{h}_l}{\mathbf{h}_l^H \Gamma_{n,l}^{-1} \mathbf{h}_l}, \tag{6.50}$$

$$\mathbf{w}_{n,l} = \frac{\lambda_{n,l}}{\lambda_{n,l} + (\tilde{\mathbf{w}}_{n,l})^H \Gamma_{n,l} \tilde{\mathbf{w}}_{n,l}} \tilde{\mathbf{w}}_{n,l}. \tag{6.51}$$

Then, the posterior pdf is given by the following equation:

$$p(s|y; \Phi, \Psi) = \prod_{l=0}^{L-1} \prod_{n=0}^{N-1} \mathcal{N}_{\mathbb{C}}\{s_{n,l}; \mu_{n,l}, \epsilon_{n,l}\}, \tag{6.52}$$

where mean $\mu_{n,l}$ and variance $\epsilon_{n,l}$ are computed by

$$\mu_{n,l} = \mathbf{w}_{n,l}^H \mathbf{x}_{n,l}, \tag{6.53}$$

$$\epsilon_{n,l} = \frac{\lambda_{n,l}(\tilde{\mathbf{w}}_{n,l})^H \Gamma_{n,l} \tilde{\mathbf{w}}_{n,l}}{\lambda_{n,l} + (\tilde{\mathbf{w}}_{n,l})^H \Gamma_{n,l} \tilde{\mathbf{w}}_{n,l}}. \tag{6.54}$$

We see that $\mathbf{w}_{n,l}$ is the gain vector of the multichannel Wiener filter [23] and $\epsilon_{n,l}$ is the associated error variance. This indicates that the MMSE estimate, $\hat{s}_{n,l} = \mu_{n,l}$, of the clean speech signal $s_{n,l}$ is a denoised version of $\mathbf{x}_{n,l}$. Thus, the clean speech signal estimator consists of a dereverberation process (6.49) followed by a denoising process (6.53), which coincides with the upper branch of the block diagram shown in Fig. 6.10.

6.3.4 Parameter Optimization

The final step in the derivation is to specify a cost function and an optimization algorithm for the model parameters. The model parameters that need to be determined are the LPCs and PRPs of the clean speech signal, the room regression matrices, the steering vectors, and the noise covariance matrices as

$$\Phi = \{\alpha_{n,1}, \cdots, \alpha_{n,P}, \nu_n\}_{0 \leq n \leq N-1}, \tag{6.55}$$

$$\Psi = \{\{G_{k,l}\}_{D_l+1 \leq k \leq D_l+K_l}, \mathbf{h}_l, \{\Gamma_{n,l}\}_{0 \leq n \leq N-1}\}_{0 \leq l \leq L-1}. \tag{6.56}$$

We again use the MLE method. Hence, we minimize the negative log likelihood function (cost function) $\mathcal{L}(\Phi, \Psi)$, which is defined as $\mathcal{L}(\Phi, \Psi) = -\log p(y; \Phi, \Psi)$, to determine the parameter values. Since $p(y; \Phi, \Psi) = \int p_{Y|S}(y|s; \Psi) p_S(s; \Phi)$ and the room acoustics and clean speech pdfs are given by (6.46) and (6.27), respectively, we obtain the following marginal pdf:

$$p(y; \Phi, \Psi) = \prod_{l=0}^{L-1} \prod_{n=0}^{N-1} \mathcal{N}_{\mathbb{C}} \left\{ \mathbf{y}_{n,l}; \sum_{k=D_l+1}^{D_l+K_l} G_{k,l}^H \mathbf{y}_{n-k,l}, \lambda_{n,l} \mathbf{h}_l \mathbf{h}_l^H + \Gamma_{n,l} \right\}. \tag{6.57}$$

The negative log likelihood function is therefore obtained by taking the negative logarithm of (6.57) as

$$\mathcal{L}(\Phi, \Psi) = \sum_{l=0}^{L-1} \sum_{n=0}^{N-1} \left\{ \log |\lambda_{n,l} \mathbf{h}_l \mathbf{h}_l^H + \Gamma_{n,l}| + \left(\mathbf{y}_{n,l} - \sum_{k=D_l+1}^{D_l+K_l} G_{k,l}^H \mathbf{y}_{n-k,l} \right)^H \right.$$

$$\left. \times \left(\lambda_{n,l} \mathbf{h}_l \mathbf{h}_l^H + \Gamma_{n,l} \right)^{-1} \left(\mathbf{y}_{n,l} - \sum_{k=D_l+1}^{D_l+K_l} G_{k,l}^H \mathbf{y}_{n-k,l} \right) \right\}. \tag{6.58}$$

As regards to the negative log likelihood function given by (6.58), we find that multiplying $\mathbf{h}_l$ by arbitrary real number c and simultaneously dividing $\{\lambda_{n,l}\}_l$ (i.e., PRP v_n in practice) by c^2 does not change the function value. To eliminate this scale ambiguity, we put a constraint on steering vector $\mathbf{h}_l$ as

$$\mathbf{h}_l^H \Gamma_{n,l}^{-1} \mathbf{h}_l = 1. \tag{6.59}$$

Unfortunately, the negative log likelihood function of (6.58) may not be minimized analytically. Thus, we optimize the parameter values based on an interative algorithm. We use a variant of the expectation-maximization (EM) algorithm [14], which converges to at least a local optimum.

Before describing this algorithm, we make two preliminary definitions. First, we assume that the noise signals are stationary over the observation period. Therefore, noise covariance matrix $\Gamma_{n,l}$ is simplified to a time-invariant covariance matrix as

$$\Gamma_{n,l} = \Gamma_l \quad \text{for } 0 \le n \le N - 1. \tag{6.60}$$

Secondly, we reclassify the parameters as follows:

$$\Theta_g = \{G_{D_l+1,l}, \cdots, G_{D_l+K_l,l}\}_{0 \le l \le L-1}, \tag{6.61}$$

$$\Theta_{-g} = \{\{\mathbf{h}_l, \Lambda_l\}_{0 \le l \le L-1}, \{\alpha_{n,1}, \cdots, \alpha_{n,P}, \nu_n\}_{0 \le n \le N-1}\}. \tag{6.62}$$

Θ_g is a set of room regression matrices while Θ_{-g} consists of all the parameters except for the room regression matrices. In the following, we use both forms of parameter set representation; Θ_g and Θ_{-g} are used in some contexts while Φ and Ψ are used in others.

We briefly summarize the concept of the EM-based parameter optimization algorithm that we use here. In this algorithm, the parameter values are updated iteratively. Suppose that we now have the parameter values, $\Theta_g[i]$ and $\Theta_{-g}[i]$, after the ith iteration has been completed. Then, the posterior pdf of the clean speech signal is obtained by substituting $\Theta_g = \Theta_g[i]$ and $\Theta_{-g} = \Theta_{-g}[i]$ (i.e., $\Phi = \Phi[i]$ and $\Psi = \Psi[i]$) into (6.52). This step is called the E-step. Now, we define an auxiliary function[12] as

$$\mathcal{Q}^{[i]}(\Theta_g, \Theta_{-g}) = -\int p_{S|Y}(s|y; \Theta_g[i], \Theta_{-g}[i]) \log p_{Y,S}(y, s; \Theta_g, \Theta_{-g}) ds. \tag{6.63}$$

Then, $\Theta_{-g}[i+1]$ is inductively obtained by minimizing this auxiliary function while keeping the value of Θ_g fixed at $\Theta_g[i]$, i.e.,

$$\Theta_{-g}[i + 1] = \underset{\Theta_{-g}}{\operatorname{argmin}} \, \mathcal{Q}^{[i]}(\Theta_g[i], \Theta_{-g}). \tag{6.64}$$

This step is called CM-step1. After CM-step1, $\Theta_g[i + 1]$ is obtained by minimizing negative log likelihood function (6.58) for a fixed Θ_{-g} value with $\Theta_{-g} = \Theta_{-g}[i + 1]$ as

$$\Theta_g[i + 1] = \underset{\Theta_g}{\operatorname{argmin}} \, \mathcal{L}(\Theta_g, \Theta_{-g}[i + 1]). \tag{6.65}$$

This step is called CM-step2. The above update process consisting of the E-step, CM-step1, and CM-step2 begins with initial values $\Theta_g[0]$ and $\Theta_{-g}[0]$, and is repeated until convergence is attained. We may readily prove that this iterative algorithm features the monotonic increase and convergence of the value of the negative log likelihood function.

---

[12] The intuitive idea of the auxiliary function is that if we possessed the joint pdf of the observed and clean speech signals, $p_{Y,S}(y, s; \Theta_g, \Theta_{-g})$, the value of Θ_{-g} could be optimized by minimizing the negative logarithm of this joint pdf with respect to Θ_{-g}. However, since the joint pdf is unavailable, we instead minimize its expectation on the clean speech signals, s, given observed signals, y, and the current parameter values, $\Theta_g[i]$ and $\Theta_{-g}[i]$. For further details of the EM algorithm and auxiliary function, see, for example, [14].

Since the formula for the E-step has already been obtained as (6.52), we derive those for CM-step1 and CM-step2 in the following.

6.3.4.1 CM-Step1

First of all, we need to embody auxiliary function $\mathcal{Q}^{[i]}(\Theta_g, \Theta_{-g})$. By definition, the auxiliary function is written as

$$
\begin{aligned}
\mathcal{Q}^{[i]}(\Theta_g, \Theta_{-g}) = \sum_{l=0}^{L-1} \sum_{n=0}^{N-1} &\Big\{ \log |\Gamma_l| + \log \lambda_{n,l} + \frac{|\mu_{n,l}[i]|^2 + \epsilon_{n,l}[i]}{\lambda_{n,l}} \\
&+ (|\mu_{n,l}[i]|^2 + \epsilon_{n,l}[i]) \mathbf{h}_l^H \Gamma_l^{-1} \mathbf{h}_l - \mathbf{h}_l^H \Gamma_l^{-1} m_{n,l}^*[i] \mathbf{x}_{n,l} \\
&- \mathbf{x}_{n,l}^H m_{n,l}[i] \Gamma_l^{-1} \mathbf{h_l} + \mathbf{x}_{n,l}^H \Gamma_l^{-1} \mathbf{x}_{n,l} \Big\},
\end{aligned}
\tag{6.66}
$$

where $\mathbf{x}_{n,l}$ and $\lambda_{n,l}$ are given by (6.49) and (6.47), respectively. $\mu_{n,l}[i]$ and $\epsilon_{n,l}[i]$ are the values of $\mu_{n,l}$ and $\epsilon_{n,l}$, respectively, obtained by substituting $\Theta_g = \Theta_g[i]$ and $\Theta_{-g} = \Theta_{-g}[i]$ into (6.53) and (6.54). Note that $\mu_{n,l}[i]$ and $\epsilon_{n,l}[i]$ have already been computed at the E-step. The Θ_{-g} value is updated by minimizing $\mathcal{Q}^{[i]}(\Theta_g[i], \Theta_{-g})$. The room regression matrices, and hence the dereverberated speech signal vector, $\mathbf{x}_{n,l}$, as well, are fixed at CM-step1. We denote the fixed value of $\mathbf{x}_{n,l}$ by $\mathbf{x}_{n,l}[i]$.

Now, let us derive minimization formulae for this auxiliary function given by (6.66). Θ_{-g} is composed of the noise covariance matrices, steering vectors, and the LPCs and PRPs of the clean speech signal. We first derive the update formula for noise covariance matrix Γ_l. By virtue of the stationarity assumption on the noise signals, for each l value, the new noise covariance matrix, Γ_l, is calculated from the dereverberated signal vector $\mathbf{x}_{n,l}[i]$ during periods where the speech is inactive. If we assume that the speech begins to be active at time frame N', we have the following update formula:

$$
\Gamma_l[i+1] = \sum_{n=0}^{N'-1} \mathbf{x}_{n,l}[i] \mathbf{x}_{n,l}[i]^H.
\tag{6.67}
$$

Next, we derive the update formula for steering vector $\mathbf{h}_l$. The updated steering vector value, $\mathbf{h}_l[i+1]$, is defined as the solution to the following optimization task:

$$
\mathbf{h}_l[i+1] = \operatorname*{argmin}_{\mathbf{h}_l} \; \mathcal{Q}^{[i]}(\Theta_g[i], \Theta_{-g}) \quad \text{subject to } \mathbf{h}_l^H \Gamma_l[i+1]^{-1} \mathbf{h}_l = 1.
$$

$$
\tag{6.68}
$$

Solving (6.68) by using the Lagrange multiplier method, we obtain the following update formula:

$$\mathbf{h}_l[i+1] = \frac{\tilde{\mathbf{h}}_l}{\tilde{\mathbf{h}}_l^H \tilde{\mathbf{h}}_l}, \tag{6.69}$$

$$\tilde{\mathbf{h}}_l = \frac{\mu_{n,l}[i]^* \mathbf{x}_{n,l}[i]}{|\mu_{n,l}[i]|^2 + \epsilon_{n,l}[i]}. \tag{6.70}$$

We see that the updated steering vector value, $\mathbf{h}_l[i+1]$, is the normalized cross correlation between the tentative estimate of the clean speech signal, $\mu_{n,l}[i]$, and that of the noisy anechoic speech signal, $\mathbf{x}_{n,l}[i]$.

Finally, let us derive the update formula for the LPCs and PRPs. For each time frame n, the updated values of the LPCs and PRPs are defined as the solution to the following optimization task:

$$\alpha_{n,1}[i+1], \cdots, \alpha_{n,P}[i+1], \nu_n[i+1] = \underset{\alpha_{n,1}, \cdots, \alpha_{n,P}, \nu_n}{\operatorname{argmin}} \; \mathcal{Q}^{[i]}(\Theta_g[i], \Theta_{-g}). \tag{6.71}$$

We can show that (6.71) leads to the following linear equations:

$$\sum_{k'=1}^{P} \alpha_{n,k'}[i+1] r_{k'-k} = r_k \quad \text{for } k = 1, \cdots, P, \tag{6.72}$$

$$\nu_n[i+1] = r_0 - \sum_{k=1}^{P} \alpha_{n,k}[i+1] r_k, \tag{6.73}$$

where r_k is the time-domain correlation coefficient corresponding to the power spectrum given by $\{|m_{n,l}[i]|^2 + \epsilon_{n,l}[i]\}_l$, i.e.,

$$r_k = \sum_{l=0}^{L-1} (|\mu_{n,l}[i]|^2 + \epsilon_{n,l}[i]) e^{jk\omega_l}. \tag{6.74}$$

We find that (6.72) and (6.73) have the same form as the Yule-Walker equation. Hence, $\{\alpha_{n,k}[i+1]\}_k$ and $\nu_n[i+1]$ are obtained via the Levinson-Durbin algorithm[13].

This update rule for the LPCs and PRPs seems to be quite natural. Indeed, if we had the clean speech signal, the LPCs and PRP at time frame n could be computed by applying the Levinson-Durbin algorithm to the power spectrum given by $\{|s_{n,l}|^2\}_l$. Because this true power spectrum is unavailable, the present update rule substitutes the expected power spectrum given by $\{|\mu_{n,l}[i]|^2 + \epsilon_{n,l}[i]\}_l$ for the true power spectrum.

---

[13] See [24] for details on the Yule-Walker equations and the Levinson-Durbin algorithm.

6.3.4.2 CM-Step2

In CM-step2, we minimize the log likelihood function (6.58) with respect to Θ_g while keeping the Θ_{-g} value fixed at $\Theta_{-g}[i+1]$. Let us define M-by-M matrix $\Lambda_{n,l}$ as[14]

$$\Lambda_{n,l} = \lambda_{n,l}\mathbf{h}_l\mathbf{h}_l^H + \Gamma_l. \tag{6.75}$$

Owing to the condition where $\Theta_{-g} = \Theta_{-g}[i+1]$, $\Lambda_{n,l}$ is fixed at CM-step2; hence, we denote the fixed value of $\Lambda_{n,l}$ by $\Lambda_{n,l}[i+1]$. By using this notation, the negative log likelihood function we wish to minimize is expressed as

$$\mathcal{L}(\Theta_g, \Theta_{-g}[i+1]) = \sum_{l=0}^{L-1}\sum_{n=0}^{N-1}\left(\mathbf{y}_{n,l} - \sum_{k=D_l+1}^{D_l+K_l} G_{k,l}^H\mathbf{y}_{n-k,l}\right)^H \Lambda_{n,l}[i+1]^{-1}$$
$$\times \left(\mathbf{y}_{n,l} - \sum_{k=D_l+1}^{D_l+K_l} G_{k,l}^H\mathbf{y}_{n-k,l}\right). \tag{6.76}$$

It is obvious that all the elements of the room regression matrices for the lth frequency bin, $\{G_{k,l}\}_k$ are dependent on each other. Thus, we define vector $\bar{\bar{\mathbf{g}}}_l$, which we call an extended room regression vector, as

$$\bar{\bar{\mathbf{g}}}_l = [(\mathbf{g}_{D_l+1,l}^1)^T, \cdots, (\mathbf{g}_{D_l+1,l}^M)^T, \cdots, (\mathbf{g}_{D_l+K_l,l}^1)^T, \cdots, (\mathbf{g}_{D_l+K_l,l}^M)^T], \tag{6.77}$$

where $\mathbf{g}_{k,l}^m$ is the mth column of $G_{k,l}$. The extended room regression vector is a row vector of M^2K_l dimensions. In response to this, we define an M-by-M^2K_l matrix $\bar{Y}_{n,l}$ consisting of the microphone signals as

$$\bar{Y}_{n,l} = \begin{bmatrix} \mathbf{y}_{n,l}^T & O & \mathbf{y}_{n-K_l+1,l}^T & O \\ & \ddots & \cdots & \ddots \\ O & \mathbf{y}_{n,l}^T & O & \mathbf{y}_{n-K_l+1,l}^T \end{bmatrix}. \tag{6.78}$$

Then, we can easily show that $\bar{\bar{\mathbf{g}}}_l[i+1]$ that minimizes (6.76) is computed by the following formula:

$$\bar{\bar{\mathbf{g}}}_l[i+1] = \left\{\left(\sum_{n=0}^{N-1} \bar{Y}_{n-D_l-1}^H \Lambda_{n,l}[i+1]^{-1}\bar{Y}_{n-D_l-1}\right)^{-1}\right.$$
$$\left.\times \left(\sum_{n=0}^{N-1} \bar{Y}_{n-D_l-1}^H \Lambda_{n,l}[i+1]^{-1}\mathbf{y}_{n,l}\right)\right\}^H. \tag{6.79}$$

All the steps in the derivation of the clean speech signal estimator described in Section 6.1.3 have now been completed. We can now summarize the combined dereverberation and denoising algorithm as shown in Table 6.2.

---

[14] $\Lambda_{n,l}$ is the covariance matrix of noisy anechoic speech signal vector $\mathbf{x}_{n,l}$.

Table 6.2 Summary of combined dereverberation and denoising algorithm.

Parameter optimization:

Initialize the parameter values, for example, as

$$\bar{\bar{\mathbf{g}}}_l[0] = [0, \cdots, 0]^T$$

$$\xi_l = \frac{1}{MN'} \sum_{n=0}^{N'-1} \mathbf{y}_{n,l}^H \mathbf{y}_{n,l}$$

$$\Gamma_l[0] = \xi_l I_M$$

$$\mathbf{h}_l[0] = \left(\frac{\xi_l}{M}\right)^{\frac{1}{2}} [1, \cdots, 1]^T \quad \text{for } 0 \le l \le L-1$$

$$\alpha_{n,1}[0], \cdots, \alpha_{n,P}[0], \nu_n[0] = \text{Levinson}(\{|y_{n,l}^1|^2\}_l) \quad \text{for } 0 \le n \le N-1$$

where Levinson($\cdot$) computes the LPCs and PRP for a given power spectrum with the Levinson-Durbin algorithm.

Update the parameter values iteratively until convergence by performing the following three steps for $i = 0, 1, 2, \cdots$.

   E-step: Compute $\mathbf{x}_{n,l}[i]$, $\mu_{n,l}[i]$, and $\epsilon_{n,l}[i]$. by using (6.49), (6.53), (6.54), (6.50), (6.51), and (6.47).

   CM-step1: Compute

$$\Gamma_l[i+1] = \sum_{n=0}^{N'-1} \mathbf{x}_{n,l}[i]\mathbf{x}_{n,l}[i]^H$$

$$\tilde{\mathbf{h}}_l = \frac{\mu_{n,l}^*[i]\mathbf{x}_{n,l}[i]}{|\mu_{n,l}[i]|^2 + \epsilon_{n,l}[i]}$$

$$\mathbf{h}_l[i+1] = \frac{\tilde{\mathbf{h}}_l}{\tilde{\mathbf{h}}_l^H \tilde{\mathbf{h}}_l} \quad \text{for } 0 \le l \le L-1$$

$$\alpha_{n,1}[i+1], \cdots, \alpha_{n,P}[i+1], \nu_n[i+1] = \text{Levinson}(\{|\mu_{n,l}[i]|^2 + \epsilon_{n,l}[i]\}_l) \quad \text{for } 0 \le n \le N-1$$

   CM-step2: For all l from 0 to $L-1$, compute

$$\lambda_{n,l}[i+1] = \frac{v_n[i+1]}{|1 - \alpha_{n,1}[i+1]e^{-j\omega_l} - \cdots - \alpha_{n,P}[i+1]e^{-jP\omega_l}|^2}$$

$$\Lambda_{n,l}[i+1] = \lambda_{n,l}[i+1]\mathbf{h}_l[i+1]\mathbf{h}_l[i+1]^H + \Gamma_l[i+1]$$

$$\bar{\bar{\mathbf{g}}}_l[i+1] = \left\{ \left(\sum_{n=0}^{N-1} \bar{Y}_{n-D_l-1}^H \Lambda_{n,l}[i+1]^{-1} \bar{Y}_{n-D_l-1} \right)^{-1} \times \left(\sum_{n=0}^{N-1} \bar{Y}_{n-D_l-1} \Lambda_{n,l}[i+1]^{-1} \mathbf{y}_{n,l} \right) \right\}^H$$

Clean speech signal estimation:

Compute the MMSE estimate of the clean speech signal given the optimized parameter values as

$$\hat{s}_{n,l} = \mu_{n,l} = \hat{\mathbf{w}}_{n,l}^H \left(\mathbf{y}_{n,l} - \sum_{k=D_l+1}^{D_l+K_l} \hat{G}_{k,l}^H \mathbf{y}_{n-k,l} \right)$$

where $\hat{\mathbf{w}}_{n,l}$ is computed by (6.51), (6.50), and (6.47).

In this table, we can see that the E-step coincides with the upper branch of the block diagram shown in Fig. 6.10 while CM-step1 and CM-step2 correspond to the Wiener filter controller and the dereverberation filter controller, respectively.

6.4 Experiments

This section describes experimental results on dereverberation and denoising to demonstrate the effectiveness of the combined dereverberation and denoising method described above. The system parameters were set as follows. The frame size and frame shift were 256 and 128 samples, respectively, for a sampling frequency of 8 kHz. The linear prediction order, P, was set at 12. The room regression orders were determined depending on frequency. Let f_l be the frequency in Hz corresponding to the lth frequency bin. Then, the room regression orders were defined as

$$
\begin{aligned}
&K_l = 5 \quad \text{for } f_l < 100; && K_l = 10 \quad \text{for } 100 \leq f_l < 200; \\
&K_l = 30 \quad \text{for } 200 \leq f_l < 1000; && K_l = 20 \quad \text{for } 1000 \leq f_l < 1500; \\
&K_l = 15 \quad \text{for } 1500 \leq f_l < 2000; && K_l = 10 \quad \text{for } 2000 \leq f_l < 3000; \\
&K_l = 5 \quad \text{for } f_l \geq 3000.
\end{aligned}
$$

We can save much computation time without degrading performance by setting the room regression orders at smaller values for higher frequency bins in this way.

For this experiment, we took 10 audio files from the ASJ-JNAS database. Each file contained the voice of one of 10 different (five male and five female) talkers. We then played each file through a loudspeaker in a room and recorded the sound with two microphones. We also played uncorrelated pink noise signals simultaneously from four loudspeakers located at different positions in the same room and recorded the sound with the same microphone setup. The captured (reverberant) speech and noise signals were finally mixed by using a computer with a signal-to-noise ratio (SNR) of 10, 15, or 20 dB. Thus, we had a total of 30 test samples. The reverberation time of the room was around 0.6 sec, and the distance between the loudspeaker for speech and the microhphone was 1.8 meters. The audio durations ranged from 3.16 to 7.17 sec.

We evaluated each test result in terms of cepstrum distance from a clean speech. A cepstrum distance of order C between two signals is defined as

$$
\text{CD} = \frac{10}{\log 10} \sqrt{2 \sum_{k=1}^{C} (c_k - \tilde{c}_k)^2} \quad \text{(dB)}, \tag{6.80}
$$

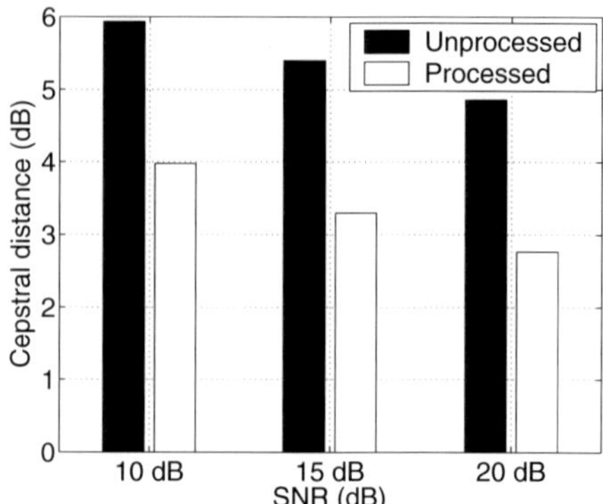

Fig. 6.12 Cepstrum distances for the combined dereverberation and denoising method with three different input SNRs: 10, 15, and 20 dB.

where c_k and c_k' are the cepstrum coefficients of the respective signals. The cepstrum order, C, was set at 12. Fig. 6.12 summarizes the cepstrum distances averaged over the 10 talkers for each input SNR. We can observe that the average cepstrum distances were improved by about 2 dB regardless of the input SNR, which indicates the effectiveness of the above method for dereverberation and denoising.

6.5 Conclusions

In this chapter, we described the weighted prediction error (WPE) method for dereverberation and the combined dereverberation and denoising method. These methods were derived by using a model based approach, which has been a dominant approach in denoising studies. Specifically, we used a multichannel autoregressive (MCAR) model to characterize the reverberation and ambient noise in a room. In contrast, we modeled a clean speech signal by using a time-varying power spectrum (TVPS) model.

As pointed out in [13], model selection is the key to the successful design of dereverberation and denoising methods. Different models lead to dereverberation and denoising methods with different characteristics. The methods based on the MCAR room acoustics model and the TVPS clean speech model would be advantageous in terms of the output speech distortion since they perform dereverberation based on linear filtering. On the other hand, from another viewpoint, for example insensitivity to acoustic environmental changes, other models and methods may be preferable. Further research is required to investigate the pros and cons of different models and methods for dereverberation and denoising.

References

1. J. Benesty, T. Gänsler, D. R. Morgan, M. M. Sondhi, and S. L. Gay, *Advances in Network and Acoustic Echo Cancellation.* Springer, 2001.
2. E. Hänsler and G. Schmidt, Eds, *Topics in Acoustic Echo and Noise Control.* Springer, 2006.
3. J. L. Flanagan, J. D. Johnston, R. Zahn, and G. W. Elko, "Computer-steered microphone arrays for sound transduction in large rooms," *Journal of the Acoustical Society of America*, vol. 78, no. 11, pp. 1508–1518, Nov. 1985.
4. J. L. Flanagan, D. A. Berkley, G. W. Elko, J. E. West, and M. M. Sondhi, "Autodirective microphone systems," *Acustica*, vol. 73, pp. 58–71, Feb. 1991.
5. M. Brandstein and D. Ward, Eds., *Microphone Arrays.* Springer, 2001.
6. Q.-G. Liu, B. Champagne, and P. Kabal, "A microphone array processing technique for speech enhancement in a reverberant space," *Speech Communication*, vol. 18, no. 4, pp. 317–334, Jun. 1996.
7. B. W. Gillespie, H. S. Malvar, and D. A. F. Florêncio, "Speech dereverberation via maximum-kurtosis subband adaptive filtering," in *Proc. IEEE ICASSP*, 2001, pp. 3701–3704.
8. K. Lebart, J. M. Boucher, and P. N. Denbigh, "A new method based on spectral subtraction for speech dereverberation," *Acta Acustica united with Acustica*, vol. 87, no. 3, pp. 359–366, May/Jun. 2001.
9. K. Eneman and M. Moonen, "Multimicrophone speech dereverberation: experimental validation," *EURASIP J. Audio, Speech, and Music Processing*, vol. 2007, Article ID 51 831, 19 pages, Apr. 2007, doi:10.1155/2007/51831.
10. A. Abramson, E. A. P. Habets, S. Gannot, and I. Cohen, "Dual-microphone speech dereverberation using GARCH modeling," in *Proc. IEEE ICASSP*, 2008, pp. 4565–4568.
11. T. Yoshioka, T. Nakatani, and M. Miyoshi, "An integrated method for blind separation and dereverberationof convolutive audio mixtures," in *Proc. EUSIPCO*, 2008.
12. ——, "Integrated speech enhancement method using noise suppression and dereverberation," *IEEE Trans. Audio, Speech, and Language Processing*, vol. 17, no. 2, pp. 231–246, Feb. 2009.
13. Y. Ephraim, "Statistical-model-based speech enhancement systems," *Proc. IEEE.* vol. 80, no 10, pp. 1526–1555, 1992.
14. C. Bishop, Ed., *Pattern Recognition and Machine Learning.* Springer, 2007.
15. T. Nakatani, T. Yoshioka, K. Kinoshita, M. Miyoshi, and B.-H. Juang, "Blind speech dereverberation with multi-channel linear prediction based on short time Fourier transform representation," in *Proc. IEEE ICASSP*, 2008, pp. 85–88.
16. H. Kuttruff, *Room Acoustics*, 4th ed. Taylor & Francis, 2000.
17. H. Attias, J. C. Platt, A. Acero, and L. Deng, "Speech denoising and dereverberation using probabilistic models," in *Advances in Neural Information Processing Systems, 13. NIPS 13.*, Nov. 30 2000-Dec. 2 2000, pp. 758–764.
18. K. Abed-Meraim, E. Moulines, and P. Loubaton, "Prediction error method for second-order blind identification," *IEEE Trans. Signal Processing*, vol. 45, no. 3, pp. 694–705, Mar. 1997.
19. K. Kinoshita, M. Delcroix, T. Nakatani, and M. Miyoshi, "Suppression of late reverberation effect on speech signal using long-term multiple-step linear prediction," *IEEE Trans. Audio, Speech, and Language Processing*, vol. 17, no. 4, pp. 534–545, May 2009.
20. A. van den Bos, "The multivariate complex normal distribution–A generalization," *IEEE Trans. Information Theory*, vol. 41, no. 2, pp. 537–539, Mar. 1995.
21. J. S. Lim and A. V. Oppenheim, "Enhancement and bandwidth compression of noisy speech," *Proc. IEEE*, vol. 67, no. 12, pp. 1586–1604, Dec. 1979.
22. S. F. Boll, "Suppression of acoustic noise in speech using spectral subtraction," *IEEE Trans. Acoustics, Speech and Signal Processing*, vol. 27, no. 2, pp. 113–120, Apr. 1979.

23. M. Feder, A. V. Oppenheim, and E. Weinstein, "Maximum likelihood noise cancellation using the em algorithm," *IEEE Trans. Acoustics, Speech and Signal Processing*, vol. 37, no. 2, pp. 204–216, Feb. 1989.
24. S. Haykin, *Adaptive Filter Theory*, 4th ed.　Prentice Hall, 2001.

Chapter 7
Codebook Approaches for Single Sensor Speech/Music Separation

Raphaël Blouet and Israel Cohen

Abstract The work presented is this chapter is an introduction to the subject of single sensor source separation dedicated to the case of speech/music audio mixtures. Approaches related in this study are all based on a full (Bayesian) probabilistic framework for both source modeling and source estimation. We first present a review of several codebook approaches for single sensor source separation as well as several attempts to enhance the algorithms. All these approaches aim at adaptively estimating the optimal time-frequency masks for each audio component within the mixture. Three strategies for source modeling are presented: Gaussian scaled mixture models, codebooks of autoregressive models, and Bayesian non-negative matrix factorization (BNMF). These models are described in details and two estimators for the time-frequency masks are presented, namely the *minimum mean-squared error* and the *maximum a posteriori*. We then propose two extensions and improvements on the BNMF method. The first one suggests to enhance discrimination between speech and music through multi-scale analysis. The second one suggests to constrain the estimation of the expansion coefficients with prior information. We finally demonstrate the improved performance of the proposed methods on mixtures of voice and music signals before conclusions and perspectives.

7.1 Introduction

The goal of blind source separation is to estimate the original sources given mixture(s) of those sources. Among other criteria, source separation systems are often categorized by the relative number of observation channels and

Raphaël Blouet
Audionamix, France, e-mail: `raphael.blouet@audionamix.com`

Israel Cohen
Technion–Israel Institute of Technology, Israel, e-mail: `icohen@ee.technion.ac.il`

I. Cohen et al. (Eds.): Speech Processing in Modern Communication, STSP 3, pp. 183–198.
springerlink.com

sources. A separation problem is over-determined if the number of channels is larger than the number of sources. It is determined if both number are equal and under-determined if the number of channels is smaller than the number of sources. Other mixing criteria such as the length and temporal variation of the mixing filter are fundamental in some applications but not relevant to the work presented here. An accurate classification of source separation system can be found in [1].

Under the determined and over-determined cases, independent component analysis (ICA) allows to separate sources assuming their statistical independence. Multichannel source separation systems can rely on the use of spatial cues even for the under-determined case [2]. Many applications however require the processing of composite signals given a unique observation channel. Single sensor audio sources separation is indeed a challenging research topic that has taken great importance in many fields including audio processing, medical imaging and communications. Attempts to solve this task were proposed in the context of the computational auditory scene analysis (CASA) [3] or binary masking technics [4]. However, audio sources received in the mixture signal generally overlap in the time-frequency plane and source separation via binary time-frequency masks often offers poor performance. Other approaches were hence proposed using various techniques and models such as dual Kalman filters [5], sparse decompositions [6], non-negative matrix factorization [7, 8, 9], support vector machines [10], and harmonic models [11]. Most of them use perceptual principles and/or statistical models for yielding the *most likely* explanation of the observed mixture given the target audio sources. These methods attempt to remove ambiguity induced by the overlap of sources in the time-frequency plane. Ambiguity can be expressed probabilistically and we here focus our study on statistical modeling of the sources. We indeed present several methods that all use a full Bayesian probabilistic framework for both modeling and estimation of the audio source. Each source model is based on a codebook that captures its spectral variability. In this work, spectral information is handled by the short-time Fourier transform (STFT) or by autoregressive models. Separation is obtained through an extension of the classical Wiener filtering based on an adaptive estimation of each source power spectral density (PSD). Sources in the mixture are indeed considered as stationary conditionally to these coefficients and the Wiener filtering is applicable. Each source PSD involves the estimation of codebook elements expansion coefficient.

This chapter is organized as follow. In Section 7.2, we review three single sensor source separation algorithms, which rely on codebooks to model audio sources. We propose in Section 7.3, an improvement over the BNMF algorithm by using a multiple-window STFT representation. This basically decomposes, in an iterative fashion, the observed signal into source and residual components for several window lengths. At each step the most relevant components allow to build a source estimate. In Section 7.4 we add several constraints to enhance the estimation of the expansion coefficients. These constraints consist adding sparse or smooth priors in the expansion coefficients estimation process. An evaluation of the presented methods is given in Section 7.5. Finally, conclusions and perspectives are discussed in Section 7.6.

7.2 Single Sensor Source Separation

In this section, we state the problem of single sensor source separation and present three codebook based methods for source separation, namely Gaussian scaled mixture models (GSMMs), codebooks of autoregressive (AR) models and Bayesian non-negative matrix factorization (BNMF). Although different, these methods share common properties: all of them use a full Bayesian probabilistic framework for both the audio source models and the estimation procedure; all of them use codebooks of stationary spectral shapes to model non-stationary signal; all of them handle separately spectral shape data and amplitude information.

7.2.1 Problem Formulation

Given an observed signal x considered as the mixture of two sources s_1 and s_2, the source separation problem is to estimate s_1 and s_2 from x. Algorithms presented in this chapter use the short-time Fourier transform (STFT) domain and we denote $X(f,t)$ the STFT of x for the frame index t and the frequency bin f. We have

$$X(f,t) = S_1(t,f) + S_2(t,f), \tag{7.1}$$

and the source separation problem consists of obtaining the estimates $\hat{S}_1(t,f)$ and $\hat{S}_2(t,f)$. Codebook approaches are widely used for single sensor source separation. These methods rely on the assumption that each source can be represented by a dictionary and usually work in two stages: (1) *Building a codebook in an offline learning step*; (2) *Given all mixture components dictionaries, finding source representation that best describes the mixture*. Building a codebook for one source implies the definition of source models. Codebook (CB) representatives of the first and second source are described as $\phi_1 = \{\phi_{1,k}\}_{k=1,...,K_1}$ and $\phi_2 = \{\phi_{2,k}\}_{k=1,...,K_2}$, where K_1 and K_2 are respectively the codebooks' lengths. The nature of the codebooks is algorithm dependent. For the GMM-based and the BNMF algorithms it contains different Gaussian parameters, while for the AR-based separation system it contains linear predictive coefficients (LPC) of the autoregressive (AR) process. As the definition of the optimal time-frequency masks estimator depends on the source models, we present a study of several criteria, namely MAP (maximum a posteriori) and MMSE (minimum mean-squared error) for the three models.

7.2.2 GSMM-Based Source Separation

The source separation technique presented in [12] has suggested the use of the Gaussian scaled mixture model (GSMM) to model the source statistics. As Gaussian mixture model (GMM) can be regarded as a codebook of zero-mean Gaussian components where each component is identified by a diagonal covariance matrix and a prior probability, the GSMM incorporates a supplementary scale parameter which aims at better taking into account non-stationarity and dynamics of the sources. Each component k of source i is identified by a diagonal covariance matrix $\Sigma_{i,k}$ and a state prior probability $\omega_{i,k}$, so that in this case we have $\phi_{i,k} = \{\Sigma_{i,k}, \omega_{i,k}\}$. The GSMM model is then simply defined by

$$p(S_i(:,t)|\{\phi_{i,k}\}_k) = \sum_{k=1}^{K_i} \omega_{i,k} \mathcal{N}(S_i(:,t)|0, a_{i,k}(t)\Sigma_{i,k}), \qquad (7.2)$$

where $\sum_{k=1}^{K_i} \omega_{i,k} = 1$, $a_{i,k}(t)$ is a time-varying amplitude factor and $S_i(:,t)$ denotes the vector of frequency coefficients of source i at frame t. Though GSMM is a straightforward extension of GMM, it is unfortunately untractable due to the added amplitude factor. Benaroya, Bimbot, and Gribonval [12] suggested to estimate these amplitude factors pairwise, in a maximum likelihood (ML) sense, as follows:

$$\gamma_{a_{1,k},a_{2,q}}(t) = P(\phi_{1,k}, \phi_{2,q}|X(:,t), a_{1,k}(t), a_{2,q}(t))$$
$$\hat{a}_{1,k}(t), \hat{a}_{2,q}(t) = \max_{a_{1,k},a_{2,q}} \{\gamma_{a_{1,k},a_{2,q}}(t)\}. \qquad (7.3)$$

The source STFTs can then be estimated either in a MAP or MMSE sense, as follows.

MAP estimator:

$$\hat{S}_i(f,t) = \frac{\hat{a}_{i,k^*}\sigma_{i,k^*}^2(f)}{\hat{a}_{1,k^*}\sigma_{1,k^*}^2(f) + \hat{a}_{2,q^*}\sigma_{2,q^*}^2(f)} X(f,t), \qquad (7.4)$$

where $(k^*, q^*) = \underset{(k,q)}{\operatorname{argmax}}\{\gamma_{\hat{a}_{1,k},\hat{a}_{2,q}}(t)\}$.

MMSE estimator:

$$\hat{S}_i(f,t) = \sum_{k,q} \gamma_{\hat{a}_{1,k},\hat{a}_{2,q}}(t) \frac{\hat{a}_{i,k}\sigma_{i,k}^2(f)}{\hat{a}_{1,k}\sigma_{1,k}^2(f) + \hat{a}_{2,q}\sigma_{2,q}^2(f)} X(f,t). \qquad (7.5)$$

Note that since the covariance matrices are assumed diagonal, separation is performed independently in each frequency bin.

7.2.3 AR-Based Source Separation

Spectral envelopes of speech signals in the STFT domain are efficiently characterized by AR models, which have been used for enhancement in [13, 14]. Many earlier methods for speech enhancement assume that the interfering signal is quasi-stationary, which restricts their usage for non-stationary environments, such as music interferences. Srinivasan et al. [13, 14] suggest to represent the speech and interference signals by using codebooks of AR processes. The predefined codebooks now contain the linear prediction coefficients of the AR processes, denoted by $\phi_1 = \{\phi_{1,k}\}_{k=1,\dots,K_1}$ and $\phi_2 = \{\phi_{2,k}\}_{k=1,\dots,K_2}$ ($\phi_{i,k}$ is now a vector of length equal to the AR order).

ML approach:

Srinivasan, Samuelsson, and Kleijn [13] proposed a source separation approach based on the ML. The goal is to find the most probable pair $\{\phi_{1,k^*}, \phi_{2,q^*}\}$ for a given observation, with

$$(k^*, q^*) = \operatorname*{argmax}_{k,q} \left\{ \max_{\lambda_{1,k}(t),\lambda_{2,q}(t)} \{p(x(:,t)|\phi_{1,k}, \phi_{2,q}; \lambda_{1,k}(t), \lambda_{2,q}(t))\} \right\},$$
$$(7.6)$$

where $x(:,t)$ denotes frame t of mixture x (this time in the time domain) and $\lambda_{1,k}(t), \lambda_{2,k}(t)$ are the frame-varying variances of the AR processes describing each source. In [13] a method is proposed to estimate the excitation variances pairwise. Like previously, once the optimal pair is found, source separation can be achieved through Wiener filtering on the given observation $x(:,t)$.

MMSE approach:

Srinivasan, Samuelsson, and Kleijn [14] also proposed an MMSE estimation approach for separation. In a Bayesian setting, the LPC and excitation variances are now considered as random variables, which can be given prior distributions to reflect *a priori* knowledge. Denoting

$$\theta = \{\phi_1, \phi_2, \{\lambda_1(t)\}_t, \{\lambda_2(t)\}_t\},$$

the MMSE estimator of θ is

$$\hat{\theta} = E[\theta|x] = \frac{1}{p(x)} \int_\theta \theta\, p(x|\theta)p(\theta)d\theta. \tag{7.7}$$

We take $p(\theta) = p(\phi_1)\, p(\phi_2)\, p(\{\lambda_1(t)\}_t), p(\{\lambda_2(t)\}_t)\}$. Then the likelihood function $p(x|\theta)$ decays rapidly when deviating from the true excitation variances [14]. This gives ground to approximating the true excitation variances by their ML estimates, and (7.7) can be rewritten as

$$\hat{\theta} = \frac{1}{p(x)} \times$$
$$\int_{\phi_1,\phi_2} [\phi_1, \phi_2]\, p(x|\phi_1, \phi_2; \hat{\lambda}_1^{ML}, \hat{\lambda}_2^{ML})p(\phi_1)p(\hat{\lambda}_1^{ML})p(\phi_2)p(\hat{\lambda}_2^{ML})d\phi_1 d\phi_2, \tag{7.8}$$

where $\hat{\lambda}_1^{ML}$ and $\hat{\lambda}_2^{ML}$ are the ML estimates of the excitation variances. We use codebook representatives as entries in the integration in (7.8).

Assuming that they are uniformly distributed, $\hat{\theta}$ is given by [14]:

$$\hat{\theta} = \frac{1}{K_1 K_2} \sum_{k=1}^{K_1} \sum_{q=1}^{K_2} \theta_{kq} \frac{p(x|\phi_{1,k}, \phi_{2,q}; \lambda_{1,k}^{ML}, \lambda_{2,q}^{ML})}{p(x)} p(\lambda_{1,k}^{ML})p(\lambda_{2,q}^{ML}), \tag{7.9}$$

where $\theta_{kq} = [\phi_{1,k}, \phi_{2,q}, \hat{\lambda}_{1,k}^{ML}, \hat{\lambda}_{2,q}^{ML}]$. Given two fixed AR codebooks, (7.9) allows an MMSE estimation of AR processes jointly associated to source 1 and source 2. Once $\hat{\theta}$ is known, we can use Wiener filtering for the separation stage.

7.2.4 Bayesian Non-Negative Matrix Factorization

The source separation technique described in [12] proposes to model each STFT frame of each source as a sum of *elementary components* modeled as zero-mean complex Gaussian distribution with known power spectral density (PSD), also referred to as *spectral shape*, and scaled by amplitude factors. Specifically, each source STFT is modeled as

$$S_i(f,t) = \sum_{k=1}^{K_i} \sqrt{a_{i,k}(t) \cdot E_{i,k}(f,t)}, \tag{7.10}$$

where $E_{i,k}(f,t) \sim \mathcal{N}_c(0, \sigma_k^2(f))$. The representatives of the codebooks are now

$$\phi_{i,k} = [\sigma_{i,k}^2(f_1), \ldots, \sigma_{i,k}^2(f_N)]^T \,.$$

This model is well adapted to the complexity of musical sources, as it explicitly represents the signal as linear combination of more simple components, with various spectral shapes.

Given the codebooks, the separation algorithm based on this model consists of two steps, as follows:

i) Compute of the amplitude parameters $\{a_{i,k}(t)\}$ in an ML sense; this is equivalent to performing a nonnegative expansion of $|X(f,t)|^2$ onto the basis formed by the union of the codebooks,
ii) Given the estimated $\{a_{i,k}(t)\}$, estimate each source in an MMSE sense through Wiener filtering:

$$\hat{S}_i(f,t) = \frac{\sum_{k=1}^{K_i} \hat{a}_{i,k}\sigma_{i,k}^2(f)}{\sum_{k=1}^{K_1} \hat{a}_{1,k}\sigma_{1,k}^2(f) + \sum_{k=1}^{K_2} \hat{a}_{2,k}\sigma_{2,k}^2(f)} X(f,t) \,. \qquad (7.11)$$

Amplitude parameters estimation.

Conditionally upon the amplitude parameters $\{a_{i,k}(t)\}_k$, elementary sources $\mathbf{s}_{i,k}(t,f)$ are (independent) zero-mean Gaussian processes with variance $a_{i,k}(t)\sigma_{i,k}^2(f)$.

The observed mixture is also a zero-mean Gaussian process with variance $\sum_{i,k} a_{i,k}(t)\sigma_{i,k}^2(f)$.

We hence have the following log-likelihood equation:

$$\log p\left(X(f,t)|\{a_{i,k}(t)\}_{i,k}\right) = -\frac{1}{2}\sum_f \left[\frac{|X(f,t)|^2}{en(f,t)} + \log(en(f,t))\right], \quad (7.12)$$

where $en(f,t) = \sum_{i,k} a_{i,k}(t)\sigma_{i,k}^2(f)$.

Amplitude parameters $\{a_{i,k}(t)\}_{i,k}$ can be estimated by setting the first derivative of the log-likelihood to zero under a non negativity constraint. As this problem has no analytic solution, we use an iterative, fixed point algorithm with multiplicative updates [15, 16, 17], yielding:

$$a_{i,k}^{(l+1)}(t) = a_{i,k}^{(l)}(t)\frac{\sum_f \sigma_{i,k}^2(f)\frac{|X(f,t)|^2}{en^{(l)}(t,f)^2}}{\sum_f \sigma_{i,k}^2(f)\frac{1}{en^{(l)}(f,t)}}, \qquad (7.13)$$

where $en^{(l)}(f,t) = \sum_{i=1,2}\sum_k a_{i,k}^{(l)}(t)\sigma_{i,k}^2(f)$.

7.2.5 Learning the Codebook

We assume that we have some clean training samples of each source. These training excerpts do not need to be identical to the source contained in the observed mixture but we assume that they are *representatives* of the source. We estimate the codebooks on the training samples according to the models previously presented.

1. Model of Section 7.2.2:
   The expectation-maximization (EM) algorithm [18] is used to estimate $\{\omega_{l,k}, \Sigma_{i,k}\}_{k=1}^{K_i}$.
2. Model of Section 7.2.3:
   A generalized Lloyd algorithm is used to learn the LPC coefficients [19].
3. Model of Section 7.2.4:
   A vector quantization algorithm is applied to the short-term power spectra of the training samples.

7.3 Multi-Window Source Separation

This section investigates the use of a multiresolution framework. It intents to enhance the single sensor source separation system presented in Section 7.2.4. This work has been published in [20] and suggests the use of multiple STFT windows in order to deal separately with long and short term elements. By that, we obtain a multi-resolution algorithm for source separation. The basic idea is to decompose, in an iterative fashion, the observed signal into source components and a residual for several window lengths. At each iteration the residual contains components that are not properly represented at the current resolution. The input signal at iteration i is then the residual generated at iteration $i - 1$. The algorithm starts with a long window sizes which is decreased throughout the iterations.

7.3.1 General Description of the Algorithm

We assume that $w_1(n), \ldots, w_N(n)$ are N windows with decreasing support lengths. We denote by X_{w_i} the STFT of x with analysis window $w_i(n)$.

We first apply the algorithm of Section 7.2.4 with the longest window $w_1(n)$. This algorithm is slightly modified as to yield a residual signal, such that

$$X_{w_1}(t, f) = S_{1,w_1}(t, f) + S_{2,w_1}(t, f) + R_{w_1}(t, f). \qquad (7.14)$$

After inverse-STFT, we iterate on $r_1(n)$ with analysis window w_2. At the end of the day, the decomposition at iteration i is

$$R_{w_{i-1}}(t,f) = S_{1,w_i}(t,f) + S_{2,w_i}(t,f) + R_{w_i}(t,f). \qquad (7.15)$$

While no residual is computed with the monoresolution approach, the multiresolution approach involves the partition of the PSDs set. This is done through a partition of the amplitude parameters indices $k \in K_1 \cup K_2$ into three different sets $Q_1(t)$, $Q_2(t)$, and $R(t)$. The set $R(t)$ contains the indices k such that the corresponding $\{a_k(t)\}_{k \in R(t)}$ are "unreliably" estimated and the set $Q_1(t)$ [resp. $Q_2(t)$] contains the indices $k \in K_1$ (rep. $k \in K_2$) of reliably estimated $a_k(t)$. More precisely, this partition is done through the computation of a confidence measure $J_k(t)$. This confidence measure should be small if the corresponding estimate of $a_k(t)$ is accurate. As will be seen in Section 7.3.2, the confidence measure that we have chosen is related to the Fisher information matrix of the likelihood of the amplitude parameters.

Note that these three sets of indices $Q_1(t)$, $Q_2(t)$, and $R(t)$ are frame dependent. Relying on similar filtering formulae than those used in the classical algorithm, we get three estimates $\hat{s}_{1,w_i}(n)$, $\hat{s}_{2,w_i}(n)$ and $\hat{r}_{w_i}(n)$ (back in the time domain). Then we can iterate on $\hat{r}_{w_i}(n)$ with a different STFT window $w_{i+1}(n)$.

Finally, we get the estimates:

$$\hat{s}_1(t) = \sum_{i=1}^{N} \hat{s}_{1,w_i}(t), \qquad (7.16)$$

$$\hat{s}_2(t) = \sum_{i=1}^{N} \hat{s}_{2,w_i}(t), \qquad (7.17)$$

$$\hat{r}(t) = \hat{r}_{w_N}(t). \qquad (7.18)$$

We expect that short components such as transients are unreliably estimated with long analysis windows and therefore fall in the residual until the window length is sufficiently small to capture them reliably.

7.3.2 Choice of a Confidence Measure

Suppose we have a confidence interval on each amplitude parameter $a_k(t)$:

$$a_k(t) \in [\hat{a}_k(t) - l_k(t); \hat{a}_k(t) + L_k(t)].$$

The quantity $J_k(t) = \frac{L_k(t) - l_k(t)}{\hat{a}_k(t)}$ can be seen as the relative confidence measure on the estimate $\hat{a}_k(t)$. $J_k(t)$ allows to set the following accept/reject rule: $J_k(t) > \lambda \, a_k(t)$ is considered as reliable and contributes to $\hat{s}_{1,w_i}(t)$ or $\hat{s}_{2,w_i}(t)$, whereas $J_k(t) \leq \lambda\, a_k(t)$ is not considered as reliable and contributes to $\hat{r}_{w_i}(t)$, where λ is an experimentally tuned threshold. Using a Taylor expansion of the opposite log-likelihood around the ML estimate, we have

$$-\log p(r_{w_i}|\{\hat{a}_k(t) + \delta a_k(t)\}_k) \approx$$
$$-\log p(r_{w_i}|\{\hat{a}_k(t)\}_k) + \frac{1}{2}[\delta a_k(t)]^T H(t)[\delta a_k(t)], \qquad (7.19)$$

where

$$H_{i,j}(t) = -\frac{\partial^2}{\partial a_i(t)\partial a_j(t)} \log p(r_{w_i}|\{\hat{a}_k(t)\}_k)\,.$$

Then taking the expectation on both sides of (7.19), we get

$$E\left(\log \frac{p(r_{w_i}|\{a_k(t)\})}{p(r_{w_i}|\{a_k(t) + \delta a_k(t)\})}|\{a_k(t)\}\right) \approx \frac{1}{2}[\delta a_k(t)]^T I(t)[\delta a_k(t)], (7.20)$$

where the left side of the equality is the Kullback-Leibler divergence and $I(t)$ is the Fisher information matrix for $a_k(t) = \hat{a}_k(t)$. This relationship is well known and is true even if $\{a_k(t)\}_k$ is not a local optimum [21]. For a given admissible error E on the Kullback-Leibler divergence, we get

$$|\delta a_k(t)| \leq \sqrt{2E} \cdot \sqrt{[I^{-1}(t)]_{k,k}}\,. \qquad (7.21)$$

Equation (7.21) defines a confidence interval on $a_k(t)$ for a given admissible error E on the objective function.

Note that we see here that the *sensitivity* of the estimated parameters to a small change of the objective function (here, the opposite log-likelihood) or a mis-specification of the objective function is related to the inverse of the Fisher information matrix. In our model, the Fisher information matrix is

$$I_{i,j}(t) = \frac{1}{2}\sum_f \frac{\sigma_i^2(f)\sigma_j^2(f)}{en(f,t)^2}\,. \qquad (7.22)$$

We have to take the inverse of $I(t)$ for all t and we get

$$J_k(t) = \frac{\sqrt{[I^{-1}(t)]_{k,k}}}{\hat{a}_k(t)}\,. \qquad (7.23)$$

7.3.3 Practical Choice of the Thresholds

It is possible to experimentally tune the threshold. However, we here suggest two practical methods to select the PSDs. The first one consists in choosing the M most reliable estimates for each frame, all other indices being kept to build the residual $R(t)$. Given $\epsilon \in [0, 1]$, the second method consists in building the residual $R(t)$ by taking the less reliable indices such that

$$\sum_{k \in R(t)} a_k(t) < \epsilon \sum_{k \in K_{1,w_i} \cup K_{2,w_i}} a_k(t) .$$

For each frame, the left sum is the estimated variance of the residual r_{w_i} while the right sum is the estimated variance of the overall decomposition $r_{w_{i-1}}$. This method insures that after N iterations, the residual variance is (approximately) lower that ϵ^N times the original signal variance.

7.4 Estimation of the Expansion Coefficients

In this section, we focus on the estimation of the expansion coefficients and investigate several tracks to enhance performances through modifications of their estimate. Improvements proposed here only concern the single sensor source separation system presented in Section 7.2.4.

ML estimation of the expansion coefficients $a_k(t)$ have high variance and show great temporal instability. This is especially true when several spectral shapes strongly overlap. Three methods aiming to reduce temporal variability of these estimates are presented. They all consist of adding a constraint on the time continuity of the envelope parameters. The first method consists of applying a plain windowed median filter to the ML estimates of the $a_k(t)$ coefficients. The second method consists of incorporating a prior density for the envelope parameters, whose logarithm is equal to $\lambda_k |a_k(t) - a_k(t-1)|$ up to a constant. We thus change the ML estimation to a MAP estimation method. This method is inspired from [7], although we use in our approach a full Bayesian framework. The third method is also based on a MAP estimate of the envelope coefficients, with GMM modeling of these coefficients.

7.4.1 Median Filter

Given the ML estimates of the envelope parameters $a_k^{ML}(t)$ and an integer J, we compute the filtered coefficients $a_k^{med}(t)$:

$$a_k^{med}(t) = \text{median}\left(\{a_k^{ML}(\tau)\}_{\tau \in [t-J,...,t+J]}\right) . \tag{7.24}$$

The use a median filter, rather than a linear filter allows to take into account some possible fast changes on the coefficients. Indeed, as the envelope coefficients may be seen as the time envelope of notes, the attack part of the note presents generally fast changes.

7.4.2 Smoothing Prior

We here model the filtering of the envelope coefficients as a smoothing prior on $a_k(t)$. Equation (7.12) gives the log-likelihood of the envelope parameters $\log p(X(f,t)\}_f|\{a_k(t)\}_{k\in K_1\cup K_2})$. We add a prior density on the envelope coefficients $a_k(t) \geq 0$ by using a Laplacian density centered on $\hat{a}_k(t-1)$. Density of a_k at time index $t-1$ then become

$$p(a_k(t)|\hat{a}_k(t-1)) = \frac{1}{2\lambda_k} \exp\left(-\lambda_k|a_k(t) - \hat{a}_k(t-1)|\right), \qquad (7.25)$$

and the log posterior distribution becomes

$$\log p(\{a_k(t)\}|\{X(f,t)\}_f, \{\hat{a}_k(t-1)\}) =$$
$$-\frac{1}{2}\sum_f \left[\frac{|X(f,t)|^2}{en(f,t)} + \log(en(f,t))\right] - \sum_k \lambda_k|a_k(t) - \hat{a}_k(t-1)|, \quad (7.26)$$

where $en(f,t) = \sum_{k\in K_1\cup K_2} a_k(t)\sigma_k^2(f) + \sigma_b^2$.

A fixed point algorithm allows to obtain the MAP estimates of the parameters. The update rule for $a_k(t)$ is closely related to non-negative matrix factorization [16], because of the non negativity constrains. At iteration $\ell+1$, the estimation is given by

$$a_k^{(\ell+1)}(t) = a_k^{(\ell)}(t) \frac{\sum_f \sigma_k^2(f)\frac{|X(t,f)|^2}{en^{(\ell)}(f,t)^2}}{\sum_f \sigma_k^2(f)\frac{1}{en^{(\ell)}(f,t)} + \lambda_k}. \qquad (7.27)$$

One issue is to estimate the hyperparameters $\{\lambda_k\}$. These parameters can be estimated from the marginal distribution $p(\{X(t,f)\}_{t,f}|\{\lambda_k\})$ or thanks to an approximate estimation procedure that simultaneously estimate the amplitude parameters $a_k(t)$ and their hyperparameters λ_k at each iteration. Relying on the exponential distribution $E(a_k(t)|\lambda_k) = \frac{1}{\lambda_k}$, we have

$$\lambda_k^{(\ell)} = \frac{T}{\sum_{t=1}^T a_k^{(\ell)}(t)}.$$

The use of the Laplacian prior allows to take into account fast changes of the envelope coefficients, for instance on the onset of a note. Note that we use here a causal estimator, as $a_k(t)$ only depends on $X(f,t)$ and $\hat{a}_k(t-1)$. We could have constructed a noncausal estimator in the same scheme. Note that the smoothing parameter λ_k depends on the elementary source index k. We should set, for instance, greater values of λ_k in case of harmonic instruments compared to percussive instruments.

7.4.3 GMM Modeling of the Amplitude Coefficients

We here propose two estimation methods for the amplitude coefficients $\{a_k\}$ that take into account prior density modeling of those coefficients. In the following subsection we indeed train a statistical model from the coefficients $\{a_k^{ML}\}$, and use those densities to re-estimate $\{a_k\}$. We propose the use of a static model that fits with an ergodic modeling of $\{a_k\}$ (GMM).

Given the ML estimates of the logarithm of the envelope parameters $\{\log\,(a_k(t))\}_{k\in[1:K]\,,t\in[1:T]}$, we train a GMM with two components using the EM algorithm. The model parameters of the GMM are $\{(w_i,\mu_i,\Sigma_i)\}_{i=1,2}$ with w_i, μ_i and Σ_i respectively the weight, the mean and the covariance matrix of the ith component.

For each spectral form, the smoothing re-estimation formula is given by

$$\hat{a}_k(t) = \alpha[p_1(t)a_{k_1} + p_2(t)a_{k_2}] + (1 - \alpha)[a_k^{ML}(t)],$$

with $p_i(t) = p(a_k(t)|\{(w_i,\mu_i,\Sigma_i)\})$, and α is an adaptation factor that must be tuned, $\alpha \in [0,1]$.

7.5 Experimental Study

7.5.1 Evaluation Criteria

We used the standard source-to-distortion ratio (SDR) and the signal-to-interference ratio (SIR) described in [22]. In short, the SDR provides an overall separation performance criterion, while the SIR only measures the level of residual interference in each estimated source. The higher are the ratios, the better is the quality of the estimation. Note that in underdetermined source separation, the SDR is usually driven by the amount of artifacts in the source estimates.

7.5.2 Experimental Setup and Results

The evaluation task consists of unmixing a mixture of speech and piano. The signals are sampled at 16 kHz and the STFT is calculated using a Hamming window of 512 samples length (32 ms) with 50% overlap between consecutive frames. For the learning step we used piano and speech segments that were 10 minutes long. The observed signals are obtained from mixtures of 25 s long test signals. We use two sets of data. The first one is used to evaluate GSMM and AR based methods, while the second one is used to

Table 7.1 SIR/SDR measures for GMM/AR based methods.

	GSMM		AR	
	MAP	MMSE	ML	MMSE
SIR	6.8	7.1	9.8	12.5
SDR	4.9	4.6	4.5	4.9

(a. Speech)

	GSMM		AR	
	MAP	MMSE	ML	MMSE
SIR	6.9	7.7	4.4	3.2
SDR	3.1	3.1	2.1	2.0

(b. Music)

Table 7.2 SIR/SDR measures for BNMF and proposed extensions.

	BNMF	3 windows BNMF	Median Filter	Smoothing prior	GMM
SIR	5.1	9.7	9.7	6.7	8.3
SDR	−1.9	−2.3	0.6	−0.5	0.7

(a. Speech)

	BNMF	3 windows BNMF	Median Filter	Smoothing prior	GMM
SIR	6.1	7.4	9.3	7.2	7.2
SDR	3.8	4.2	5.2	4.6	4.7

(b. Music)

evaluate the AF method and a proposed extension. The first data set consists of speech segments taken from the TIMIT database and piano segments acquired through the web. The second data set consists of speech segments taken from the BREF database and piano segments that are extracted from solo piano recordings. Results are shown in Tables 7.1 and 7.2.

When observing the simulation results, one can see that no single algorithm is superior for all criteria. However, the AR/MMSE performs well when separating the speech. Another observation is that the AR has low SIR results for the piano source; this can be explained by the fact that AR model is not adequate for representing the piano signal. The multi-window approach slightly improves the SDR and the SIR on the music part (around 0.4 dB) but the improvement is not clear in the speech component case. The proposed extensions of BNMF source separation methods (median filter, temporal smoothing, and GMM methods) perform better than the baseline system for all SDR and SIR values.

7.6 Conclusions

We have presented three codebook approaches for single channel source separation. Each codebook underlies different models for the sources, i.e addresses different features of the sources. The experimental separation results show that AR-based model efficiently captures speech features, while the BNMF model is good at representing music because of its additive nature (a complex music signal is represented as a sum of simpler elementary components). On the other hand, the GSMM assumes in its conception that the audio signal is *exclusively* in one state *or* another, which intuitively does not best explain music. The separation results presented here also tend to confirm this fact. Extensions of the BNMF approach have been proposed, which allow a slight performance improvement. It is worthwhile noting that the above methods rely on the assumptions that sources are continuously active in all time frames. This is generally incorrect for audio signals, and we will try in our future work to use source presence probability estimation in the separation process. The separation algorithms define the posterior probabilities and gain factors of each pair based on the entire frequency range. This causes numerical instabilities and does not take into consideration local features of the sources, e.g., for speech signals the lower frequencies may contain most of the energy. Another aspect of our future work will consist in adding perceptual frequency weighting in the expansion coefficient estimation.

Acknowledgements This work has been supported by the European Commission's IST program under project Memories. Thanks to Elie Benaroya and Cédric Févotte for their contributions to this work.

References

1. E. Vincent, M. G. Jafari, S. A. Abdallah, M. D. Plumbley, and M. E. Davies, "Blind audio source separation," Centre for Digital Music, Queen Mary University of London, Technical Report C4DM-TR-05-01, 2005.
2. O. Yilmaz and S. Richard, "Blind separation of speech mixtures via time-frequency masking," *IEEE Trans. Signal Processing*, vol. 52, no. 7, pp. 1830–1847, 2004.
3. D. L. Wang and G. J. Brown, "Separation of speech from interfering sounds based on oscillatory correlation," *IEEE Trans. Neural Networks*, vol. 10, pp. 684–697, 1999.
4. S. T. Roweis, "One microphone source separation," *Advances in Neural Information Processing Systems*, vol. 13, pp. 793–799, 2000.
5. E. Wan and A. Nelson, "Neural dual extended Kalman filtering: Applications in speech enhancement and monaural blind signal separation," in *Proc. IEEE Workshop On Neural Networks And Signal Processing*, 1997.
6. B. A. Pearlmutter and A. M. Zador, "Monaural source separation using spectral cues," in *Proc. Int. Congress on Independent Component Analysis and Blind Signal Separation*, 2004.
7. T. Virtanen, "Sound source separation using sparse coding with temporal continuity objective," in *Proc. Int. Computer Music Conference*, 2003.
8. P. Smaragdis, "Non-negative matrix factor deconvolution; extraction of multiple sound sources from monophonic inputs," in *Proc. Int. Congress on Independent Component Analysis and Blind Signal Separation*, 2004.
9. M. Kim and S. Choi, "Monaural music source separation: Nonnegativity, sparseness, and shift-invariance," in *Proc. Int. Congress on Independent Component Analysis and Blind Signal Separation*, 2006, pp. 617–624.
10. S. Hochreiter and M. C. Mozer, "Monaural separation and classification of nonlinear transformed and mixed independent signals: an SVM perspective," in *Proc. Int. Congress on Independent Component Analysis and Blind Signal Separation*, 2001.

11. E. Vincent and M. D. Plumbley, "Single-channel mixture decomposition using Bayesian harmonic models," in *Proc. Int. Congress on Independent Component Analysis and Blind Signal Separation*, 2006.

12. L. Benaroya, F. Bimbot, and R. Gribonval, "Audio source separation with a single sensor," *IEEE Trans. Audio, Speech, Language Processing*, vol. 14, no. 1, pp. 191–199, Jan. 2006.

13. S. Srinivasan, J. Samuelsson, and W. B. Kleijn, "Codebook driven short-term predictor parameter estimation for speech enhancement," *IEEE Trans. Audio, Speech, Language Processing*, vol. 14, no. 1, pp. 163–176, 2006.

14. ——, "Codebook-based Bayesian speech enhancement," in *Proc. Int. Conf. on Acoustics, Speech and Signal Processing (ICASSP)*, 2005.

15. D. D. Lee and H. S. Seung, "Algorithms for non-negative matrix factorization," *Advanced Neural Information Proceecing Systems*, vol. 13, pp. 556–562, 2001.

16. P. O. Hoyer, "Non-negative sparse coding," in *Proc. IEEE Workshop on Neural Networks for Signal Processing*, 2002.

17. L. Benaroya, L. M. Donagh, F. Bimbot, and R. Gribonval, "Non negative sparse representation for Wiener based source separation with a single sensor," in *Proc. Int. Conf. on Acoustics, Speech and Signal Processing (ICASSP)*, Apr. 2003, pp. 613–616.

18. A. P. Dempster, N. M. Laird, and D. B. Rubin, "Maximum likelihood from incomplete data via the EM algorithm," *Journal of the Royal Statistical Society*, vol. 39, no. 1, pp. 1–38, 1977.

19. K. K. Paliwal and W. B. Kleijn, *Digital Communications*, 4th ed. New York: McGraw Hill, 2001.

20. L. Benaroya, R. Blouet, C. Févotte, and I. Cohen, "Single sensor source separation based on Wiener filtering and multiple window STFT," in *Proc. Int. Workshop Acoust. Echo Noise Control (IWAENC)*, Paris, France, Sep. 2006, pp. 1–4, paper no. 52.

21. C. Arndt, *Information Measures: Information and Its Description in Science and Engineering*. Springer-Verlag, 2001.

22. R. Gribonval, L. Benaroya, E. Vincent, and C. Févotte, "Proposals for performance measurement in source separation," in *Proc. of ICA*, Nara, Japan, 2003.

Chapter 8
Microphone Arrays: Fundamental Concepts

Jacek P. Dmochowski and Jacob Benesty

Abstract Microphone array beamforming is concerned with the extraction of a desired acoustic signal from noisy microphone measurements. The microphone array problem is a more difficult one than that of classical sensor array applications due to several reasons: the speech signal is naturally analog and wideband. Moreover, the acoustic channel exhibits strong multipath components and long reverberation times. However, the conventional metrics utilized to evaluate signal enhancement performance do not necessarily reflect these differences. In this chapter, we attempt to reformulate the objectives of microphone array processing in a unique manner, one which explicitly accounts for the broadband signal and the reverberant channel. A distinction is made between wideband and narrowband metrics. The relationships between broadband performance measures and the corresponding component narrowband measures are analyzed. The broadband metrics presented here provide measures which, when optimized, hopefully lead to beamformer designs tailored to the specific nature of the microphone array environment.

8.1 Introduction

Microphone arrays are becoming increasingly more common in the acquisition and de-noising of received acoustic signals. Additional microphones allow us to apply spatiotemporal filtering methods which are, at least in theory, significantly more powerful in their ability to rid the received signal of the unwanted additive noise than conventional temporal filtering techniques which simply emphasize certain temporal frequencies while de-emphasizing others.

Jacek P. Dmochowski
City College of New York, NY, USA, e-mail: `jdmochowski@ccny.cuny.edu`

Jacob Benesty
INRS-EMT, QC, Canada, e-mail: `benesty@emt.inrs.ca`

I. Cohen et al. (Eds.): Speech Processing in Modern Communication, STSP 3, pp. 199–223.
springerlink.com   © Springer-Verlag Berlin Heidelberg 2010

It may be argued that the advantage of multiple microphones has not been fully realized in practice. In this chapter, we attempt to shed light on the fundamental problems and goals of microphone array beamforming by studying the metrics by which performance is measured.

The initial microphone array designs [1], [2] are formulated around rather lofty expectations. For example, the minimum variance distortionless response (MVDR) [3] beamformer has long been studied in the microphone array context. Notice that the MVDR beamformer attempts to perform both dereverberation *and* noise reduction simultaneously in each frequency bin. In a reverberant environment with unknown impulse responses, acoustic dereverberation is a challenging problem in of itself; constraining the frequency-domain solution to achieve perfect dereverberation while at the same time reducing additive noise is ambitious. It may be speculated that the coupling of dereverberation and noise reduction in adaptive beamformer designs leads to poor performance in practice [4].

This chapter attempts to clearly identify the challenges and define the metrics involved in the microphone array beamforming problem. Instead of attempting to develop new beamformer designs, we focus on clarifying the goals of microphone arrays, which will then hopefully lead to the development of more powerful beamformers that are tailored to the distinct nature of the environment.

8.2 Signal Model

Consider the conventional signal model in which an N-element microphone array captures a convolved source signal in some noise field. The received signals at the time instant t are expressed as [2], [5], [6]

$$y_n(t) = g_n(t) * s(t) + v_n(t) \tag{8.1}$$
$$= x_n(t) + v_n(t), \ n = 1, 2, \ldots, N,$$

where $g_n(t)$ is the impulse response from the unknown source $s(t)$ to the nth microphone, $*$ stands for linear convolution, and $v_n(t)$ is the additive noise at microphone n. We assume that the signals $x_n(t)$ and $v_n(t)$ are uncorrelated and zero mean. By definition, $x_n(t)$ is coherent across the array. The noise signals $v_n(t)$ are typically only partially (if at all) coherent across the array. All previous signals are considered to be real and broadband.

Conventionally, beamforming formulations have attempted to recover $s(t)$ from the microphone measurements $y_n(t)$, $n = 1, \ldots, N$. This involves two processes: dereverberation and noise reduction. In this chapter, the desired signal is instead designated by the clean (but convolved) signal received at microphone 1, namely $x_1(t)$. The problem statement may be posed as follows: given N mixtures of two uncorrelated signals $x_n(t)$ and $v_n(t)$, our aim is to preserve $x_1(t)$ while minimizing the contribution of the noise terms $v_n(t)$ in the array output. While the array processing does not attempt to perform any inversion of the acoustic channels g_n, (single-channel) dereverberation techniques may be applied to the beamformer output in a post-processing fashion. Such techniques are not considered in this chapter, however.

The main objective of this chapter is to properly define all relevant measures that aid us in recovering the desired signal $x_1(t)$, to analyze the signal components at the beamformer output, and to clarify the most important concepts in microphone arrays.

In the frequency domain, (8.1) can be rewritten as

$$Y_n(f) = G_n(f)S(f) + V_n(f) \tag{8.2}$$
$$= X_n(f) + V_n(f), \ n = 1, 2, \ldots, N,$$

where $Y_n(f)$, $G_n(f)$, $S(f)$, $X_n(f) = G_n(f)S(f)$, and $V_n(f)$ are the frequency-domain representations of $y_n(t)$, $g_n(t)$, $s(t)$, $x_n(t)$, and $v_n(t)$, respectively, at temporal frequency f, and the time-domain signal

$$a(t) = \int_{-\infty}^{\infty} A(f)e^{j2\pi ft} df \tag{8.3}$$

is the inverse Fourier transform of $A(f)$.

The N microphone signals in the frequency domain are better summarized in a vector notation as

$$\mathbf{y}(f) = \mathbf{g}(f)S(f) + \mathbf{v}(f)$$
$$= \mathbf{x}(f) + \mathbf{v}(f)$$
$$= \mathbf{d}(f)X_1(f) + \mathbf{v}(f), \tag{8.4}$$

where

$$\mathbf{y}(f) = \left[Y_1(f) \ Y_2(f) \ \cdots \ Y_N(f) \right]^T,$$
$$\mathbf{x}(f) = \left[X_1(f) \ X_2(f) \ \cdots \ X_N(f) \right]^T,$$
$$= S(f) \left[G_1(f) \ G_2(f) \ \cdots \ G_N(f) \right]^T$$
$$= S(f)\mathbf{g}(f),$$
$$\mathbf{v}(f) = \left[V_1(f) \ V_2(f) \ \cdots \ V_N(f) \right]^T,$$
$$\mathbf{d}(f) = \left[1 \ \frac{G_2(f)}{G_1(f)} \ \cdots \ \frac{G_N(f)}{G_1(f)} \right]^T$$
$$= \frac{\mathbf{g}(f)}{G_1(f)},$$

and superscript ^{T} denotes transpose of a vector or a matrix. The vector $\mathbf{d}(f)$ is termed the *steering vector* or *direction vector* since it determines the

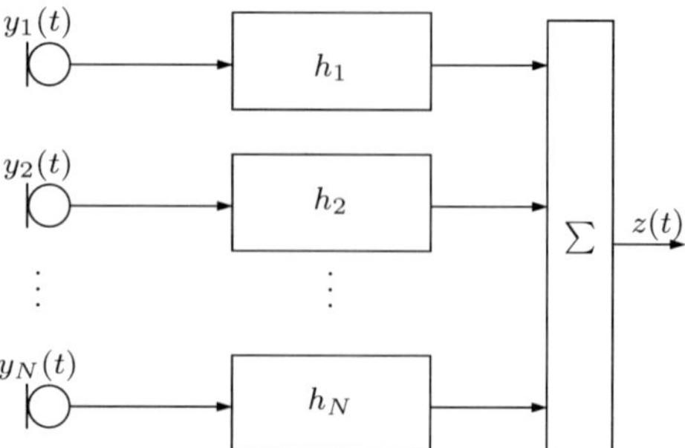

Fig. 8.1 Structure of a broadband beamformer, where h_n, $n = 1, 2, \ldots, N$, are finite impulse response (FIR) filters.

direction of the desired signal $X_1(f)$ [7], [8]. This definition is a generalization of the classical steering vector to a reverberant (multipath) environment. Indeed, the acoustic impulse responses ratios from a broadband source to the aperture convey information about the position of the source.

8.3 Array Model

Usually, the array processing or beamforming is performed by applying a temporal filter to each microphone signal and summing the filtered signals (see Fig. 8.1). In the frequency domain, this is equivalent to adding a complex weight to the output of each sensor and summing across the aperture:

$$\begin{aligned}
Z(f) &= \mathbf{h}^H(f)\mathbf{y}(f) \\
&= \mathbf{h}^H(f)\left[\mathbf{d}(f)X_1(f) + \mathbf{v}(f)\right] \\
&= X_{1,\mathrm{f}}(f) + V_{\mathrm{rn}}(f),
\end{aligned} \tag{8.5}$$

where $Z(f)$ is the beamformer output signal,

$$\mathbf{h}(f) = \begin{bmatrix} H_1(f) & H_2(f) & \cdots & H_N(f) \end{bmatrix}^T$$

is the beamforming weight vector which is suitable for performing spatial filtering at frequency f, superscript ^{H} denotes transpose conjugation of a vector or a matrix, $X_{1,\mathrm{f}}(f) = \mathbf{h}^H(f)\mathbf{d}(f)X_1(f)$ is the filtered desired signal, and $V_{\mathrm{rn}}(f) = \mathbf{h}^H(f)\mathbf{v}(f)$ is the residual noise.

In the time domain, the beamformer output signal is

$$z(t) = \int_{-\infty}^{\infty} Z(f)e^{j2\pi ft}df$$

$$= \int_{-\infty}^{\infty} X_{1,\mathrm{f}}(f)e^{j2\pi ft}df + \int_{-\infty}^{\infty} V_{\mathrm{rn}}(f)e^{j2\pi ft}df$$

$$= x_{1,\mathrm{f}}(t) + v_{\mathrm{rn}}(t). \tag{8.6}$$

8.4 Signal-to-Noise Ratio

One of the most important measures in all aspects of speech enhancement is the *signal-to-noise ratio (SNR)*. The SNR is a second-order measure which quantifies the level of noise present relative to the level of the desired signal.

Since the processing of the array signals may be done either in the time- or frequency-domain, the SNR becomes domain-dependent. To that end, one must differentiate between the narrowband (i.e., at a single frequency) SNR and the broadband SNR (i.e., occurring across the entire frequency range). In any acoustic application, the broadband SNR is the more appropriate metric; however, since the array signals are often decomposed into narrowband bins and processed locally, the narrowband SNRs may be taken into account during the algorithmic processing.

We begin by defining the *broadband input SNR* as the ratio of the power of the time-domain desired signal over the power of the time-domain noise at the reference microphone, i.e.,

$$\mathrm{iSNR} = \frac{E\left[x_1^2(t)\right]}{E\left[v_1^2(t)\right]}$$

$$= \frac{\int_{-\infty}^{\infty} \phi_{x_1}(f)df}{\int_{-\infty}^{\infty} \phi_{v_1}(f)df}, \tag{8.7}$$

where the component *narrowband input SNR* is written as

$$\mathrm{iSNR}(f) = \frac{\phi_{x_1}(f)}{\phi_{v_1}(f)}, \tag{8.8}$$

where

$$\phi_a(f) = E\left[|A(f)|^2\right] \tag{8.9}$$

denotes the power spectral density (PSD) of the wide sense stationary (WSS) process $a(t)$ at temporal frequency f.

To quantify the level of noise remaining in the beamformer output signal, $z(t)$, we define the *broadband output SNR* as the ratio of the power of the filtered desired signal over the power of the residual noise, i.e.,

$$\mathrm{oSNR}(\mathbf{h}) = \frac{E\left[x_{1,\mathrm{f}}^2(t)\right]}{E\left[v_{\mathrm{rn}}^2(t)\right]}$$

$$= \frac{\int_{-\infty}^{\infty} \phi_{x_1}(f) \left|\mathbf{h}^H(f)\mathbf{d}(f)\right|^2 df}{\int_{-\infty}^{\infty} \mathbf{h}^H(f)\mathbf{\Phi}_v(f)\mathbf{h}(f)df}, \tag{8.10}$$

where $\mathbf{\Phi}_v(f) = E\left[\mathbf{v}(f)\mathbf{v}^H(f)\right]$ is the PSD matrix of the noise signals at the array.

The *narrowband output SNR* is given by

$$\mathrm{oSNR}\left[\mathbf{h}(f)\right] = \frac{\phi_{x_1}(f) \left|\mathbf{h}^H(f)\mathbf{d}(f)\right|^2}{\mathbf{h}^H(f)\mathbf{\Phi}_v(f)\mathbf{h}(f)}. \tag{8.11}$$

In the particular case where we only have one microphone (no spatial processing), we get

$$\mathrm{oSNR}\left[\mathbf{h}(f)\right] = \mathrm{iSNR}\left(f\right). \tag{8.12}$$

Notice that the broadband input and output SNRs cannot be expressed as an integral of their narrowband counterparts:

$$\mathrm{iSNR} \neq \int_{-\infty}^{\infty} \mathrm{iSNR}(f)df,$$

$$\mathrm{oSNR}\left(\mathbf{h}\right) \neq \int_{-\infty}^{\infty} \mathrm{oSNR}\left[\mathbf{h}(f)\right] df. \tag{8.13}$$

It is also important to understand that for all cases, the SNR has some limitations as a measure of beamforming "goodness." The measure considers signal power without taking into account distortion in the desired signal. As a result, additional measures need to be defined, as shown in upcoming sections.

8.5 Array Gain

The role of the beamformer is to produce a signal whose SNR is higher than that which was received. To that end, the *array gain* is defined as the ratio of the output SNR (after beamforming) over the input SNR (at the reference microphone) [1]. This leads to the following definitions:

- the *broadband array gain,*

$$\mathcal{A}(\mathbf{h}) = \frac{\mathrm{oSNR}(\mathbf{h})}{\mathrm{iSNR}} \tag{8.14}$$

$$= \frac{\int_{-\infty}^{\infty} \phi_{x_1}(f) \left| \mathbf{h}^H(f)\mathbf{d}(f) \right|^2 df}{\int_{-\infty}^{\infty} \phi_{x_1}(f) df} \frac{\int_{-\infty}^{\infty} \phi_{v_1}(f) df}{\int_{-\infty}^{\infty} \mathbf{h}^H(f)\boldsymbol{\Phi}_v(f)\mathbf{h}(f) df},$$

- and the *narrowband array gain,*

$$\mathcal{A}\left[\mathbf{h}(f)\right] = \frac{\mathrm{oSNR}\left[\mathbf{h}(f)\right]}{\mathrm{iSNR}(f)}$$

$$= \frac{\left| \mathbf{h}^T(f)\mathbf{d}(f) \right|^2}{\mathbf{h}^T(f)\boldsymbol{\Gamma}_v(f)\mathbf{h}(f)}, \tag{8.15}$$

where

$$\boldsymbol{\Gamma}_v(f) = \phi_{v_1}^{-1}(f)\boldsymbol{\Phi}_v(f) \tag{8.16}$$

is the spatial pseudo-coherence matrix of the noise. By inspection,

$$\mathcal{A}(\mathbf{h}) \neq \int_{-\infty}^{\infty} \mathcal{A}\left[\mathbf{h}(f)\right] df. \tag{8.17}$$

Assume that the noise is temporally and spatially white with variance σ_v^2 at all microphones; in this case, the pseudo-coherence matrix simplifies to

$$\boldsymbol{\Gamma}_v(f) = \mathbf{I}_N, \tag{8.18}$$

where $\mathbf{I}_N$ is the N-by-N identity matrix. As a result, the narrowband array gain simplifies to

$$\mathcal{A}\left[\mathbf{h}(f)\right] = \frac{\left| \mathbf{h}^H(f)\mathbf{d}(f) \right|^2}{\mathbf{h}^H(f)\mathbf{h}(f)}. \tag{8.19}$$

Using the Cauchy-Schwartz inequality, it is easy to obtain

$$\mathcal{A}\left[\mathbf{h}(f)\right] \leq \|\mathbf{d}(f)\|_2^2, \ \forall \ \mathbf{h}(f). \tag{8.20}$$

We deduce from (8.20) that the narrowband array gain never exceeds the square of the 2-norm of the steering vector $\mathbf{d}(f)$. For example, if the elements of $\mathbf{d}(f)$ are given by anechoic plane wave propagation

$$\mathbf{d}(f) = \begin{bmatrix} 1 & e^{-j2\pi f \tau_{12}} & \cdots & e^{-j2\pi f \tau_{1N}} \end{bmatrix}, \tag{8.21}$$

where τ_{1n} is the relative delay between the reference microphone and microphone n, then it follows that

$$\mathcal{A}\left[\mathbf{h}(f)\right] \leq \|\mathbf{d}(f)\|_2^2$$
$$\leq N, \tag{8.22}$$

and the array gain is upper-bounded by the number of microphones. It is important to observe that the time-domain array gain is generally different from the narrowband array gain given at each frequency.

8.6 Noise Rejection and Desired Signal Cancellation

The array gain fails to capture the presence of desired signal distortion introduced by the beamforming process. Thus, this section introduces two sub-measures which treat signal distortion and noise reduction individually.

The *noise-reduction factor* [9], [10] or *noise-rejection factor* [11] quantifies the amount of noise being rejected by the beamformer. This quantity is defined as the ratio of the power of the noise at the reference microphone over the power of the noise remaining at the beamformer output. We provide the following definitions:

- the *broadband noise-rejection factor*,

$$\xi_{\mathrm{nr}}(\mathbf{h}) = \frac{\int_{-\infty}^{\infty} \phi_{v_1}(f)df}{\int_{-\infty}^{\infty} \mathbf{h}^H(f)\mathbf{\Phi}_v(f)\mathbf{h}(f)df}, \tag{8.23}$$

- and the *narrowband noise-rejection factor*,

$$\xi_{\mathrm{nr}}\left[\mathbf{h}(f)\right] = \frac{\phi_{v_1}(f)}{\mathbf{h}^H(f)\mathbf{\Phi}_v(f)\mathbf{h}(f)}$$
$$= \frac{1}{\mathbf{h}^H(f)\mathbf{\Gamma}_v(f)\mathbf{h}(f)}. \tag{8.24}$$

The broadband noise-rejection factor is expected to be lower bounded by 1; otherwise, the beamformer amplifies the noise received at the microphones. The higher the value of the noise-rejection factor, the more the noise is rejected.

In practice, most beamforming algorithms distort the desired signal. In order to quantify the level of this distortion, we define the *desired-signal-reduction factor* [5] or *desired-signal-cancellation factor* [11] as the ratio of the variance of the desired signal at the reference microphone over the variance of the filtered desired signal at the beamformer output. It is easy to deduce the following mathematical definitions:

- the *broadband desired-signal-cancellation factor*,

$$\xi_{\text{dsc}}(\mathbf{h}) = \frac{\int_{-\infty}^{\infty} \phi_{x_1}(f)df}{\int_{-\infty}^{\infty} \phi_{x_1}(f)\left|\mathbf{h}^H(f)\mathbf{d}(f)\right|^2 df} \tag{8.25}$$

- and the *narrowband desired-signal-cancellation factor*,

$$\xi_{\text{dsc}}\left[\mathbf{h}(f)\right] = \frac{1}{\left|\mathbf{h}^H(f)\mathbf{d}(f)\right|^2}. \tag{8.26}$$

Once again, note that

$$\xi_{\text{nr}}(\mathbf{h}) \neq \int_{-\infty}^{\infty} \xi_{\text{nr}}\left[\mathbf{h}(f)\right] df,$$

$$\xi_{\text{dsc}}(\mathbf{h}) \neq \int_{-\infty}^{\infty} \xi_{\text{dsc}}\left[\mathbf{h}(f)\right] df. \tag{8.27}$$

Another key observation is that the design of broadband beamformers that do not cancel the broadband desired signal requires the constraint

$$\mathbf{h}^H(f)\mathbf{d}(f) = 1, \forall f. \tag{8.28}$$

Thus, the desired-signal-cancellation factor is equal to 1 if there is no cancellation and expected to be greater than 1 when cancellation happens.

Lastly, by making the appropriate substitutions, one can derive the following relationships between the array gain, noise-rejection factor, and desired-signal-cancellation factor:

$$\mathcal{A}(\mathbf{h}) = \frac{\xi_{\text{nr}}(\mathbf{h})}{\xi_{\text{dsc}}(\mathbf{h})},$$

$$\mathcal{A}\left[\mathbf{h}(f)\right] = \frac{\xi_{\text{nr}}\left[\mathbf{h}(f)\right]}{\xi_{\text{dsc}}\left[\mathbf{h}(f)\right]}. \tag{8.29}$$

8.7 Beampattern

The *beampattern* is a convenient way to represent the response of the beamformer to the signal $x_1(t)$ as a function of the steering vector $\mathbf{d}(f)$ (or equivalently, the location of the source), assuming the absence of any noise or interference. This steering vector spans the ratios of acoustic impulse responses from any point in space to the array of sensors. Formally, the beampattern is defined as the ratio of the variance of the beamformer output when the source impinges with a steering vector $\mathbf{d}(f)$ to the variance of the desired signal $x_1(t)$. From this definition, we deduce

- the *broadband beampattern*,

$$\mathcal{B}(\mathbf{d}) = \frac{\int_{-\infty}^{\infty} \phi_{x_1}(f) \left| \mathbf{h}^H(f)\mathbf{d}(f) \right|^2 df}{\int_{-\infty}^{\infty} \phi_{x_1}(f) df}, \qquad (8.30)$$

- and the *narrowband beampattern*,

$$\mathcal{B}\left[\mathbf{d}(f)\right] = \left| \mathbf{h}^H(f)\mathbf{d}(f) \right|^2. \qquad (8.31)$$

It is interesting to point out that the broadband beampattern is a linear combination of narrowband beampatterns:

$$\mathcal{B}(\mathbf{d}) = \frac{\int_{-\infty}^{\infty} \phi_{x_1}(f)\mathcal{B}\left[\mathbf{d}(f)\right] df}{\int_{-\infty}^{\infty} \phi_{x_1}(f) df}, \qquad (8.32)$$

as the denominator is simply a scaling factor. The contribution of each narrowband beampattern to the overall broadband beampattern is proportional to the power of the desired signal at that frequency.

It is also interesting to observe the following relations:

$$\mathcal{B}(\mathbf{d}) = \frac{1}{\xi_{\mathrm{dsc}}(\mathbf{h})},$$

$$\mathcal{B}\left[\mathbf{d}(f)\right] = \frac{1}{\xi_{\mathrm{dsc}}\left[\mathbf{h}(f)\right]}.$$

When the weights of the beamformer are chosen in such a way that there is no cancellation, the value of the beampattern is 1 in the direction of the source.

8.7.1 Anechoic Plane Wave Model

Consider the case of a far-field source impinging on the array in an anechoic environment. In that case, the transfer function from the source to each sensor is given by a phase-shift (neglecting any attenuation of the signal which is uniform across the array for a far-field source):

$$\mathbf{g}_{\mathrm{a}}(f) = \left[e^{-j2\pi f\tau_1} \; e^{-j2\pi f\tau_2} \; \cdots \; e^{-j2\pi f\tau_N} \right]^T, \qquad (8.33)$$

where τ_n is the propagation time from the source location to sensor n. The steering vector follows as

$$\begin{aligned}
\mathbf{d}_{\mathrm{a}}(f,\zeta) &= \left[1 \; e^{-j2\pi f(\tau_2-\tau_1)} \; \cdots \; e^{-j2\pi f(\tau_N-\tau_1)} \right]^T \\
&= \left[1 \; e^{-j2\pi f\tau_{12}} \; \cdots \; e^{-j2\pi f\tau_{1N}} \right]^T,
\end{aligned} \qquad (8.34)$$

where $\tau_{1n} = \tau_n - \tau_1$. Moreover, the steering vector is now parameterized by the direction-of-arrival (DOA):

$$\boldsymbol{\zeta} = \begin{bmatrix} \sin\phi\cos\theta & \sin\phi\sin\theta & \cos\phi \end{bmatrix}^T, \tag{8.35}$$

where ϕ and θ are the incoming wave's elevation and azimuth angles, respectively. The relationship between the relative delays and the plane wave's DOA follows from the solution to the wave equation [1]:

$$\tau_{1n} = \frac{1}{c}\boldsymbol{\zeta}^T \left(\mathbf{r}_n - \mathbf{r}_1\right), \ n = 1, 2, \ldots, N, \tag{8.36}$$

where $\mathbf{r}_n$ is the location of the nth microphone. For a uniform linear array (ULA), $\mathbf{r}_n = \begin{bmatrix} nd\ 0\ 0 \end{bmatrix}^T$ where d is the spacing between adjacent microphones; thus, one obtains

$$\tau_{1n} = \frac{(n-1)d}{c} \sin\phi\cos\theta, \ n = 1, 2, \ldots, N, \tag{8.37}$$

where c is the speed of sound propagation.

A conventional delay-and-sum beamformer (DSB) steered to DOA $\boldsymbol{\zeta}_o$ selects its weights according to

$$\mathbf{h}_{\mathrm{dsb}}(f) = \frac{1}{N}\mathbf{d}_{\mathrm{a}}(f, \boldsymbol{\zeta}_o). \tag{8.38}$$

This weighting time-aligns the signal component arriving from DOA $\boldsymbol{\zeta}_o$. As a result, the desired signal is coherently summed, while all other DOAs are incoherently added. The resulting narrowband beampattern of a DSB in an anechoic environment is given by

$$\begin{aligned}
\mathcal{B}\left[\mathbf{h}_{\mathrm{dsb}}(f)\right] &= \left| \frac{1}{N}\mathbf{d}_{\mathrm{a}}^H (f, \boldsymbol{\zeta}_o)\mathbf{d}_{\mathrm{a}}(f, \boldsymbol{\zeta}) \right|^2 \\
&= \frac{1}{N^2} \left| \sum_{n=0}^{N-1} e^{j2\pi f \frac{nd}{c}(\cos\theta_o - \cos\theta)} \right|^2 \\
&= \frac{1}{N^2} \left| \frac{1 - e^{j2\pi f N \frac{d}{c}(\cos\theta_o - \cos\theta)}}{1 - e^{j2\pi f \frac{d}{c}(\cos\theta_o - \cos\theta)}} \right|^2,
\end{aligned} \tag{8.39}$$

where it has been assumed that $\phi = \phi_o = \frac{\pi}{2}$ (i.e., the source and sensors lie on a plane) for simplicity.

When processing a narrow frequency range centered around frequency f, $\mathcal{B}\left[\mathbf{h}_{\mathrm{dsb}}(f)\right]$ depicts the spatial filtering capabilities of the resulting narrowband beamformer. For a wideband characterization, the broadband beampattern of the DSB is given by

$$\mathcal{B}\left(\mathbf{h}_{\mathrm{dsb}}\right) = \frac{1}{N^2} \frac{\int_{-\infty}^{\infty} \phi_{x_1}(f) \left| \frac{1-e^{j2\pi f N \frac{d}{c}(\cos\theta_o - \cos\theta)}}{1-e^{j2\pi f \frac{d}{c}(\cos\theta_o - \cos\theta)}} \right|^2 df}{\int_{-\infty}^{\infty} \phi_{x_1}(f) df}. \tag{8.40}$$

8.8 Directivity

Acoustic settings frequently have a myriad of noise sources present. In order to model this situation, a spherically isotropic or "diffuse" noise field is one in which the noise power is constant and equal at all spatial frequencies (i.e., directions) [1], [12]. When designing beamformers, one would like to be able to quantify the ability of the beamformer to attenuate such a noise field. To that end, the *directivity* factor is classically defined as the array gain of a (narrowband) beamformer in an isotropic noise field. In the narrowband case, this is equivalent to the ratio of the beampattern in the direction of the source over the resulting residual noise power. Thus, we define

- the *broadband directivity factor*,

$$\mathcal{D} = \frac{\mathcal{B}(\mathbf{d})}{\int_{-\infty}^{\infty} \mathbf{h}^H(f)\mathbf{\Gamma}_{\mathrm{diff}}(f)\mathbf{h}(f)df} \tag{8.41}$$

- and the *narrowband directivity factor*,

$$\mathcal{D}(f) = \frac{\mathcal{B}\left[\mathbf{d}(f)\right]}{\mathbf{h}^H(f)\mathbf{\Gamma}_{\mathrm{diff}}(f)\mathbf{h}(f)}, \tag{8.42}$$

where

$$\begin{aligned} \left[\mathbf{\Gamma}_{\mathrm{diff}}(f)\right]_{nm} &= \frac{\sin 2\pi f(m-n)dc^{-1}}{2\pi f(m-n)dc^{-1}} \\ &= \operatorname{sinc}\left[2\pi f(m-n)dc^{-1}\right] \end{aligned} \tag{8.43}$$

is the coherence matrix of a diffuse noise field [13].

The classical directivity index [11], [12] is simply

$$\mathcal{DI}(f) = 10\log_{10}\mathcal{D}(f). \tag{8.44}$$

8.8.1 Superdirective Beamforming

As the name suggests, a superdirective beamformer is one which is designed to optimize the beamformer's directivity; to that end, consider a beamformer which maximizes the directivity while constraining the beampattern in the direction of the source to unity. This leads to the following optimization problem for the beamforming weights $\mathbf{h}(f)$:

$$\mathbf{h}_{\mathrm{sdb}}(f) = \arg\min_{\mathbf{h}(f)} \mathbf{h}^H(f)\boldsymbol{\Gamma}_{\mathrm{diff}}(f)\mathbf{h}(f) \ \text{ subject to } \mathbf{h}^H(f)\mathbf{d}(f) = 1.$$

$$(8.45)$$

The solution to (8.45) is written as

$$\mathbf{h}_{\mathrm{sdb}}(f) = \frac{\boldsymbol{\Gamma}_{\mathrm{diff}}^{-1}(f)\mathbf{d}(f)}{\mathbf{d}^H(f)\boldsymbol{\Gamma}_{\mathrm{diff}}^{-1}(f)\mathbf{d}(f)}. \tag{8.46}$$

Like the DSB, the superdirective beamformer of (8.46) is a fixed beamformer – that is, it is not data-dependent, as it *assumes* a particular coherence structure for the noise field and requires knowledge of the source DOA.

8.9 White Noise Gain

Notice also that the optimization of (8.45) may be performed for any noise field – it is not limited to a diffuse noise field. In the case of a spatially white noise field, the pseudo-coherence matrix is

$$\boldsymbol{\Gamma}_v(f) = \mathbf{I}_N, \tag{8.47}$$

and the beamformer that maximizes the so-called *white noise gain* (WNG) is found by substituting (8.47) into (8.46):

$$\begin{aligned}\mathbf{h}_{\mathrm{wng}}(f) &= \frac{\mathbf{d}(f)}{\mathbf{d}^H(f)\mathbf{d}(f)} \\ &= \frac{1}{N}\mathbf{d}(f),\end{aligned} \tag{8.48}$$

which is indeed the DSB.

The narrowband WNG is formally defined as the array gain with a spatially white noise field:

$$\begin{aligned}\mathcal{W}\left[\mathbf{h}(f)\right] &= \frac{|\mathbf{h}(f)\mathbf{d}(f)|^2}{\mathbf{h}^H(f)\mathbf{h}(f)} \\ &= \frac{\mathcal{B}\left[\mathbf{d}(f)\right]}{\mathbf{h}^H(f)\mathbf{h}(f)}.\end{aligned} \tag{8.49}$$

Analogously, we define the broadband WNG as

$$\mathcal{W}(\mathbf{h}) = \frac{\mathcal{B}(\mathbf{d})}{\int_{-\infty}^{\infty} \mathbf{h}^H(f)\mathbf{h}(f)df}. \tag{8.50}$$

8.10 Spatial Aliasing

The phenomenon of aliasing is classically viewed as an artifact of performing spectral analysis on a sampled signal. Sampling introduces a periodicity into the Fourier transform; if the bandwidth of the signal exceeds half of the sampling frequency, the spectral replicas overlap, leading to a distortion in the observed spectrum.

Spatial aliasing is analogous to its temporal counterpart: in order to reconstruct a spatial sinusoid from a set of uniformly-spaced discrete spatial samples, the spatial sampling period must be less than half of the sinusoid's wavelength. This principle has long been applied to microphone arrays in the following sense: the spacing between adjacent microphone elements should be less than half of the wavelength corresponding to the highest temporal frequency of interest. Since microphone arrays are concerned with the naturally wideband speech (i.e. the highest frequency of interest is in the order of 4 kHz), the resulting arrays are quite small in size.

Notice that the spatial sampling theorem is formulated with respect to a temporally narrowband signal. On the other hand, microphone arrays sample a temporally broadband signal; one may view the wideband nature of sound as diversity. In this section, it is shown that this diversity allows us to increase the microphone array spacing beyond that allowed by the Nyquist theorem without suffering any aliasing artifacts.

Denote the value of the sound field by the four-dimensional function $s(\mathbf{x}, t) = s(x, y, z, t)$, where $\mathbf{x} = \begin{bmatrix} x & y & z \end{bmatrix}^T$ denotes the observation point in Cartesian co-ordinates. One may express this function as a multidimensional inverse Fourier transform:

$$s(\mathbf{x}, t) = \frac{1}{(2\pi)^4} \int_{-\infty}^{\infty} \int_{-\infty}^{\infty} S(\mathbf{k}, \omega) e^{j\left(\omega t - \mathbf{k}^T \mathbf{x}\right)} d\mathbf{k} d\omega, \qquad (8.51)$$

where $\omega = 2\pi f$ is the angular temporal frequency,

$$\mathbf{k} = \begin{bmatrix} k_x & k_y & k_z \end{bmatrix}^T$$

is the angular spatial frequency vector, and $S(\mathbf{k}, \omega)$ are the coefficients of the basis functions $e^{j\left(\omega t - \mathbf{k}^T \mathbf{x}\right)}$, which are termed *monochromatic plane waves*, as the value of each basis function at a fixed time instant is constant along any plane of the form $\mathbf{k}^T \mathbf{x} = K$, where K is some constant. Thus, any sound field may be represented as a linear combination of propagating narrowband plane waves. It is also interesting to note that according to the physical constraints posed by the wave equation to propagating waves, the following relationship exists among spatial and temporal frequencies [1]:

$$\mathbf{k} = \frac{\omega}{c} \zeta. \qquad (8.52)$$

Thus, the spatial frequency vector points in the direction of propagation $\boldsymbol{\zeta}$, while its magnitude is linearly related to the temporal frequency.

The values of the weighting coefficients follow from the multidimensional Fourier transform:

$$S(\mathbf{k}, \omega) = \int_{-\infty}^{\infty} \int_{-\infty}^{\infty} s(\mathbf{x}, t) e^{-j\left(\omega t - \mathbf{k}^T \mathbf{x}\right)} d\mathbf{x} dt. \tag{8.53}$$

Notice that each coefficient may be written as

$$S(\mathbf{k}, \omega) = \int_{-\infty}^{\infty} \left[\int_{-\infty}^{\infty} s(\mathbf{x}, t) e^{-j\omega t} dt \right] e^{j\mathbf{k}^T \mathbf{x}} d\mathbf{x}$$

$$= \int_{-\infty}^{\infty} S_{\mathbf{x}}(\omega) e^{j\mathbf{k}^T \mathbf{x}} d\mathbf{x}, \tag{8.54}$$

where

$$S_{\mathbf{x}}(\omega) = \int_{-\infty}^{\infty} s(\mathbf{x}, t) e^{-j\omega t} dt$$

is the temporal Fourier transform of the signal observed at position $\mathbf{x}$. Thus, multidimensional spectrum of a space-time field is equal to the spatial Fourier transform of the temporal Fourier coefficients across space. From the duality of the Fourier transform, one may also write

$$S_{\mathbf{x}}(\omega) = \frac{1}{(2\pi)^3} \int_{-\infty}^{\infty} S(\mathbf{k}, \omega) e^{-j\mathbf{k}^T \mathbf{x}} d\mathbf{k},$$

which expresses the Fourier transform of the signal observed at an observation point $\mathbf{x}$ as a Fourier integral through a three-dimensional slice of the space-time frequency.

In microphone array applications, the goal is to estimate the temporal signal propagating from a certain DOA; this task may be related to the estimation of the space-time Fourier coefficients $S(\mathbf{k}, \omega)$. To see this, consider forming a wideband signal by integrating the space-time Fourier transform across temporal frequency, while only retaining the portion that propagates from the desired DOA $\boldsymbol{\zeta}_o$:

$$s_{\boldsymbol{\zeta}_o}(t) \triangleq \frac{1}{2\pi} \int_{-\infty}^{\infty} S(\mathbf{k}_o, \omega) e^{j\omega t} d\omega, \tag{8.55}$$

where $\mathbf{k}_o = \frac{\omega}{c} \boldsymbol{\zeta}_o$ is the spatial frequency vector whose direction is that of the desired plane wave.

By substituting (8.54) into (8.55), one obtains the following expression for the resulting broadband beam:

$$s_{\zeta_o}(t) = \frac{1}{2\pi} \int_{-\infty}^{\infty} \left[\int_{-\infty}^{\infty} S_{\mathbf{x}}(\omega) e^{j\mathbf{k}_o{}^T \mathbf{x}} d\mathbf{x} \right] e^{j\omega t} d\omega$$

$$= \int_{-\infty}^{\infty} \left[\frac{1}{2\pi} \int_{-\infty}^{\infty} S_{\mathbf{x}}(\omega) e^{j\omega\left(t + \frac{1}{c}\zeta_o{}^T \mathbf{x}\right)} d\omega \right] d\mathbf{x}$$

$$= \int_{-\infty}^{\infty} s\left(\mathbf{x}, t + \frac{1}{c}\zeta_o{}^T \mathbf{x}\right) d\mathbf{x}. \tag{8.56}$$

It is evident from (8.56) that the broadband beam is formed by integrating the time-delayed (or advanced) space-time field across space. This operation is indeed the limiting case of the DSB [1] as the number of microphones tends to infinity.

When the space-time field $s(\mathbf{x}, t)$ consists of a plane wave propagating from DOA ζ, one may write

$$s(\mathbf{x}, t) = s\left(t - \frac{1}{c}\zeta^T \mathbf{x}\right). \tag{8.57}$$

This simplifies the expression for the resulting broadband beam,

$$s_{\zeta_o}(t) = \int_{-\infty}^{\infty} s\left(\mathbf{x}, t + \frac{1}{c}\zeta_o{}^T \mathbf{x}\right) d\mathbf{x}$$

$$= \int_{-\infty}^{\infty} s\left[t + \frac{1}{c}\left(\zeta_o - \zeta\right)^T \mathbf{x}\right] d\mathbf{x}. \tag{8.58}$$

Now that we have an expression for the broadband beam, we can analyze the effect of estimating this signal using a discrete spatial aperture. The general expression for the discrete-space broadband beam follows from (8.56) as

$$s_{\zeta_o}^{\mathrm{d}}(t) = \sum_{n=-\infty}^{\infty} s\left(\mathbf{x}_n, t + \frac{1}{c}\zeta_o^T \mathbf{x}_n\right), \tag{8.59}$$

where $\mathbf{x}_n$ is the nth spatial sample. When the space-time field consists of a single plane wave, the discrete beam simplifies to

$$s_{\zeta_o}^{\mathrm{d}}(t) = \sum_{n=-\infty}^{\infty} s\left[t + \frac{1}{c}\left(\zeta_o - \zeta\right)^T \mathbf{x}_n\right]. \tag{8.60}$$

We now delve into analyzing the effect of the spatial sampling frequency on the resulting broadband beams.

8.10.1 Monochromatic Signal

Consider first a narrowband plane wave:

$$s(\mathbf{x}, t) = e^{j\omega\left(t - \frac{1}{c}\boldsymbol{\zeta}^T \mathbf{x}\right)}. \tag{8.61}$$

Substituting (8.61) into (8.60) results in

$$s_{\boldsymbol{\zeta}_\mathrm{o}}^\mathrm{d}(t) = A\left(\boldsymbol{\zeta}_\mathrm{o}, \boldsymbol{\zeta}\right) e^{j\omega t}, \tag{8.62}$$

where

$$A\left(\boldsymbol{\zeta}_\mathrm{o}, \boldsymbol{\zeta}\right) = \sum_{n=-\infty}^{\infty} e^{j\frac{\omega}{c}\left(\boldsymbol{\zeta}_\mathrm{o} - \boldsymbol{\zeta}\right)^T \mathbf{x}_n}$$

and

$$s_{\boldsymbol{\zeta}_\mathrm{o}}^\mathrm{d}(0) = A\left(\boldsymbol{\zeta}_\mathrm{o}, \boldsymbol{\zeta}\right).$$

Thus, the beam is a complex weighted version of the desired signal $s(t) = e^{j\omega t}$. It is instructive to analyze the values of the complex amplitude $A\left(\boldsymbol{\zeta}_\mathrm{o}, \boldsymbol{\zeta}\right)$ in terms of the spatial sampling rate.

To ease the analysis, assume that the spatial sampling is in the x-direction only and with a sampling period of d:

$$\mathbf{x}_n = \begin{bmatrix} nd & 0 & 0 \end{bmatrix}^T. \tag{8.63}$$

As a result,

$$A\left(\theta_\mathrm{o}, \theta\right) = \sum_{n=-\infty}^{\infty} e^{j\omega \frac{nd}{c}\left(\sin \phi_\mathrm{o} \cos \theta_\mathrm{o} - \sin \phi \cos \theta\right)}. \tag{8.64}$$

Moreover, assume that the source lies on the x-y plane and that we are only concerned with the azimuthal component of the spatial spectrum; that is, $\phi_o = \phi = \frac{\pi}{2}$. In that case

$$A\left(\theta_\mathrm{o}, \theta\right) = \sum_{n=-\infty}^{\infty} e^{j\omega \frac{nd}{c}\left(\cos \theta_\mathrm{o} - \cos \theta\right)}. \tag{8.65}$$

Recall the following property of an infinite summation of complex exponentials:

$$\sum_{n=-\infty}^{\infty} e^{j\omega nT} = \frac{1}{T} \sum_{k=-\infty}^{\infty} \delta\left(\frac{\omega}{2\pi} - \frac{k}{T}\right)$$

$$= \frac{2\pi}{T} \sum_{k=-\infty}^{\infty} \delta\left(\omega - \frac{2\pi k}{T}\right), \tag{8.66}$$

where $\delta(\cdot)$ is the delta-Dirac function. Substituting (8.66) into (8.65),

$$A(\theta_\mathrm{o},\theta) = \frac{2\pi}{\frac{d}{c}(\cos\theta_\mathrm{o} - \cos\theta)} \sum_{k=-\infty}^{\infty} \delta\left[\omega - \frac{2\pi k}{\frac{d}{c}(\cos\theta_\mathrm{o} - \cos\theta)}\right]. \tag{8.67}$$

Consider now the conditions for the argument of the delta Dirac function in (8.67) to equal zero. This requires

$$\omega = \frac{2\pi k}{\frac{d}{c}(\cos\theta_\mathrm{o} - \cos\theta)}, \tag{8.68}$$

which is equivalent to

$$d(\cos\theta_\mathrm{o} - \cos\theta) = k\lambda, \tag{8.69}$$

where $\lambda = 2\pi\frac{c}{\omega}$ is the wavelength. For $k = 0$, (8.69) holds if $\cos\theta_\mathrm{o} = \cos\theta$, meaning that

$$s_{\theta_\mathrm{o}}^\mathrm{d}(t) = A(\theta_\mathrm{o},\theta)\, e^{j\omega t}$$

$$= \infty \quad \text{for } \cos\theta_\mathrm{o} = \cos\theta, \tag{8.70}$$

which is the desired result; indeed, this is the true (i.e., non-aliased) spectral peak. Note that for $\cos\theta = \cos\theta_o$, the factor $A(\theta_o,\theta)$ is infinite since the analysis assumes that the number of microphones $N \to \infty$.

Given the result of (8.70), a rigorous definition of spatial aliasing in broadband applications may be proposed: aliasing occurs whenever

$$\exists\, \theta_\mathrm{o} \neq \theta \quad \text{such that } s_{\theta_\mathrm{o}}^\mathrm{d}(t) = \infty. \tag{8.71}$$

In other words, spatial aliasing occurs when the discrete-space broadband beam $s_{\zeta_\mathrm{o}}^\mathrm{d}(t)$ tends to infinity even though the steered DOA θ_o does not match the true DOA θ.

The steered range for a ULA is $0 \leq \theta \leq \pi$ and the cosine function is one-to-one over this interval. Thus

$$\cos\theta_\mathrm{o} = \cos\theta \Rightarrow \theta_\mathrm{o} = \theta, \quad 0 \leq \theta_\mathrm{o}, \theta \leq \pi. \tag{8.72}$$

It is now straightforward to determine the aliasing conditions. Under the assumption of a narrowband signal, the beam $s_{\theta_\mathrm{o}}^\mathrm{d}(t)$ tends to infinity if there exists an integer $k \in \mathbb{Z}$ such that

$$\omega = \frac{2\pi k}{\frac{d}{c}\left(\cos\theta_{\mathrm{o}} - \cos\theta\right)}, \tag{8.73}$$

or

$$\frac{d}{\lambda} = \frac{k}{\cos\theta_{\mathrm{o}} - \cos\theta}. \tag{8.74}$$

Take $k = 1$; over the range $0 \leq \theta_1 \leq 2\pi$,

$$|\cos\theta_1 - \cos\theta_0| \leq 2, \tag{8.75}$$

meaning that to prevent aliasing, one needs to ensure that

$$\frac{d}{\lambda} < \frac{1}{\cos\theta_{\mathrm{o}} - \cos\theta}, \tag{8.76}$$

or

$$d < \frac{\lambda}{2}, \tag{8.77}$$

which is indeed the classical narrowband aliasing criterion. Note that for $|k| > 1$, the condition (8.77) also prevents (8.73).

8.10.2 Broadband Signal

Consider now a wideband signal with arbitrary temporal frequency content $S(\omega)$:

$$s(t) = \int_{-\infty}^{\infty} S(\omega) e^{j\omega t} d\omega. \tag{8.78}$$

Assuming a one-dimensional sampling scheme and considering only spatial frequencies with $\phi_{\mathrm{o}} = \phi = \frac{\pi}{2}$, the continuous version of the broadband beam corresponding to this signal follows from (8.58) as

$$s_{\theta_{\mathrm{o}}}(t) = \int_{-\infty}^{\infty}\int_{-\infty}^{\infty} S(\omega) e^{j\omega\left[t + \frac{x}{c}(\cos\theta_{\mathrm{o}} - \cos\theta)\right]} d\omega dx. \tag{8.79}$$

By setting $x = nd$ and replacing the integral with an infinite summation, we obtain the discrete version of (8.79):

$$s_{\theta_o}^{d}(t) = \sum_{n=-\infty}^{\infty} \int_{-\infty}^{\infty} S(\omega)e^{j\omega\left[t+\frac{nd}{c}(\cos\theta_1-\cos\theta)\right]}d\omega \qquad (8.80)$$

$$= \int_{-\infty}^{\infty} S(\omega)e^{j\omega t}\left[\sum_{n=-\infty}^{\infty} e^{j\omega\frac{nd}{c}(\cos\theta_o-\cos\theta)}\right]d\omega$$

$$= \int_{-\infty}^{\infty} S(\omega)e^{j\omega t}\frac{2\pi}{\frac{d}{c}(\cos\theta_o-\cos\theta)}\sum_{k=-\infty}^{\infty}\delta\left[\omega-\frac{2\pi k}{\frac{d}{c}(\cos\theta_o-\cos\theta)}\right]d\omega$$

$$= \frac{2\pi}{\frac{d}{c}(\cos\theta_o-\cos\theta)}\sum_{k=-\infty}^{\infty} S\left[\frac{2\pi k}{\frac{d}{c}(\cos\theta_o-\cos\theta)}\right]e^{j\frac{2\pi k}{\frac{d}{c}(\cos\theta_o-\cos\theta)}t}.$$

Examining (8.80), it follows that the discrete-space broadband beam for an arbitrary wideband signal takes the form of a series of weighted complex exponentials. For any temporal signal which obeys

$$\sum_{k=-\infty}^{\infty} S\left[\frac{2\pi k}{\frac{d}{c}(\cos\theta_o-\cos\theta)}\right]e^{j\frac{2\pi k}{\frac{d}{c}(\cos\theta_o-\cos\theta)}t} < \infty, \quad \cos\theta_o \neq \cos\theta, \quad (8.81)$$

one can state that the beam exhibits an infinite peak only when the scaling factor

$$\frac{2\pi}{\frac{d}{c}(\cos\theta_o-\cos\theta)} = \infty, \qquad (8.82)$$

which implies $\theta_o = \theta$. Thus, for wideband signals with spectra of the form (8.81), under the definition of (8.71), spatial aliasing does not result, regardless of the spatial sampling period d.

The condition of (8.81) refers to signals which are band-limited and not dominated by a strong harmonic component. The presence of such harmonic components at integer multiples of $\frac{2\pi}{\frac{d}{c}(\cos\theta_o-\cos\theta)}$ may drive the broadband beam to infinity at DOAs not matching the true DOA.

8.11 Mean-Squared Error

The ultimate aim of a beamformer is to reproduce the desired signal, free of any noise or interference, in the array output. To that end, the *mean-squared error (MSE)* is a key measure for designing optimal beamforming algorithms.

Let us first write the time-domain (broadband) error signal between the beamformer output signal and the desired signal, i.e.,

$$
\begin{aligned}
e(t) &= z(t) - x_1(t) \\
&= x_{1,\mathrm{f}}(t) - x_1(t) + v_{\mathrm{rn}}(t) \\
&= \int_{-\infty}^{\infty} \left[\mathbf{h}^H(f)\mathbf{d}(f) - 1 \right] X_1(f) e^{j2\pi ft} df + \int_{-\infty}^{\infty} \mathbf{h}^H(f)\mathbf{v}(f) e^{j2\pi ft} df \\
&= e_{x_1}(t) + e_v(t),
\end{aligned}
\tag{8.83}
$$

where

$$
\begin{aligned}
e_{x_1}(t) &= \int_{-\infty}^{\infty} \left[\mathbf{h}^H(f)\mathbf{d}(f) - 1 \right] X_1(f) e^{j2\pi ft} df \\
&= \int_{-\infty}^{\infty} \mathcal{E}_{x_1}(f) e^{j2\pi ft} df,
\end{aligned}
\tag{8.84}
$$

is the desired signal distortion, and

$$
\begin{aligned}
e_v(t) &= \int_{-\infty}^{\infty} \mathbf{h}^H(f)\mathbf{v}(f) e^{j2\pi ft} df \\
&= \int_{-\infty}^{\infty} \mathcal{E}_v(f) e^{j2\pi ft} df
\end{aligned}
\tag{8.85}
$$

represents the broadband residual noise.

The variance of the time-domain error signal is the *broadband MSE*:

$$
\begin{aligned}
J(\mathbf{h}) &= E\left[e^2(t) \right] \\
&= E\left[e_{x_1}^2(t) \right] + E\left[e_v^2(t) \right] \\
&= \int_{-\infty}^{\infty} \phi_{x_1}(f) \left| \mathbf{h}^H(f)\mathbf{d}(f) - 1 \right|^2 df + \int_{-\infty}^{\infty} \mathbf{h}^H(f)\boldsymbol{\Phi}_v(f)\mathbf{h}(f) df.
\end{aligned}
\tag{8.86}
$$

For the particular filter $\mathbf{h}(f) = \mathbf{i}$, $\forall\, f$, where

$$
\mathbf{i} = \begin{bmatrix} 1 & 0 & \dots & 0 \end{bmatrix}^T,
\tag{8.87}
$$

we obtain

$$
J(\mathbf{i}) = \int_{-\infty}^{\infty} \phi_{v_1}(f) df.
\tag{8.88}
$$

Therefore, we define the broadband normalized MSE (NMSE) as

$$
\tilde{J}(\mathbf{h}) = \frac{J(\mathbf{h})}{J(\mathbf{i})},
\tag{8.89}
$$

which can be rewritten as

$$
\tilde{J}(\mathbf{h}) = \mathrm{iSNR} \cdot v_{\mathrm{dsd}}(\mathbf{h}) + \frac{1}{\xi_{\mathrm{nr}}(\mathbf{h})},
\tag{8.90}
$$

where

$$v_{\mathrm{dsd}}(\mathbf{h}) = \frac{\int_{-\infty}^{\infty} \phi_{x_1}(f) \left| \mathbf{h}^H(f)\mathbf{d}(f) - 1 \right|^2 df}{\int_{-\infty}^{\infty} \phi_{x_1}(f) df}, \tag{8.91}$$

is the *broadband desired-signal-distortion index* [14].

From the broadband MSE we can deduce the *narrowband MSE*

$$J\left[\mathbf{h}(f)\right] = E\left[\left|\mathcal{E}_{x_1}(f)\right|^2\right] + E\left[\left|\mathcal{E}_v(f)\right|^2\right] \tag{8.92}$$

$$= \phi_{x_1}(f)\left|\mathbf{h}^H(f)\mathbf{d}(f) - 1\right|^2 + \mathbf{h}^H(f)\mathbf{\Phi}_v(f)\mathbf{h}(f).$$

We can also deduce the narrowband NMSE:

$$\tilde{J}\left[\mathbf{h}(f)\right] = \mathrm{iSNR}\left[\mathbf{h}(f)\right] \cdot v_{\mathrm{dsd}}\left[\mathbf{h}(f)\right] + \frac{1}{\xi_{\mathrm{nr}}\left[\mathbf{h}(f)\right]}, \tag{8.93}$$

where

$$v_{\mathrm{dsd}}\left[\mathbf{h}(f)\right] = \left|\mathbf{h}^H(f)\mathbf{d}(f) - 1\right|^2, \tag{8.94}$$

is the *narrowband desired-signal-distortion index* [14].

Note that the broadband MSE is a linear combination of the underlying narrowband MSEs:

$$J(\mathbf{h}) = \int_{-\infty}^{\infty} J\left[\mathbf{h}(f)\right] df. \tag{8.95}$$

8.11.1 Wiener Filter

Intuitively, we would like to derive a beamformer which minimizes the MSE at every frequency. This is the essence of the multichannel Wiener filter. The conventional minimum MSE (MMSE) minimizes the narrowband MSE:

$$\mathbf{h}_{\mathrm{W}}(f) = \arg \max_{\mathbf{h}(f)} J\left[\mathbf{h}(f)\right]. \tag{8.96}$$

Taking the gradient of $J\left[\mathbf{h}(f)\right]$ with respect to $\mathbf{h}^H(f)$ results in

$$\nabla_{\mathbf{h}^H(f)} J\left[\mathbf{h}(f)\right] = \phi_{x_1}(f)\left[\mathbf{d}(f)\mathbf{d}^H(f)\mathbf{h}(f) - \mathbf{d}(f)\right] + \mathbf{\Phi}_v(f)\mathbf{h}(f). \tag{8.97}$$

Setting (8.97) to zero and solving for $\mathbf{h}(f)$ reveals the conventional (narrowband) MMSE solution:

$$\mathbf{h}_{\mathrm{W}}(f) = \phi_{x_1}(f)\mathbf{\Phi}_y^{-1}(f)\mathbf{d}(f), \tag{8.98}$$

where

$$\mathbf{\Phi}_y(f) = E\left[\mathbf{y}(f)\mathbf{y}^H(f)\right]$$
$$= \phi_{x_1}(f)\mathbf{d}(f)\mathbf{d}^H(f) + \mathbf{\Phi}_v(f) \tag{8.99}$$

is the PSD matrix of the array measurements. The filter $\mathbf{h}_{\mathrm{W}}(f)$ minimizes the difference between array output and desired signal at the single frequency f.

Since the broadband MSE is a linear combination of the narrowband MSEs, minimizing the MSE at every frequency guarantees the minimization of the broadband MSE. Thus, applying the narrowband solution $\mathbf{h}_{\mathrm{W}}(f)$ at every component frequency results in the broadband MMSE solution.

8.11.2 Minimum Variance Distortionless Response

The celebrated minimum variance distortionless response (MVDR) beamformer proposed by Capon [3], [15] is also easily derived from the narrowband MSE. Indeed, minimizing $E\left[|\mathcal{E}_v(f)|^2\right]$ with the constraint that $E\left[|\mathcal{E}_{x_1}(f)|^2\right] = 0$ [or $\mathbf{h}^H(f)\mathbf{d}(f) - 1 = 0$], we obtain the classical MVDR filter:

$$\mathbf{h}_{\mathrm{MVDR}}(f) = \frac{\mathbf{\Phi}_v^{-1}(f)\mathbf{d}(f)}{\mathbf{d}^H(f)\mathbf{\Phi}_v^{-1}(f)\mathbf{d}(f)}. \tag{8.100}$$

Using the fact that $\mathbf{\Phi}_x(f) = E\left[\mathbf{x}(f)\mathbf{x}^H(f)\right] = \phi_{x_1}(f)\mathbf{d}(f)\mathbf{d}^H(f)$, the explicit dependence of the above filter on the steering vector is eliminated to obtain the following forms [5]:

$$\mathbf{h}_{\mathrm{MVDR}}(f) = \frac{\mathbf{\Phi}_v^{-1}(f)\mathbf{\Phi}_x(f)}{\mathrm{tr}\left[\mathbf{\Phi}_v^{-1}(f)\mathbf{\Phi}_x(f)\right]}\mathbf{i}$$
$$= \frac{\mathbf{\Phi}_v^{-1}(f)\mathbf{\Phi}_y(f) - \mathbf{I}_N}{\mathrm{tr}\left[\mathbf{\Phi}_v^{-1}(f)\mathbf{\Phi}_y(f)\right] - N}\mathbf{i}, \tag{8.101}$$

where $\mathrm{tr}[\cdot]$ denotes the trace of a square matrix.

The MVDR beamformer rejects the maximum level of noise allowable without distorting the desired signal at each frequency; however, the level broadband noise rejection is unclear. On the other hand, since the constraint is verified at all frequencies, the MVDR filter guarantees zero desired signal distortion at every frequency.

8.12 Conclusions

This chapter has reformulated the objectives of microphone arrays taking into account the wideband nature of the speech signal and the reverberant properties of acoustic environments. The SNR, array gain, noise-reduction factor, desired-signal-cancellation factor, beampattern, directivity factor, WNG, and MSE were defined in both narrowband and broadband contexts. Additionally, an analysis of spatial aliasing with broadband signals revealed that the spatial Nyquist criterion may be relaxed in microphone array applications. To this point in time, microphone array designs have been mainly focused on optimizing narrowband measures at each frequency bin. The broadband criteria presented in this chapter will hopefully serve as the metrics which future beamformer designs will focus on.

References

1. D. H. Johnson and D. E. Dudgeon, *Array Signal Processing–Concepts and Techniques*. Englewood Cliffs, NJ: Prentice-Hall, 1993.
2. M. Brandstein and D. B. Ward, Eds., *Microphone Arrays: Signal Processing Techniques and Applications*. Berlin, Germany: Springer-Verlag, 2001.
3. J. Capon, "High resolution frequency-wavenumber spectrum analysis," *Proc. IEEE*, vol. 57, pp. 1408–1418, Aug. 1969.
4. J. Benesty, J. Chen, Y. Huang, and J. Dmochowski, "On microphone-array beamforming from a MIMO acoustic signal processing perspective," *IEEE Trans. Audio, Speech, Language Processing*, vol. 15, pp. 1053–1065, Mar. 2007.
5. J. Benesty, J. Chen, and Y. Huang, *Microphone Array Signal Processing*. Berlin, Germany: Springer-Verlag, 2008.
6. S. Gannot and I. Cohen, "Adaptive beamforming and postfiltering," in *Springer Handbook of Speech Processing*, J. Benesty, M. M. Sondhi, and Y. Huang, Eds., Berlin, Germany: Springer-Verlag, 2008, Chapter 47, pp. 945–978.
7. B. D. Van Veen and K. M. Buckley, "Beamforming: a versatile approach to spatial filtering," *IEEE Acoust., Speech, Signal Process. Mag.*, vol. 5, pp. 4–24, Apr. 1988.
8. W. Herbordt and W. Kellermann, "Adaptive beamforming for audio signal acquisition," in *Adaptive Signal Processing: Applications to Real-World Problems*, J. Benesty and Y. Huang, Eds., Berlin, Germany: Springer-Verlag, 2003, Chapter 6, pp. 155–194.
9. J. Benesty, J. Chen, Y. Huang, and S. Doclo, "Study of the Wiener filter for noise reduction," in *Speech Enhancement*, J. Benesty, S. Makino, and J. Chen, Eds., Berlin, Germany: Springer-Verlag, 2005, Chapter 2, pp. 9–41.
10. J. Chen, J. Benesty, Y. Huang, and S. Doclo, "New insights into the noise reduction Wiener filter," *IEEE Trans. Audio, Speech, Language Process.*, vol. 14, pp. 1218–1234, July 2006.
11. W. Herbordt, *Combination of Robust Adaptive Beamforming with Acoustic Echo Cancellation for Acoustic Human/Machine Interfaces*. PhD Thesis, Erlangen-Nuremberg University, Germany, 2004.
12. G. W. Elko and J. Meyer, "Microphone arrays," in *Springer Handbook of Speech Processing*, J. Benesty, M. M. Sondhi, and Y. Huang, Eds., Berlin, Germany: Springer-Verlag, 2008, Chapter 48, pp. 1021–1041.

13. A. Spriet, *Adaptive Filtering Techniques for Noise Reduction and Acoustic Feedback Cancellation in Hearing Aids*. PhD Thesis, Katholieke Universiteit Leuven, Belgium, 2004.
14. J. Benesty, J. Chen, Y. Huang, and I. Cohen, *Noise Reduction in Speech Processing*. Berlin, Germany: Springer-Verlag, 2009.
15. R. T. Lacoss, "Data adaptive spectral analysis methods," *Geophysics*, vol. 36, pp. 661–675, Aug. 1971.

Chapter 9
The MVDR Beamformer for Speech Enhancement

Emanuël A. P. Habets, Jacob Benesty, Sharon Gannot, and Israel Cohen

Abstract [1]The minimum variance distortionless response (MVDR) beamformer is widely studied in the area of speech enhancement and can be used for both speech dereverberation and noise reduction. This chapter summarizes some new insights into the MVDR beamformer. Specifically, the local and global behaviors of the MVDR beamformer are analyzed, different forms of the MVDR beamformer and relations between the MVDR and other optimal beamformers are discussed. In addition, the tradeoff between dereverberation and noise reduction is analyzed. This analysis is done for a mixture of coherent and non-coherent noise fields and entirely non-coherent noise fields. It is shown that maximum noise reduction is achieved when the MVDR beamformer is used for noise reduction only. The amount of noise reduction that is sacrificed when complete dereverberation is required depends on the direct-to-reverberation ratio of the acoustic impulse response between the source and the reference microphone. The performance evaluation demonstrates the tradeoff between dereverberation and noise reduction.

Emanuël A. P. Habets
Imperial College London, UK, e-mail: `e.habets@imperial.ac.uk`

Jacob Benesty
INRS-EMT, QC, Canada, e-mail: `benesty@emt.inrs.ca`

Sharon Gannot
Bar-Ilan University, Israel, e-mail: `gannot@eng.biu.ac.il`

Israel Cohen
Technion–Israel Institute of Technology, Israel, e-mail: `icohen@ee.technion.ac.il`

[1] This work was supported by the Israel Science Foundation under Grant 1085/05.

I. Cohen et al. (Eds.): Speech Processing in Modern Communication, STSP 3, pp. 225–254.
springerlink.com   © Springer-Verlag Berlin Heidelberg 2010

9.1 Introduction

Distant or hands-free audio acquisition is required in many applications such as audio-bridging and teleconferencing. Microphone arrays are often used for the acquisition and consist of sets of microphone sensors that are arranged in specific patterns. The received sensor signals usually consist of a desired sound signal, coherent and non-coherent interferences. The received signals are processed in order to extract the desired sound, or in other words to suppress the interferences. In the last four decades many algorithms have been proposed to process the received sensor signals [1, 2].

For single-channel noise reduction, the Wiener filter can be considered as one of the most fundamental approaches (see for example [3] and the references therein). The Wiener filter produces a minimum mean-squared error (MMSE) estimate of the desired speech component received by the microphone. Doclo and Moonen [4], proposed a multichannel Wiener Filter (MWF) technique that produces an MMSE estimate of the desired speech component in one of the microphone signals. In [5], the optimization criterion of the MWF was modified to take the allowable speech distortion into account, resulting in the speech-distortion-weighted MWF (SDW-MWF). Another interesting solution is provided by the minimum variance distortionless response (MVDR) beamformer, also known as Capon beamformer [6], which minimizes the output power of the beamformer under a single linear constraint on the response of the array towards the desired signal. The idea of combining multiple inputs in a statistically optimum manner under the constraint of no signal distortion can be attributed to Darlington [7]. Several researchers developed beamformers in which additional linear constraints were imposed (e.g., Er and Cantoni [8]). These beamformers are known as linear constraint minimum variance (LCMV) beamformers, of which the MVDR beamformer is a special case. In [9], Frost proposed an adaptive scheme of the MVDR beamformer, which is based on a constrained least-mean-square (LMS) type adaptation. Kaneda et al. [10] proposed a noise reduction system for speech signals, termed AMNOR, which adopts a soft-constraint that controls the tradeoff between speech distortion and noise reduction. To avoid the constrained adaptation of the MVDR beamformer, Griffiths and Jim [11] proposed the generalized sidelobe canceler (GSC) structure, which separates the output power minimization and the application of the constraint. While Griffiths and Jim only considered one constraint (i.e., MVDR beamformer) it was later shown in [12] that the GSC structure can also be used in the case of multiple constraints (i.e., LCMV beamformer). The original GSC structure is based on the assumption that the different sensors receive a delayed version of the desired signal. The GSC structure was re-derived in the frequency-domain, and extended to deal with general acoustic transfer functions (ATFs) by Affes and Grenier [13] and later by Gannot et al. [14]. The frequency-domain version in [14], which takes into account the reverberant nature of the enclosure, was termed the transfer-function generalized sidelobe canceler (TF-GSC).

In theory the LCMV beamformer can achieve perfect dereverberation and noise cancellation when the ATFs between all sources (including interferences) and the microphones are known [15]. Using the MVDR beamformer we can achieve perfect reverberation cancellation when the ATFs between the desired source and the microphones are known. In the last three decades various methods have been developed to blindly identify the ATFs, more details can be found in [16] and the references therein and in [17]. Blind estimation of the ATFs is however beyond the scope of this chapter in which we assume that the ATFs between the source and the sensors are known. In earlier works [15], it was observed that there is a tradeoff between the amount of speech dereverberation and noise reduction. Only recently this tradeoff was analyzed by Habets et al. in [18].

In this chapter we study the MVDR beamformer in room acoustics and with broadband signals. First, we analyze the local and global behaviors [1] of the MVDR beamformer. Secondly, we derive several different forms of the MVDR filter and discuss the relations between the MVDR beamformer and other optimal beamformers. Finally, we analyze the tradeoff between noise and reverberation reduction. The local and global behaviors, as well as the tradeoff, are analyzed for different noise fields, viz. a mixture of coherent and non-coherent noise fields and entirely non-coherent noise fields.

The chapter is organized as follows: in Section 9.2 the array model is formulated and the notation used in this chapter is introduced. In Section 9.3 we start by formulating the SDW-MWF in the frequency domain. We then show that the MWF as well as the MVDR filter are special cases of the SDW-MWF. In Section 9.4 we define different performance measures that will be used in our analysis. In Section 9.5 we analyze the performance of the MVDR beamformer. The performance evaluation that demonstrates the tradeoff between reverberation and noise reduction is presented in Section 9.6. Finally, conclusions are provided in Section 9.7.

9.2 Problem Formulation

Consider the conventional signal model in which an N-element sensor array captures a convolved desired signal (speech source) in some noise field. The received signals are expressed as [19, 1]

$$y_n(k) = g_n * s(k) + v_n(k) \tag{9.1}$$
$$= x_n(k) + v_n(k), \ n = 1, 2, \ldots, N,$$

where k is the discrete-time index, g_n is the impulse response from the unknown (desired) source $s(k)$ to the nth microphone, $*$ stands for convolution, and $v_n(k)$ is the noise at microphone n. We assume that the signals $x_n(k)$ and $v_n(k)$ are uncorrelated and zero mean. All signals considered in this work are broadband. In this chapter, without loss of generality, we consider the first microphone ($n = 1$) as the reference microphone. Our main objective is then to study the recovering of any one of the signals $x_1(k)$ (noise reduction only), $s(k)$ (total dereverberation and noise reduction), or a filtered version of $s(k)$ with the MVDR beamformer. Obviously, we can recover the reverberant component at one of the other microphones $x_2(k), \ldots, x_N(k)$. When we desire noise reduction only the largest amount of noise reduction is attained by using the reference microphone with the highest signal to noise ratio [1].

In the frequency domain, (9.1) can be rewritten as

$$Y_n(\omega) = G_n(\omega)S(\omega) + V_n(\omega) \tag{9.2}$$
$$= X_n(\omega) + V_n(\omega), \ n = 1, 2, \ldots, N,$$

where $Y_n(\omega)$, $G_n(\omega)$, $S(\omega)$, $X_n(\omega) = G_n(\omega)S(\omega)$, and $V_n(\omega)$ are the discrete-time Fourier transforms (DTFTs) of $y_n(k)$, g_n, $s(k)$, $x_n(k)$, and $v_n(k)$, respectively, at angular frequency ω ($-\pi < \omega \le \pi$) and j is the imaginary unit ($j^2 = -1$). We recall that the DTFT and the inverse transform [20] are

$$A(\omega) = \sum_{k=-\infty}^{\infty} a(k)e^{-j\omega k}, \tag{9.3}$$

$$a(k) = \frac{1}{2\pi} \int_{-\pi}^{\pi} A(\omega)e^{j\omega k}d\omega. \tag{9.4}$$

The N microphone signals in the frequency domain are better summarized in a vector notation as

$$\mathbf{y}(\omega) = \mathbf{g}(\omega)S(\omega) + \mathbf{v}(\omega) \tag{9.5}$$
$$= \mathbf{x}(\omega) + \mathbf{v}(\omega),$$

where

$$\mathbf{y}(\omega) = \left[Y_1(\omega) \ Y_2(\omega) \ \cdots \ Y_N(\omega) \right]^T,$$
$$\mathbf{x}(\omega) = \left[X_1(\omega) \ X_2(\omega) \ \cdots \ X_N(\omega) \right]^T,$$
$$= S(\omega) \left[G_1(\omega) \ G_2(\omega) \ \cdots \ G_N(\omega) \right]^T$$
$$= S(\omega)\mathbf{g}(\omega),$$
$$\mathbf{v}(\omega) = \left[V_1(\omega) \ V_2(\omega) \ \cdots \ V_N(\omega) \right]^T,$$

and superscript ^{T} denotes transpose of a vector or a matrix.

Using the power spectral density (PSD) of the received signal and the fact that $x_n(k)$ and $v_n(k)$ are uncorrelated, we get

$$\phi_{y_n y_n}(\omega) = \phi_{x_n x_n}(\omega) + \phi_{v_n v_n}(\omega) \qquad (9.6)$$

$$= |G_n(\omega)|^2 \, \phi_{ss}(\omega) + \phi_{v_n v_n}(\omega), \;\; n = 1, 2, \ldots, N,$$

where $\phi_{y_n y_n}(\omega)$, $\phi_{x_n x_n}(\omega)$, $\phi_{ss}(\omega)$, and $\phi_{v_n v_n}(\omega)$ are the PSDs of the nth sensor input signal, the nth sensor reverberant speech signal, the desired signal, and the nth sensor noise signal, respectively.

The array processing, or beamforming, is then performed by applying a complex weight to each sensor and summing across the aperture:

$$Z(\omega) = \mathbf{h}^H(\omega)\mathbf{y}(\omega)$$

$$= \mathbf{h}^H(\omega)\left[\mathbf{g}(\omega)S(\omega) + \mathbf{v}(\omega)\right], \qquad (9.7)$$

where $Z(\omega)$ is the beamformer output,

$$\mathbf{h}(\omega) = \left[\, H_1(\omega)\; H_2(\omega)\; \cdots\; H_N(\omega) \,\right]^T$$

is the beamforming weight vector which is suitable for performing spatial filtering at frequency ω, and superscript ^{H} denotes transpose conjugation of a vector or a matrix.

The PSD of the beamformer output is given by

$$\phi_{zz}(\omega) = \mathbf{h}^H(\omega)\Phi_{xx}(\omega)\mathbf{h}(\omega) + \mathbf{h}^H(\omega)\Phi_{vv}(\omega)\mathbf{h}(\omega), \qquad (9.8)$$

where

$$\Phi_{xx}(\omega) = E\left[\mathbf{x}(\omega)\mathbf{x}^H(\omega)\right]$$

$$= \phi_{ss}(\omega)\mathbf{g}(\omega)\mathbf{g}^H(\omega) \qquad (9.9)$$

is the rank-one PSD matrix of the convolved speech signals with $E(\cdot)$ denoting mathematical expectation, and

$$\Phi_{vv}(\omega) = E\left[\mathbf{v}(\omega)\mathbf{v}^H(\omega)\right] \qquad (9.10)$$

is the PSD matrix of the noise field. In the sequel we assume that the noise is not fully coherent at the microphones so that $\Phi_{vv}(\omega)$ is a full-rank matrix and its inverse exists.

Now, we define a parameterized desired signal, which we denote by $Q(\omega)S(\omega)$, where $Q(\omega)$ refers to a complex scaling factor that defines the nature of our desired signal. Let $G_1^{\mathrm{d}}(\omega)$ denote the DTFT of the direct path response from the desired source to the first microphone. By setting $Q(\omega) = G_1^{\mathrm{d}}(\omega)$, we are stating that we desire both noise reduction and complete dereverberation. By setting $Q(\omega) = G_1(\omega)$, we are stating that we only desire noise reduction or in other words we desire to recover the reference sensor signal $X_1(\omega) = G_1(\omega)S(\omega)$. In the following, we use the factor $Q(\omega)$ in the definitions of performance measures and in the derivation of the various beamformers.

9.3 From Speech Distortion Weighted Multichannel Wiener Filter to Minimum Variance Distortionless Response Filter

In this section, we first formulate the SDW-MWF in the context of room acoustics. We then focus on a special case of the SDW-MWF, namely the celebrated MVDR beamformer proposed by Capon [6]. It is then shown that the SDW-MWF can be decomposed into an MVDR beamfomer and a speech distortion weighted single-channel Wiener filter. Finally, we show that the MVDR beamformer and the maximum SNR beamfomer are equivalent.

9.3.1 Speech Distortion Weighted Multichannel Wiener Filter

Let us define the error signal between the output of the beamformer and the parameterized desired signal at frequency ω:

$$\mathcal{E}(\omega) = Z(\omega) - Q(\omega)S(\omega)$$
$$= \underbrace{\left[\mathbf{h}^H(\omega)\mathbf{g}(\omega) - Q(\omega)\right]S(\omega)}_{\mathcal{E}_{\tilde{s}}(\omega)} + \underbrace{\mathbf{h}^H(\omega)\mathbf{v}(\omega)}_{\mathcal{E}_v(\omega)}, \qquad (9.11)$$

where $\tilde{S}(\omega) = Q(\omega)\,S(\omega)$. The first term $\mathcal{E}_{\tilde{s}}(\omega)$ denotes the residual desired signal at the output of the beamformer and the second term $\mathcal{E}_v(\omega)$ denotes the residual noise signal at the output of the beamformer. The mean-squared error (MSE) is given by

$$J\left[\mathbf{h}(\omega)\right] = E\left[|\mathcal{E}(\omega)|^2\right]$$
$$= E\left[|\mathcal{E}_{\tilde{s}}(\omega)|^2\right] + E\left[|\mathcal{E}_v(\omega)|^2\right]$$
$$= \left|\mathbf{h}^H(\omega)\mathbf{g}(\omega) - Q(\omega)\right|^2 \phi_{ss}(\omega) + \mathbf{h}^H(\omega)\Phi_{vv}(\omega)\mathbf{h}(\omega). \quad (9.12)$$

The objective of the MWF is to provide an MMSE estimate of either the clean speech source signal, the (reverberant) speech component in one of the microphone signals, or a reference signal. Therefore, the MWF inevitably introduces some speech distortion. To control the tradeoff between speech distortion and noise reduction the SDW-MWF was proposed [5, 21]. The objective of the SDW-MWF can be described as follows[2]

$$\underset{\mathbf{h}(\omega)}{\operatorname{argmin}} E\left[|\mathcal{E}_v(\omega)|^2\right] \quad \text{subject to} \quad E\left[|\mathcal{E}_{\tilde{s}}(\omega)|^2\right] \leq \sigma^2(\omega), \qquad (9.13)$$

where $\sigma^2(\omega)$ defines the maximum local power of the residual desired signal. Since the maximum local power of the residual desired signal is upper-bounded by $|Q(\omega)|^2 \phi_{ss}(\omega)$ we have $0 \leq \sigma^2(\omega) \leq |Q(\omega)|^2 \phi_{ss}(\omega)$. The solution of (9.13) can be found using the Karush-Kuhn-Tucker necessary conditions for constrained minimization [22]. Specifically, $\mathbf{h}(\omega)$ is a feasible point if it satisfies the gradient equation of the Lagrangian

$$\mathcal{L}[\mathbf{h}(\omega), \lambda(\omega)] = E\left[|\mathcal{E}_v(\omega)|^2\right] + \lambda(\omega)\left(E\left[|\mathcal{E}_{\tilde{s}}(\omega)|^2\right] - \sigma^2(\omega)\right), \qquad (9.14)$$

where $\lambda(\omega)$ denotes the Lagrange multiplier for angular frequency ω and

$$\lambda(\omega)\left(E\left[|\mathcal{E}_{\tilde{s}}(\omega)|^2\right] - \sigma^2(\omega)\right) = 0, \quad \lambda(\omega) \geq 0 \text{ and } \omega \in (-\pi, \pi]. \qquad (9.15)$$

The SDW-MWF can now be obtained by setting the derivative of (9.14) with respect to $\mathbf{h}(\omega)$ to zero:

$$\mathbf{h}_{\text{SDW}-\text{MWF}}(\omega, \lambda) = Q^*(\omega)\left[\Phi_{xx}(\omega) + \lambda^{-1}(\omega)\Phi_{vv}(\omega)\right]^{-1}\phi_{ss}(\omega)\mathbf{g}(\omega), \quad (9.16)$$

where superscript * denotes complex conjugation. Using the Woodbury's identity (also known as the matrix inversion lemma) we can write (9.16) as

$$\mathbf{h}_{\text{SDW}-\text{MWF}}(\omega, \lambda) = Q^*(\omega)\frac{\phi_{ss}(\omega)\Phi_{vv}^{-1}(\omega)\mathbf{g}(\omega)}{\lambda^{-1}(\omega) + \phi_{ss}(\omega)\mathbf{g}^H(\omega)\Phi_{vv}^{-1}(\omega)\mathbf{g}(\omega)}. \qquad (9.17)$$

In order to satisfy (9.15) we require that

$$\sigma^2(\omega) = E\left[|\mathcal{E}_{\tilde{s}}(\omega)|^2\right]. \qquad (9.18)$$

Therefore, the Lagrange multiplier $\lambda(\omega)$ must satisfy

---

[2] The employed optimization problem differs from the one used in [5, 21]. However, it should be noted that the solutions are mathematically equivalent. The advantage of the employed optimization problem is that it is directly related to the MVDR beamformer as will be shown in the following section.

$$\sigma^2(\omega) = \left| \mathbf{h}_{\mathrm{SDW-MWF}}^H(\omega, \lambda)\mathbf{g}(\omega) - Q(\omega) \right|^2 \phi_{ss}(\omega) \qquad (9.19)$$

$$= \mathbf{h}_{\mathrm{SDW-MWF}}^H(\omega, \lambda)\Phi_{xx}(\omega)\mathbf{h}_{\mathrm{SDW-MWF}}(\omega, \lambda)$$

$$- \mathbf{h}_{\mathrm{SDW-MWF}}^H(\omega, \lambda)\mathbf{g}(\omega)Q^*(\omega)\phi_{ss}(\omega)$$

$$- \mathbf{g}^H(\omega)\mathbf{h}_{\mathrm{SDW-MWF}}(\omega, \lambda)Q(\omega)\phi_{ss}(\omega)$$

$$+ |Q(\omega)|^2 \phi_{ss}(\omega).$$

Using (9.19), it is possible to find the Lagrange multiplier that results in a specific maximum local power of the residual desired signal. It can be shown that λ monotonically increases when σ^2 decreases. When $\sigma^2(\omega) = |Q(\omega)|^2 \phi_{ss}(\omega)$ we are stating that we allow maximum speech distortion. In order to satisfy (9.19), $\mathbf{h}_{\mathrm{SDW-MWF}}(\omega, \lambda)$ should be equal to $[\,0\ 0\ \cdots\ 0\,]^T$, which is obtain when $\lambda(\omega)$ approaches zero. Consequently, we obtain maximum noise reduction and maximum speech distortion. Another interesting solution is obtained when $\lambda(\omega) = 1$, in this case $\mathbf{h}_{\mathrm{SDW-MWF}}(\omega, \lambda)$ is equal to the non-causal multichannel Wiener filter.

For the particular case, $Q(\omega) = G_1(\omega)$, where we only want to reduce the level of the noise (no dereverberation at all), we can eliminate the explicit dependence of (9.17) on the acoustic transfer functions. Specifically, by using the fact that $\phi_{ss}(\omega)\mathbf{g}^H(\omega)\Phi_{vv}^{-1}(\omega)\mathbf{g}(\omega)$ is equal to $\mathrm{tr}\left[\Phi_{vv}^{-1}(\omega)\Phi_{xx}(\omega)\right]$ we obtain the following forms:

$$\mathbf{h}_{\mathrm{SDW-MWF}}(\omega, \lambda) = \frac{\Phi_{vv}^{-1}(\omega)\Phi_{xx}(\omega)}{\lambda^{-1}(\omega) + \mathrm{tr}\left[\Phi_{vv}^{-1}(\omega)\Phi_{xx}(\omega)\right]}\mathbf{u}$$

$$= \frac{\Phi_{vv}^{-1}(\omega)\Phi_{yy}(\omega) - \mathbf{I}}{\lambda^{-1}(\omega) + \mathrm{tr}\left[\Phi_{vv}^{-1}(\omega)\Phi_{yy}(\omega)\right] - N}\mathbf{u}, \qquad (9.20)$$

where $\mathrm{tr}(\cdot)$ denotes the trace of a matrix, $\mathbf{I}$ is the $N \times N$ identity matrix,

$$\Phi_{yy}(\omega) = E\left[\mathbf{y}(\omega)\mathbf{y}^H(\omega)\right] \qquad (9.21)$$

is the PSD matrix of the microphone signals, and $\mathbf{u} = [\,1\ 0\ \cdots\ 0\ 0\,]^T$ is a vector of length N.

9.3.2 Minimum Variance Distortionless Response Filter

The MVDR filter can be found by minimizing the local power of the residual desired signal at the output of the beamformer. This can be achieved by setting the maximum local power of the residual desired signal $\sigma^2(\omega)$ in (9.13) equal to zero. We then obtain the following optimalization problem

$$\mathbf{h}_{\mathrm{MVDR}}(\omega) = \underset{\mathbf{h}(\omega)}{\mathrm{argmin}}\, E\left[|\mathcal{E}_v(\omega)|^2\right] \quad \text{subject to} \quad E\left[|\mathcal{E}_{\tilde{s}}(\omega)|^2\right] = 0, \quad (9.22)$$

Alternatively, we can use the MSE in (9.12) to derive the MVDR filter, which is conceived by providing a fixed gain [in our case modelled by $Q(\omega)$] to the signal while utilizing the remaining degrees of freedom to minimize the contribution of the noise and interference [second term of the right-hand side of (9.12)] to the array output. The latter optimization problem can be formulated as

$$\mathbf{h}_{\mathrm{MVDR}}(\omega) = \underset{\mathbf{h}(\omega)}{\mathrm{argmin}}\, E\left[|\mathcal{E}_v(\omega)|^2\right] \quad \text{subject to} \quad \mathbf{h}^H(\omega)\mathbf{g}(\omega) = Q(\omega). \quad (9.23)$$

Since $E\left[|\mathcal{E}_{\tilde{s}}(\omega)|^2\right] = \left|\mathbf{h}^H(\omega)\mathbf{g}(\omega) - Q(\omega)\right|^2 \phi_{ss}(\omega) = 0$ for $\mathbf{h}^H(\omega)\mathbf{g}(\omega) = Q(\omega)$ we obtain the same solution for both optimization problems, i.e.,

$$\mathbf{h}_{\mathrm{MVDR}}(\omega) = Q^*(\omega)\frac{\Phi_{vv}^{-1}(\omega)\mathbf{g}(\omega)}{\mathbf{g}^H(\omega)\Phi_{vv}^{-1}(\omega)\mathbf{g}(\omega)}. \quad (9.24)$$

The MVDR filter can also be obtained from the SDW-MWF defined in (9.17) by finding the Lagrange multiplier $\lambda(\omega)$ that satisfies (9.15) for $\sigma^2(\omega) = 0$. To satisfy (9.15) we require that the local power of the residual desired signal at the output of the beamformer, $E\left[|\mathcal{E}_{\tilde{s}}(\omega)|^2\right]$, is equal to zero. From (9.19) it can be shown directly that $\sigma^2(\omega) = 0$ when $\mathbf{h}_{\mathrm{SDW-MWF}}^H(\omega, \lambda)\mathbf{g}(\omega) = Q(\omega)$. Using (9.17) the latter expression can be written as

$$Q(\omega)\frac{\phi_{ss}(\omega)\mathbf{g}^H(\omega)\Phi_{vv}^{-1}(\omega)\mathbf{g}(\omega)}{\lambda^{-1}(\omega) + \phi_{ss}(\omega)\mathbf{g}^H(\omega)\Phi_{vv}^{-1}(\omega)\mathbf{g}(\omega)} = Q(\omega). \quad (9.25)$$

Hence, when $\lambda(\omega)$ goes to infinity the left and right hand sides of (9.25) are equal. Consequently, we have

$$\lim_{\lambda(\omega)\to\infty} \mathbf{h}_{\mathrm{SDW-MWF}}(\omega, \lambda) = \mathbf{h}_{\mathrm{MVDR}}(\omega). \quad (9.26)$$

We can get rid of the explicit dependence on the acoustic transfer functions $\{G_2(\omega), \ldots, G_M(\omega)\}$ of the MVDR filter (9.24) by multiplying the numerator and denominator in (9.24) by $\phi_{ss}(\omega)$ and using the fact that $\mathbf{g}^H(\omega)\Phi_{vv}^{-1}(\omega)\mathbf{g}(\omega)$ is equal to $\mathrm{tr}\left[\Phi_{vv}^{-1}(\omega)\mathbf{g}(\omega)\mathbf{g}^H(\omega)\right]$ to obtain the following form [18]:

$$\mathbf{h}_{\mathrm{MVDR}}(\omega) = \frac{Q^*(\omega)}{G_1^*(\omega)}\frac{\Phi_{vv}^{-1}(\omega)\Phi_{xx}(\omega)}{\mathrm{tr}\left[\Phi_{vv}^{-1}(\omega)\Phi_{xx}(\omega)\right]}\mathbf{u}. \quad (9.27)$$

Basically, we only need $G_1(\omega)$ to achieve dereverberation and noise reduction. It should however be noted that $\mathbf{h}_{\mathrm{MVDR}}(\omega)$ is a non-causal filter.

Using the Woodbury's identity, another important form of the MVDR filter is derived [18]:

$$\mathbf{h}_{\mathrm{MVDR}}(\omega) = C(\omega)\Phi_{yy}^{-1}(\omega)\Phi_{xx}(\omega)\mathbf{u}, \tag{9.28}$$

where

$$C(\omega) = \frac{Q^*(\omega)}{G_1^*(\omega)} \left\{ 1 + \frac{1}{\mathrm{tr}\left[\Phi_{vv}^{-1}(\omega)\Phi_{xx}(\omega)\right]} \right\}. \tag{9.29}$$

For the particular case, $Q(\omega) = G_1(\omega)$, where we only want to reduce the level of the noise (no dereverberation at all), we can get rid of the explicit dependence of the MVDR filter on all acoustic transfer functions to obtain the following forms [1]:

$$\begin{aligned}
\mathbf{h}_{\mathrm{MVDR}}(\omega) &= \frac{\Phi_{vv}^{-1}(\omega)\Phi_{xx}(\omega)}{\mathrm{tr}\left[\Phi_{vv}^{-1}(\omega)\Phi_{xx}(\omega)\right]}\mathbf{u} \\
&= \frac{\Phi_{vv}^{-1}(\omega)\Phi_{yy}(\omega) - \mathbf{I}}{\mathrm{tr}\left[\Phi_{vv}^{-1}(\omega)\Phi_{yy}(\omega)\right] - N}\mathbf{u}.
\end{aligned} \tag{9.30}$$

Hence, noise reduction can be achieved without explicitly estimating the acoustic transfer functions.

9.3.3 Decomposition of the Speech Distortion Weighted Multichannel Wiener Filter

Using (9.17) and (9.24) the SDW-MWF can be decomposed into an MVDR beamformer and a speech distortion weighted single-channel Wiener filter, i.e.,

$$\mathbf{h}_{\mathrm{SDW-MWF}}(\omega, \lambda) = \mathbf{h}_{\mathrm{MVDR}}(\omega) \cdot h_{\mathrm{SDW\text{-}WF}}(\omega, \lambda), \tag{9.31}$$

where

$$\begin{aligned}
h_{\mathrm{SDW\text{-}WF}}(\omega, \lambda) &= \frac{\phi_{\tilde{s}\tilde{s}}(\omega)}{\phi_{\tilde{s}\tilde{s}}(\omega) + \lambda^{-1}(\omega)\,\mathbf{g}^H(\omega)\Phi_{vv}(\omega)\mathbf{g}(\omega)} \\
&= \frac{\phi_{\tilde{s}\tilde{s}}(\omega)}{\phi_{\tilde{s}\tilde{s}}(\omega) + \lambda^{-1}(\omega)\,\mathbf{h}_{\mathrm{MVDR}}^H(\omega)\Phi_{vv}(\omega)\mathbf{h}_{\mathrm{MVDR}}(\omega)},
\end{aligned} \tag{9.32}$$

and $\phi_{\tilde{s}\tilde{s}}(\omega) = |Q(\omega)|^2 \phi_{ss}(\omega)$. Indeed, for $\lambda(\omega) \to \infty$ (i.e., no speech distortion) the (single-channel) speech distortion weighted Wiener filter $h_{\mathrm{SDW\text{-}WF}}(\omega, \lambda) = 1$ for all ω.

9.3.4 Equivalence of MVDR and Maximum SNR Beamformer

It is interesting to show the equivalence between the MVDR filter (9.24) and the maximum SNR (MSNR) beamformer [23], which is obtained from

$$\mathbf{h}_{\mathrm{MSNR}}(\omega) = \underset{\mathbf{h}(\omega)}{\mathrm{argmax}} \; \frac{|\mathbf{h}^H(\omega)\mathbf{g}(\omega)|^2 \phi_{ss}(\omega)}{\mathbf{h}^H(\omega)\Phi_{vv}(\omega)\mathbf{h}(\omega)}. \tag{9.33}$$

The well-known solution to (9.33) is the (colored noise) matched filter

$$\mathbf{h}(\omega) \propto \Phi_{vv}^{-1}(\omega)\mathbf{g}(\omega).$$

If the array response is constrained to fulfil $\mathbf{h}^H(\omega)\mathbf{g}(\omega) = Q(\omega)$ we have

$$\mathbf{h}_{\mathrm{MSNR}}(\omega) = Q^*(\omega)\frac{\Phi_{vv}^{-1}(\omega)\mathbf{g}(\omega)}{\mathbf{g}^H(\omega)\Phi_{vv}^{-1}(\omega)\mathbf{g}(\omega)}. \tag{9.34}$$

This solution is identical to the solution of the MVDR filter (9.24).

9.4 Performance Measures

In this section, we present some very useful measures that will help us better understand how noise reduction and speech dereverberation work with the MVDR beamformer in a real room acoustic environment.

To be consistent with prior works we define the local input signal-to-noise ratio (SNR) with respect to the the parameterized desired signal [given by $Q(\omega)S(\omega)$] and the noise signal received by the first microphone, i.e.,

$$\mathrm{iSNR}\left[Q(\omega)\right] = \frac{|Q(\omega)|^2 \phi_{ss}(\omega)}{\phi_{v_1 v_1}(\omega)}, \;\; \omega \in (-\pi, \pi], \tag{9.35}$$

where $\phi_{v_1 v_1}(\omega)$ is the PSD of the noise signal $v_1(\omega)$. The global input SNR is given by

$$\mathrm{iSNR}(Q) = \frac{\int_{-\pi}^{\pi} |Q(\omega)|^2 \phi_{ss}(\omega)d\omega}{\int_{-\pi}^{\pi} \phi_{v_1 v_1}(\omega)d\omega}. \tag{9.36}$$

After applying the MVDR on the received signals, given by (9.8), the local output SNR is

$$\mathrm{oSNR}\left[\mathbf{h}_{\mathrm{MVDR}}(\omega)\right] = \frac{\left|\mathbf{h}_{\mathrm{MVDR}}^{H}(\omega)\mathbf{g}(\omega)\right|^{2}\phi_{ss}(\omega)}{\mathbf{h}_{\mathrm{MVDR}}^{H}(\omega)\Phi_{vv}(\omega)\mathbf{h}_{\mathrm{MVDR}}(\omega)}$$

$$= \frac{\left|Q(\omega)\right|^{2}\phi_{ss}(\omega)}{\mathbf{h}_{\mathrm{MVDR}}^{H}(\omega)\Phi_{vv}(\omega)\mathbf{h}_{\mathrm{MVDR}}(\omega)}. \tag{9.37}$$

By substituting (9.24) in (9.37) it can be shown that

$$\mathrm{oSNR}\left[\mathbf{h}_{\mathrm{MVDR}}(\omega)\right] = \frac{\left|Q(\omega)\right|^{2}\phi_{ss}(\omega)}{\left|Q(\omega)\right|^{2}\frac{\mathbf{g}^{H}(\omega)\Phi_{vv}^{-1}(\omega)}{\mathbf{g}^{H}(\omega)\Phi_{vv}^{-1}(\omega)\mathbf{g}(\omega)}\Phi_{vv}(\omega)\frac{\Phi_{vv}^{-1}(\omega)\mathbf{g}(\omega)}{\mathbf{g}^{H}(\omega)\Phi_{vv}^{-1}(\omega)\mathbf{g}(\omega)}}$$

$$= \phi_{ss}(\omega)\mathbf{g}^{H}(\omega)\Phi_{vv}^{-1}(\omega)\mathbf{g}(\omega)$$

$$= \mathrm{tr}\left[\Phi_{vv}^{-1}(\omega)\Phi_{xx}(\omega)\right], \ \omega \in (-\pi, \pi]. \tag{9.38}$$

It is extremely important to observe that the desired response $Q(\omega)$ has no impact on the resulting local output SNR (but has an impact on the local input SNR). The global output SNR with the MVDR filter is

$$\mathrm{oSNR}\left(\mathbf{h}_{\mathrm{MVDR}}\right) = \frac{\int_{-\pi}^{\pi}\left|\mathbf{h}_{\mathrm{MVDR}}^{H}(\omega)\mathbf{g}(\omega)\right|^{2}\phi_{ss}(\omega)d\omega}{\int_{-\pi}^{\pi}\mathbf{h}_{\mathrm{MVDR}}^{H}(\omega)\Phi_{vv}(\omega)\mathbf{h}_{\mathrm{MVDR}}(\omega)d\omega}$$

$$= \frac{\int_{-\pi}^{\pi}\left|Q(\omega)\right|^{2}\phi_{ss}(\omega)d\omega}{\int_{-\pi}^{\pi}\left|Q(\omega)\right|^{2}\left[\mathbf{g}^{H}(\omega)\Phi_{vv}^{-1}(\omega)\mathbf{g}(\omega)\right]^{-1}d\omega}$$

$$= \frac{\int_{-\pi}^{\pi}\left|Q(\omega)\right|^{2}\phi_{ss}(\omega)d\omega}{\int_{-\pi}^{\pi}\mathrm{oSNR}^{-1}\left[\mathbf{h}_{\mathrm{MVDR}}(\omega)\right]\left|Q(\omega)\right|^{2}\phi_{ss}(\omega)d\omega}. \tag{9.39}$$

Contrary to the local output SNR, the global output SNR depends strongly on the complex scaling factor $Q(\omega)$.

Another important measure is the level of noise reduction achieved through beamforming. Therefore, we define the local noise-reduction factor as the ratio of the PSD of the original noise at the reference microphone over the PSD of the residual noise:

$$\xi_{\mathrm{nr}}\left[\mathbf{h}(\omega)\right] = \frac{\phi_{v_{1}v_{1}}(\omega)}{\mathbf{h}^{H}(\omega)\Phi_{vv}(\omega)\mathbf{h}(\omega)} \tag{9.40}$$

$$= \frac{\mathrm{oSNR}\left[\mathbf{h}(\omega)\right]}{\mathrm{iSNR}\left[Q(\omega)\right]} \cdot \frac{\left|Q(\omega)\right|^{2}}{\left|\mathbf{h}^{H}(\omega)\mathbf{g}(\omega)\right|^{2}}, \ \omega \in (-\pi, \pi].$$

We see that $\xi_{\mathrm{nr}}\left[\mathbf{h}(\omega)\right]$ is the product of two terms. The first one is the ratio of the output SNR over the input SNR at frequency ω while the second term represents the local distortion introduced by the beamformer $\mathbf{h}(\omega)$. For the MVDR beamformer we have $\left|\mathbf{h}_{\mathrm{MVDR}}^{H}(\omega)\mathbf{g}(\omega)\right|^{2} = \left|Q(\omega)\right|^{2}$. Therefore we can further simplify (9.40):

$$\xi_{\mathrm{nr}}\left[\mathbf{h}_{\mathrm{MVDR}}(\omega)\right] = \frac{\mathrm{oSNR}\left[\mathbf{h}_{\mathrm{MVDR}}(\omega)\right]}{\mathrm{iSNR}\left[Q(\omega)\right]}, \ \omega \in (-\pi, \pi]. \tag{9.41}$$

In this case the local noise-reduction factor tells us exactly how much the output SNR is improved (or not) compared to the input SNR.

Integrating across the entire frequency range in the numerator and denominator of (9.40) yields the global noise-reduction factor:

$$\begin{aligned}
\xi_{\mathrm{nr}}(\mathbf{h}) &= \frac{\int_{-\pi}^{\pi} \phi_{v_1 v_1}(\omega)d\omega}{\int_{-\pi}^{\pi} \mathbf{h}^H(\omega)\boldsymbol{\Phi}_{vv}(\omega)\mathbf{h}(\omega)d\omega} \\
&= \frac{\mathrm{oSNR}(\mathbf{h})}{\mathrm{iSNR}(Q)} \cdot \frac{\int_{-\pi}^{\pi} \left|Q(\omega)\right|^2 \phi_{ss}(\omega)d\omega}{\int_{-\pi}^{\pi} \left|\mathbf{h}^H(\omega)\mathbf{g}(\omega)\right|^2 \phi_{ss}(\omega)d\omega}.
\end{aligned} \tag{9.42}$$

The global noise-reduction factor is also the product of two terms. While the first one is the ratio of the global output SNR over the global input SNR, the second term is the global speech distortion due the beamformer. For the MVDR beamformer the global noise-reduction factor further simplifies to

$$\xi_{\mathrm{nr}}(\mathbf{h}_{\mathrm{MVDR}}) = \frac{\mathrm{oSNR}(\mathbf{h}_{\mathrm{MVDR}})}{\mathrm{iSNR}(Q)}. \tag{9.43}$$

9.5 Performance Analysis

In this section we analyze the performance of the MVDR beamformer and the tradeoff between the amount of speech dereverberation and noise reduction. When comparing the noise-reduction factor of different MVDR beamformers (with different objectives) it is of great importance that the comparison is conducted in a fair way. In Subsection 9.5.1 we will discuss this issue and propose a viable comparison method. In Subsections 9.5.2 and 9.5.3, we analyze the local and global behaviors of the output SNR and the noise-reduction factor obtained by the MVDR beamformer, respectively. In addition, we analyze the tradeoff between dereverberation and noise reduction. In Subsections 9.5.4 and 9.5.5 we analyze the MVDR performance in non-coherent noise fields and mixed coherent and non-coherent noise fields, respectively.

9.5.1 On the Comparison of Different MVDR Beamformers

One of the main objectives of this work is to compare MVDR beamformers with different constraints. When we desire noise-reduction only, the constraint of the MVDR beamformer is given by $\mathbf{h}^H(\omega)\mathbf{g}(\omega) = G_1(\omega)$. When we

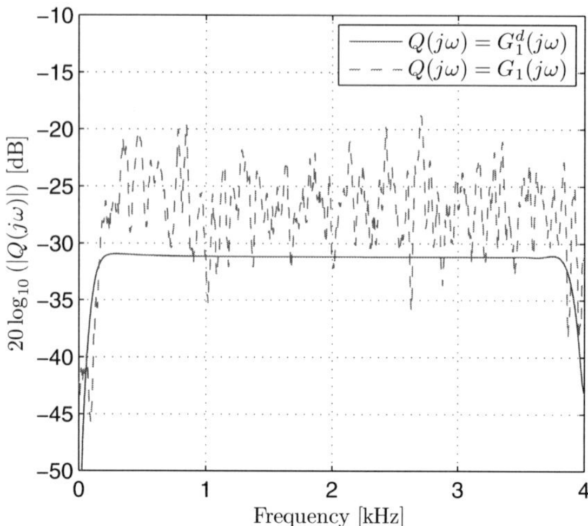

Fig. 9.1 Magnitude of the transfer functions $Q(\omega) = \{G_1(\omega), G_1^{\mathrm{d}}(\omega)\}$ (reverberation time $T_{60} = 0.5$ s, source-receiver distance $D = 2.5$ m).

desire complete dereverberation and noise reduction we can use the constraint $\mathbf{h}^H(\omega)\mathbf{g}(\omega) = G_1^{\mathrm{d}}(\omega)$, where $G_1^{\mathrm{d}}(\omega)$ denotes the transfer function of the direct path response from the source to the first microphone. In Fig. 9.1 the magnitude of the transfer functions $G_1(\omega)$ and $G_1^{\mathrm{d}}(\omega)$ are shown. The transfer function $G_1(\omega)$ was generated using the image-method [24], the distance between the source and the microphone was 2.5 m and the reverberation time was 500 ms. The transfer function $G_1^{\mathrm{d}}(\omega)$ was obtained by considering only the direct path. As expected from a physical point of view, we can see that the energy of $G_1(\omega)$ is larger than the energy of $G_1^{\mathrm{d}}(\omega)$. In addition we observe that for very few frequencies $|G_1(\omega)|^2$ is smaller than $|G_1^{\mathrm{d}}(\omega)|^2$. Evidently, the power of the desired signal $G_1^{\mathrm{d}}(\omega)S(\omega)$ is always smaller than the power of the desired signal $G_1(\omega)S(\omega)$.

Now let us first look at an illustrative example. Obviously, by choosing any constraint $Q(\omega, \gamma) = \gamma \cdot G_1^{\mathrm{d}}(\omega)$ ($\gamma > 0 \ \wedge \ \gamma \in \mathbb{R}$) we desire both noise reduction and complete dereverberation. Now let us define the MVDR filter with the constraint $Q(\omega, \gamma)$ by $\mathbf{h}_{\mathrm{MVDR}}(\omega, \gamma)$. Using (9.24) it can be shown that $\mathbf{h}_{\mathrm{MVDR}}(\omega, \gamma)$ is equal to $\gamma\,\mathbf{h}_{\mathrm{MVDR}}(\omega)$, i.e., by scaling the desired signal we scale the MVDR filter. Consequently, we have also scaled the noise signal at the output. When we would directly calculate the noise-reduction factor of the beamformers $\mathbf{h}_{\mathrm{MVDR}}(\omega)$ and $\mathbf{h}_{\mathrm{MVDR}}(\omega, \gamma)$ using (9.41) we obtain different results, i.e.,

$$\xi_{\mathrm{nr}}\left[\mathbf{h}_{\mathrm{MVDR}}(\omega)\right] \neq \xi_{\mathrm{nr}}\left[\mathbf{h}_{\mathrm{MVDR}}(\omega, \gamma)\right] \quad \text{for} \ \ \gamma \neq 1. \tag{9.44}$$

This can also be explained by the fact that the local output SNRs of all MVDR beamformers $\mathbf{h}_{\text{MVDR}}(\omega, \gamma)$ are equal because the local output SNR [as defined in (9.37)] is independent of γ while the local input SNR [as defined in (9.35)] is dependent on γ. A similar problem occurs when we like to compare the noise-reduction factor of MVDR beamformers with completely different constraints because the power of the reverberant signal is much larger than the power of the direct sound signal. This abnormality can be corrected by normalizing the power of the output signal, which can be achieved my normalizing the MVDR filter. Fundamentally, the definition of the MVDR beamformer depends on $Q(\omega)$. Therefore, the choice of different desired signals [given by $Q(\omega)S(\omega)$] is part of the (local and global) input SNR definitions. Basically we can apply any normalization provided that the power of the desired signals at the output of the beamformer is equal. However, to obtain a meaningful output power and to be consistent with earlier works, we propose to make the power of the desired signal at the output equal to the power of the signal that would be obtained when using the constraint $\mathbf{h}^H(\omega)\mathbf{g}(\omega) = G_1(\omega)$. The global normalization factor $\eta(Q, G_1)$ is therefore given by

$$\eta(Q, G_1) = \sqrt{\frac{\int_{-\pi}^{\pi} |G_1(\omega)|^2 \, \phi_{ss}(\omega) \, d\omega}{\int_{-\pi}^{\pi} |Q(\omega)|^2 \, \phi_{ss}(\omega) \, d\omega}}, \tag{9.45}$$

which can either be applied to the output signal of the beamformer or the filter $\mathbf{h}_{\text{MVDR}}(\omega)$.

9.5.2 Local Analyzes

The most important goal of a beamforming algorithm is to improve the local SNR after filtering. Therefore, we must design the beamforming weight vectors, $\mathbf{h}(\omega)$, $\omega \in (-\pi, \pi]$, in such a way that oSNR $[\mathbf{h}(\omega)] \geq$ iSNR $[Q(\omega)]$. We next give an interesting property that will give more insights into the local SNR behavior of the MVDR beamformer.

Property 9.1. With the MVDR filter given in (9.24), the local output SNR times $|Q(\omega)|^2$ is always greater than or equal to the local input SNR times $|G_1(\omega)|^2$, i.e.,

$$|Q(\omega)|^2 \cdot \text{oSNR}\left[\mathbf{h}_{\text{MVDR}}(\omega)\right] \geq |G_1(\omega)|^2 \cdot \text{iSNR}\left[Q(\omega)\right], \quad \forall \omega, \tag{9.46}$$

which can also be expressed using (9.35) as

$$\text{oSNR}\left[\mathbf{h}_{\text{MVDR}}(\omega)\right] \geq \frac{|G_1(\omega)|^2 \, \phi_{ss}(\omega)}{\phi_{v_1 v_1}(\omega)}, \quad \forall \omega. \tag{9.47}$$

Proof. See Appendix.

This property proofs that the local output SNR obtained using the MVDR filter will always be equal or larger than the ratio of the reverberant desired signal power and the noise power received by the reference microphone (in this case the first microphone).

The normalized local noise-reduction factor is defined as

$$
\begin{aligned}
\tilde{\xi}_{\mathrm{nr}}\left[\mathbf{h}_{\mathrm{MVDR}}(\omega)\right] &= \xi_{\mathrm{nr}}\left[\eta(Q, G_1)\,\mathbf{h}_{\mathrm{MVDR}}(\omega)\right] \\
&= \frac{1}{\eta^2(Q, G_1)}\,\frac{\mathrm{oSNR}\left[\mathbf{h}_{\mathrm{MVDR}}(\omega)\right]}{\mathrm{iSNR}\left[Q(\omega)\right]} \\
&= \frac{1}{\eta^2(Q, G_1)\,|Q(\omega)|^2}\cdot\frac{\mathrm{oSNR}\left[\mathbf{h}_{\mathrm{MVDR}}(\omega)\right]\cdot\phi_{v_1 v_1}(\omega)}{\phi_{ss}(\omega)} \\
&= \frac{1}{\zeta[Q(\omega), G_1(\omega)]}\cdot\frac{\mathrm{oSNR}\left[\mathbf{h}_{\mathrm{MVDR}}(\omega)\right]\cdot\phi_{v_1 v_1}(\omega)}{\phi_{ss}(\omega)},
\end{aligned}
\qquad (9.48)
$$

where $\zeta[Q(\omega), G_1(\omega)] = \eta^2(Q, G_1)\,|Q(\omega)|^2$. Indeed, for different MVDR beamformers the noise-reduction factor varies due to $\zeta[Q(\omega), G_1(\omega)]$, since the local output SNR, $\phi_{v_1 v_1}(\omega)$, and $\phi_{ss}(\omega)$ do not depend on $Q(\omega)$.

Since $\zeta[Q(\omega), G_1(\omega)] = \zeta[\gamma\,Q(\omega), G_1(\omega)]$ ($\gamma > 0$) the normalized local noise-reduction factor is independent of the global scaling factor γ.

To gain more insight into the local behavior of $\zeta[Q(\omega), G_1(\omega)]$ we analyzed several acoustic transfer functions. To simplify the following discussion we assume that the power spectral density $\phi_{ss}(\omega) = 1$ for all ω. Let us decompose the transfer function $G_1(\omega)$ into two parts. The first part $G_1^{\mathrm{d}}(\omega)$ is the DTFT the direct path, while the second part $G_1^{\mathrm{r}}(\omega)$ is the DTFT of the reverberant part. Now let us define the desired response as

$$
Q(\omega, \alpha) = G_1^{\mathrm{d}}(\omega) + \alpha\,G_1^{\mathrm{r}}(\omega),
\qquad (9.49)
$$

where the parameter $0 \leq \alpha \leq 1$ controls the direct-to-reverberation ratio (DRR) of the desired response. In Fig. 9.2(a) we plotted $\zeta[Q(\omega, \alpha), G_1(\omega)]$ for $\alpha = \{0, 0.2, 1\}$. Due to the normalization the energy of $\zeta[Q(\omega, \alpha), G_1(\omega)]$ (and therefore its mean value) does not depend on α. Locally we can see that the deviation with respect to $|G_1^{\mathrm{d}}(\omega)|^2$ increases when α increases (i.e., when the DRR decreases). In Fig. 9.2(b) we plotted the histogram of $\zeta[Q(\omega, \alpha), G_1(\omega)]$ for $\alpha = \{0, 0.2, 1\}$. First, we observe that the probability that $\zeta[Q(\omega, \alpha), G_1(\omega)]$ is smaller than its mean value decreases when α decreases (i.e., when the DRR increases). Secondly, we observe that the distribution is stretched out towards negative values when α increases. Hence, when the desired speech signal contains less reverberation it is more likely that $\zeta[Q(\omega, \alpha), G_1(\omega)]$ will increase and that the local noise-reduction factor will decrease. Therefore, it is likely that the highest local noise reduction is achieved when we desire only noise reduction, i.e., for $Q(\omega) = G_1(\omega)$.

Using Property 9.1 we deduce a lower bound for the normalized local noise-reduction factor:

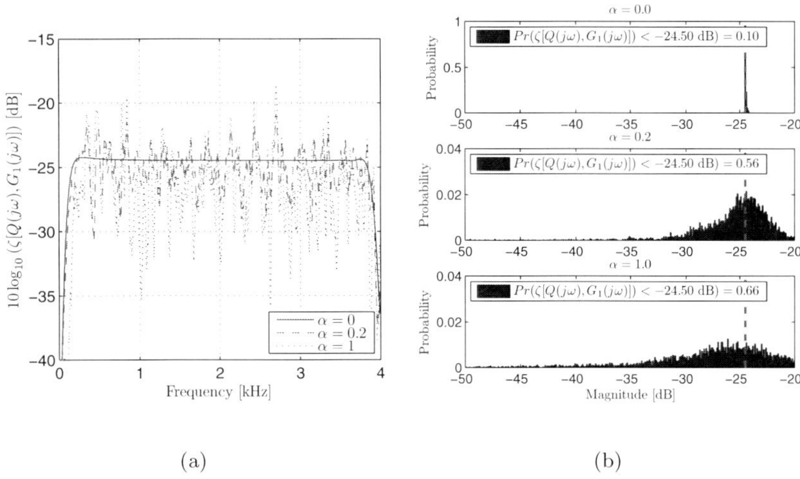

$$\begin{array}{cc} \text{(a)} & \text{(b)} \end{array}$$

Fig. 9.2 a) The normalized transfer functions $\zeta[Q(\omega,\alpha),G_1(\omega)]$ with $Q(\omega,\alpha) = G_1^{\mathrm{d}}(\omega) + \alpha\,G_1^{\mathrm{r}}(\omega)$ for $\alpha = \{0, 0.2, 1\}$, b) the histograms of $10\log_{10}(\zeta[Q(\omega,\alpha),G_1(\omega)])$.

$$\tilde{\xi}_{\mathrm{nr}}\left[\mathbf{h}_{\mathrm{MVDR}}(\omega)\right] \geq \frac{1}{\eta^2(Q,G_1)\,|Q(\omega)|^2}\,|G_1(\omega)|^2. \tag{9.50}$$

For $Q(\omega) = G_1(\omega)$ we obtain

$$\tilde{\xi}_{\mathrm{nr}}\left[\mathbf{h}_{\mathrm{MVDR}}(\omega)\right] \geq 1. \tag{9.51}$$

Expression (9.51) proves that there is always noise-reduction when we desire only noise reduction. However, in other situations we cannot guarantee that there is noise reduction.

9.5.3 Global Analyzes

Using (9.43), (9.39), and (9.36) we deduce the normalized global noise-reduction factor:

$$\tilde{\xi}_{\mathrm{nr}}(\mathbf{h}_{\mathrm{MVDR}}) = \xi_{\mathrm{nr}}(\eta(Q, G_1)\,\mathbf{h}_{\mathrm{MVDR}})$$

$$= \frac{1}{\eta^2(Q, G_1)}\,\frac{\mathrm{oSNR}(\mathbf{h}_{\mathrm{MVDR}})}{\mathrm{iSNR}(Q)}$$

$$= \frac{1}{\eta^2(Q, G_1)}\,\frac{\int_{-\pi}^{\pi}\phi_{v_1 v_1}(\omega)d\omega}{\int_{-\pi}^{\pi}\mathrm{oSNR}^{-1}\left[\mathbf{h}_{\mathrm{MVDR}}(\omega)\right]\left|Q(\omega)\right|^2\phi_{ss}(\omega)d\omega}$$

$$= \frac{\int_{-\pi}^{\pi}\phi_{v_1 v_1}(\omega)d\omega}{\int_{-\pi}^{\pi}\mathrm{oSNR}^{-1}\left[\mathbf{h}_{\mathrm{MVDR}}(\omega)\right]\zeta[Q(\omega), G_1(\omega)]\,\phi_{ss}(\omega)d\omega}. \tag{9.52}$$

This normalized global noise-reduction factor behaves, with respect to $Q(\omega)$, similarly to its local counterpart. It can be verified that the normalized global noise-reduction factor for $\gamma \cdot Q(\omega)$ is independent of γ. Due to the complexity of (9.52) it is difficult to predict the exact behavior of the normalized global noise-reduction factor. From the analyzes in the previous subsection we do know that the distribution of $\zeta[Q(\omega), G_1(\omega)]$ is stretched out towards zero when the DRR decreases. Hence, for each frequency it is likely that $\zeta[Q(\omega), G_1(\omega)]$ will decrease when the DRR decreases. Consequently, we expect that the normalized global noise-reduction factor will always increase when the DRR decreases. The expected behavior of the normalized global noise-reduction factor is consistent with the results presented in Section 9.6.

9.5.4 Non-Coherent Noise Field

Let us assume that the noise field is non-coherent, also known as spatially white. In case the noise variance at each microphone is equal to $\sigma_{\mathrm{nc}}^2(\omega)$ the noise covariance matrix $\Phi_{vv}(\omega)$ simplifies to $\sigma_{\mathrm{nc}}^2(\omega)\mathbf{I}$. In the latter case the MVDR beamformer simplifies to

$$\mathbf{h}_{\mathrm{MVDR}}(\omega) = Q^*(\omega)\frac{\mathbf{g}(\omega)}{\|\mathbf{g}(\omega)\|^2}, \tag{9.53}$$

where $\|\mathbf{g}(\omega)\|^2 = \mathbf{g}^H(\omega)\mathbf{g}(\omega)$. For $Q(\omega) = G_1(\omega)$ this is the well-known matched beamformer [25], which generalizes the delay-and-sum beamformer. The normalized local noise-reduction factor can be deduced by substituting $\sigma_{\mathrm{nc}}^2(\omega)\mathbf{I}$ in (9.48), and result in

$$\tilde{\xi}_{\mathrm{nr}}\left[\mathbf{h}_{\mathrm{MVDR}}(\omega)\right] = \frac{1}{\zeta[Q(\omega), G_1(\omega)]}\,\|\mathbf{g}(\omega)\|^2. \tag{9.54}$$

When $Q(\omega) = G_1(\omega)$ the normalization factor $\eta(Q, G_1)$ equals one, the normalized noise-reduction factor then becomes

$$\tilde{\xi}_{\mathrm{nr}}\left[\mathbf{h}_{\mathrm{MVDR}}(\omega)\right] = \frac{\|\mathbf{g}(\omega)\|^2}{|G_1(\omega)|^2}$$

$$= \left(1 + \sum_{n=2}^{N} \frac{|G_n(\omega)|^2}{|G_1(\omega)|^2}\right). \tag{9.55}$$

As we expected from (9.51), the normalized noise-reduction factor is always larger than 1 when $Q(\omega) = G_1(\omega)$. However, in other situations we cannot guarantee that there is noise reduction.

The normalized global noise-reduction factor is given by

$$\tilde{\xi}_{\mathrm{nr}}\left(\mathbf{h}_{\mathrm{MVDR}}\right) = \frac{1}{\eta^2(Q,G_1)} \frac{\int_{-\pi}^{\pi} \sigma_{\mathrm{nc}}^{-2}(\omega)\,\|\mathbf{g}(\omega)\|^2\,\phi_{ss}(\omega)\,d\omega}{\int_{-\pi}^{\pi} \sigma_{\mathrm{nc}}^{-2}(\omega)\,|Q(\omega)|^2\,\phi_{ss}(\omega)\,d\omega}$$

$$= \frac{\int_{-\pi}^{\pi} \sigma_{\mathrm{nc}}^{-2}(\omega)\,\|\mathbf{g}(\omega)\|^2\,\phi_{ss}(\omega)\,d\omega}{\int_{-\pi}^{\pi} \sigma_{\mathrm{nc}}^{-2}(\omega)\,\zeta[Q(\omega),G_1(\omega)]\,\phi_{ss}(\omega)\,d\omega}. \tag{9.56}$$

In an anechoic environment where the source is positioned in the far-field of the array, $G_n(\omega)$ are steering vectors and $|Q(\omega)|^2 = |G_n(\omega)|^2$, $\forall n$. In this case the normalized global noise-reduction factor simplifies to

$$\tilde{\xi}_{\mathrm{nr}}\left(\mathbf{h}_{\mathrm{MVDR}}\right) = N. \tag{9.57}$$

The latter results in consistent with earlier works and shows that the noise-reduction factor only depends on the number of microphones. When the PSD matrices of the noise and microphone signals are known we can compute the MVDR filter using (9.30), i.e., we do not require any *a prior* knowledge of the direction of arrival.

9.5.5 Coherent plus Non-Coherent Noise Field

Let $\mathbf{d}(\omega) = [\,D_1(\omega)\,D_2(\omega)\,\cdots\,D_N(\omega)\,]^T$ denote the ATFs between a noise source and the array. The noise covariance matrix can be written as

$$\Phi_{vv}(\omega) = \sigma_{\mathrm{c}}^2(\omega)\mathbf{d}(\omega)\mathbf{d}^H(\omega) + \sigma_{\mathrm{nc}}^2(\omega)\mathbf{I}. \tag{9.58}$$

Using Woodbury's identity the MVDR beamformer becomes

$$\mathbf{h}_{\mathrm{MVDR}}(\omega) = Q^*(\omega)\,\frac{\left(\mathbf{I} - \dfrac{\mathbf{d}(\omega)\mathbf{d}^H(\omega)}{\frac{\sigma_{\mathrm{nc}}^2(\omega)}{\sigma_{\mathrm{c}}^2(\omega)} + \mathbf{d}^H(\omega)\mathbf{d}(\omega)}\right)\mathbf{g}(\omega)}{\mathbf{g}^H(\omega)\left(\mathbf{I} - \dfrac{\mathbf{d}(\omega)\mathbf{d}^H(\omega)}{\frac{\sigma_{\mathrm{nc}}^2(\omega)}{\sigma_{\mathrm{c}}^2(\omega)} + \mathbf{d}^H(\omega)\mathbf{d}(\omega)}\right)\mathbf{g}(\omega)}. \tag{9.59}$$

The normalized local noise-reduction factor is given by [18]

$$\tilde{\xi}_{\mathrm{nr}}\left[\mathbf{h}_{\mathrm{MVDR}}(\omega)\right] = C(\omega)\left(\mathbf{g}^H(\omega)\mathbf{g}(\omega) - \frac{|\mathbf{g}^H(\omega)\mathbf{d}(\omega)|^2}{\frac{\sigma_{\mathrm{nc}}^2(\omega)}{\sigma_{\mathrm{c}}^2(\omega)} + \mathbf{d}^H(\omega)\mathbf{d}(\omega)}\right), \qquad (9.60)$$

where

$$C(\omega) = \frac{1}{\zeta[Q(\omega), G_1(\omega)]}\left(1 + \frac{\sigma_{\mathrm{c}}^2(\omega)}{\sigma_{\mathrm{nc}}^2(\omega)}|D_1(\omega)|^2\right). \qquad (9.61)$$

The noise reduction depends on $\zeta[Q(\omega), G_1(\omega)]$, the ratio between the variance of the non-coherent and coherent, and on the inner product of $\mathbf{d}(\omega)$ and $\mathbf{g}(\omega)$ [26].

Obviously, the noise covariance matrix $\Phi_{vv}(\omega)$ needs to be full-rank. However, from a theoretical point of view we can analyze the residual coherent noise at the output of the MVDR beamformer, given by $\mathbf{h}_{\mathrm{MVDR}}^H(\omega)\mathbf{d}(\omega)\sigma_{\mathrm{c}}(\omega)$, when the ratio $\frac{\sigma_{\mathrm{nc}}^2(\omega)}{\sigma_{\mathrm{c}}^2(\omega)}$ approaches zero, i.e., the noise field becomes more and more coherent. Provided that $\mathbf{d}(\omega) \neq \mathbf{g}(\omega)$ the coherent noise at the output of the beamformer is given by

$$\lim_{\frac{\sigma_{\mathrm{nc}}^2(\omega)}{\sigma_{\mathrm{c}}^2(\omega)} \to 0} \mathbf{h}_{\mathrm{MVDR}}^H(\omega)\mathbf{d}(\omega)\sigma_{\mathrm{c}}(\omega) = 0.$$

For $\mathbf{d}(\omega) = \mathbf{g}(\omega)$ there is a contradiction, since the desired signal and the coherent noise signal come from the same point.

9.6 Performance Evaluation

In this section, we evaluate the performance of the MVDR beamformer in room acoustics. We will demonstrate the tradeoff between speech dereverberation and noise reduction by computing the normalized noise-reduction factor in various scenarios. A linear microphone array was used with 2 to 8 microphones and an inter-microphone distance of 5 cm. The room size is $5 \times 4 \times 6$ m (length$\times$width$\times$height), the reverberation time of the enclosure varies between 0.2 to 0.4 s. All room impulse responses are generated using the image-method proposed by Allen and Berkley [24] with some necessary modifications that ensure proper inter-microphone phase delays as proposed by Peterson [27]. The distance between the desired source and the first microphone varies from 1 to 3 m. The desired source consists of speech like noise (USASI). The noise consists of a simple AR(1) process (autoregressive process of order one) that was created by filtering a stationary zero-mean Gaussian sequences with a linear time-invariant filter. We used non-coherent noise, a mixture of non-coherent noise and a coherent noise source, and diffuse noise.

In order to study the tradeoff more carefully we need to control the amount of reverberation reduction. Here we propose to control the amount of reverberation reduction by changing the DRR of the desired response $Q(\omega)$. As proposed in Section 9.5.1, we control the DRR using the parameter α $(0 \leq \alpha \leq 1)$. The complex scaling factor $Q(\omega, \alpha)$ is calculated using (9.49). When the desired response equals $Q(\omega, 0) = G_1^{\mathrm{d}}(\omega)$ we desire both noise reduction and complete dereverberation. However, when the desired response equals $Q(\omega, 1) = G_1(\omega)$ we desire only noise reduction.

9.6.1 Influence of the Number of Microphones

In this section we study the influence of the number of microphones used. The reverberation time was set to $T_{60} = 0.3$ s and the distance between the source and the first microphone was $D = 2$ m. The noise field is non-coherent and the global input SNR [for $Q(\omega, 0) = G_1^{\mathrm{d}}(\omega)$] was iSNR = 5 dB. In this experiment 2, 4, or 8 microphones were used. In Fig. 9.3 the normalized global noise-reduction factor is shown for $0 \leq \alpha \leq 1$. Firstly, we observe that there is a tradeoff between speech dereverberation and noise reduction. The largest amount of noise reduction is achieved for $\alpha = 1$, i.e., when no dereverberation is performed. While a smaller amount of noise reduction is achieved for $\alpha = 0$, i.e., when complete dereverberation is performed. In the case of two microphones $(N = 2)$, we amplify the noise when we desire to complete dereverberate the speech signal. Secondly, we observe that the amount of noise reduction increases with approximately 3 dB if we double the number of microphones. Finally, we observe that the tradeoff becomes less evident when more microphones are used. When more microphones are available the degrees of freedom of the MVDR beamformer increases. In such a case the MVDR beamformer is apparently able to perform speech dereverberation without significantly sacrificing the amount of noise reduction.

9.6.2 Influence of the Reverberation Time

In this section we study the influence of the reverberation time. The distance between the source and the first microphone was set to $D = 4$ m. The noise field is non-coherent and the global input SNR [for $Q(\omega) = G_1^{\mathrm{d}}(\omega)$] was iSNR = 5 dB. In this experiment four microphones were used, and the reverberation time was set to $T_{60} = \{0.2, 0.3, 0.4\}$ s. The DRR ratio of the desired response $Q(\omega)$ is shown in Fig. 9.4(a). In Fig. 9.4(b) the normalized global noise-reduction factor is shown for $0 \leq \alpha \leq 1$. Again, we observe that there is a tradeoff between speech dereverberation and noise reduction. This experiment also shows that almost no noise reduction is sacrificed when we desire to

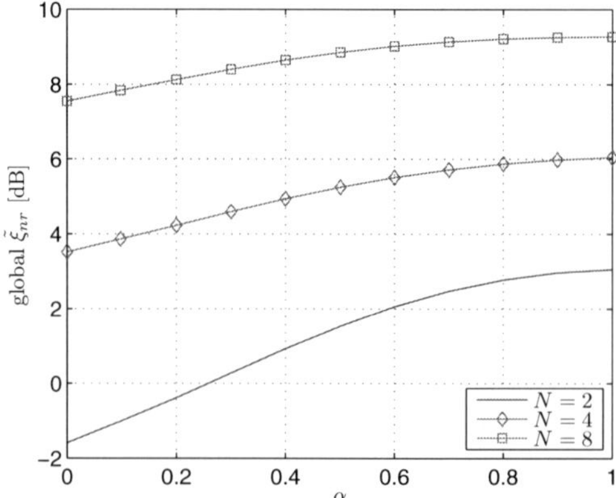

Fig. 9.3 The normalized global noise-reduction factor obtained using $N = \{2, 4, 8\}$ ($T_{60} = 0.3$, $D = 2$ m, non-coherent noise iSNR = 5 dB).

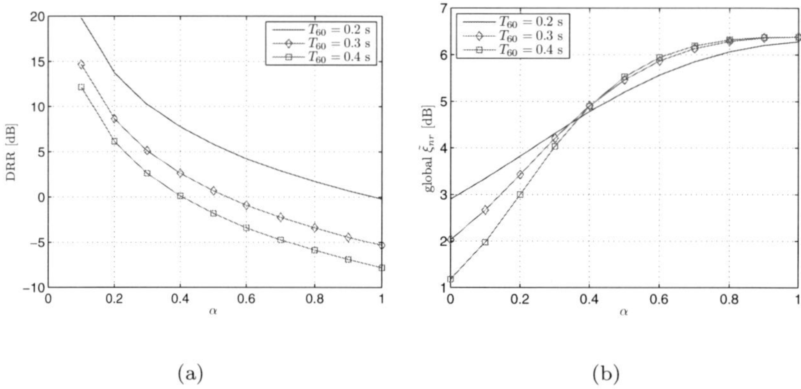

(a) (b)

Fig. 9.4 a) The DRR of $Q(\omega, \alpha)$ for $T_{60} = \{0.2, 0.3, 0.4\}$ s. b) The normalized global noise-reduction factor obtained using $T_{60} = \{0.2, 0.3, 0.4\}$ s ($N = 4$, $D = 4$ m, non-coherent noise iSNR = 5 dB).

increase the DRR to approximately -5 dB for $T_{60} \leq 0.3$ s . In other words, as long as the reverberant part of the signal is dominant (DRR≤ -5 dB) we can reduce reverberation and noise without sacrificing too much noise reduction. However, when the DRR is increased further (DRR> -5 dB) the noise-reduction decreases.

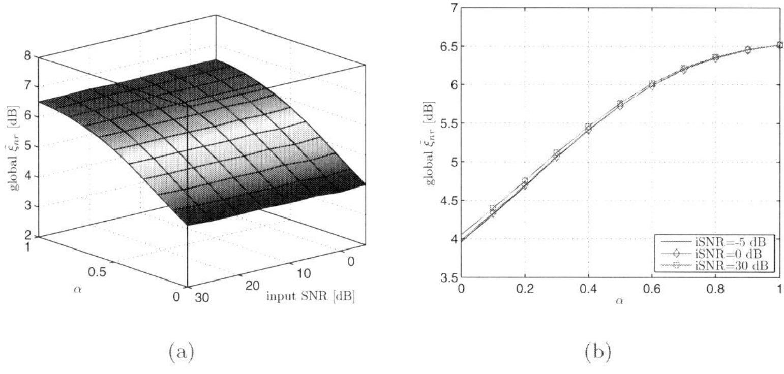

(a)　　　　　　　　　　　　　　　　(b)

Fig. 9.5 The normalized global noise-reduction factor obtained using non-coherent noise iSNR = $\{-5,\ldots,30\}$ dB ($T_{60} = 0.3$ s, $N = 4$, $D = 2$ m).

9.6.3 Influence of the Noise Field

In this section we evaluate the normalized noise-reduction factor in various noise fields and study the tradeoff between noise reduction and dereverberation.

9.6.3.1 Non-Coherent Noise Field

In this section we study the amount of noise reduction in a non-coherent noise field with different input SNRs. The distance between the source and the first microphone was set to $D = 2$ m. In this experiment four microphones were used, and the reverberation time was set to $T_{60} = 0.3$ s. In Fig. 9.5(a) the normalized global noise-reduction factor is shown for $0 \leq \alpha \leq 1$ and different input SNRs ranging from -5 dB to 30 dB. In Fig. 9.5(b) the normalized global noise-reduction factor is shown for $0 \leq \alpha \leq 1$ and input SNRs of -5, 0, and 30 dB. We observe the tradeoff between speech dereverberation and noise reduction as before. As expected from (9.56), for a non-coherent noise field the normalized global noise-reduction factor is independent of the input SNR.

In Fig. 9.6, we depicted the normalized global noise-reduction factor for $\alpha = 0$ (i.e., complete dereverberation and noise reduction) and $\alpha = 1$ (i.e., noise reduction only) for different distances. It should be noted that the DRR is not monotonically decreasing with the distance. Therefore, the noise-reduction factor is not monotonically decreasing with the distance. Here four microphones were used and the reverberation time equals 0.3 s. When we

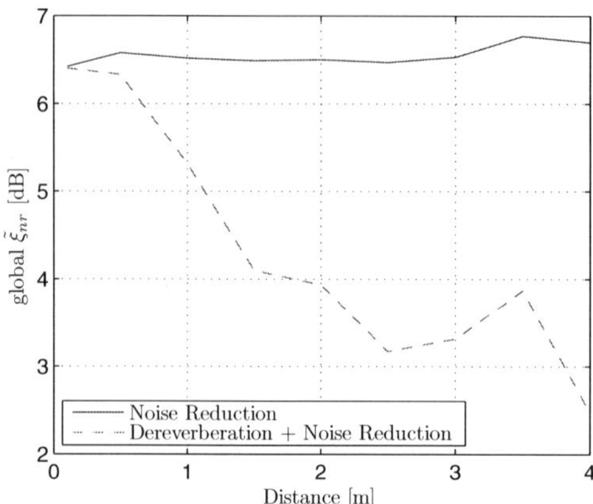

Fig. 9.6 The normalized global noise-reduction factor for one specific source trajectory obtained using $D = \{0.1, 0.5, 1, \ldots, 4\}$ m ($T_{60} = 0.3$ s, $N = 4$, non-coherent noise iSNR = 5 dB).

desire only noise reduction, the noise reduction is independent of the distance between the source and the first microphone. However, when we desire both dereverberation and noise reduction we see that the normalized global noise-reduction factor decreases rapidly. At a distance of 4 m we sacrificed approximately 4 dB noise reduction.

9.6.3.2 Coherent and Non-Coherent Noise Field

In this section we study the amount of noise reduction in a coherent plus non-coherent noise field with different input SNRs. The input SNR (iSNR$_{\mathrm{nc}}$) of the non-coherent noise is 20 dB. The distance between the source and the first microphone was set to $D = 2$ m. In this experiment four microphones were used, and the reverberation time was set to $T_{60} = 0.3$ s. In Fig. 9.7(a) the normalized global noise-reduction factor is shown for $0 \leq \alpha \leq 1$ and for input SNR (iSNR$_{\mathrm{c}}$) of the coherent noise source that ranges from -5 dB to 30 dB. In Fig. 9.7(b) the normalized global noise-reduction factor is shown for $0 \leq \alpha \leq 1$ and input SNRs of -5, 0, and 30 dB. We observe the tradeoff between speech dereverberation and noise reduction as before. In addition, we see that the noise reduction in a coherent noise field is much larger than the noise reduction in a non-coherent noise field.

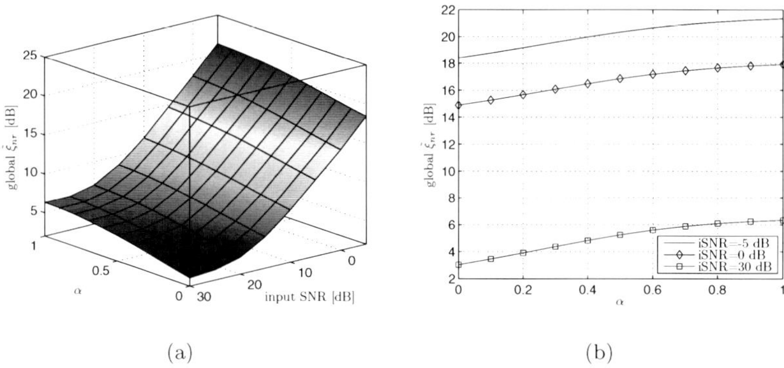

(a)   (b)

Fig. 9.7 The normalized global noise-reduction factor obtained using a coherent plus non-coherent noise $\mathrm{iSNR_c} = \{-5,\ldots,30\}$ dB ($\mathrm{iSNR_{nc}} = 20$ dB, $T_{60} = 0.3$ s, $N = 4$, $D = 2$ m).

9.6.4 Example Using Speech Signals

Finally, we show an example obtained using a real speech sampled at 16 kHz. The speech sample was taken from the APLAWD speech database [28]. For this example non-coherent noise was used and the input SNR ($\mathrm{iSNR_{nc}}$) was 10 dB. The distance between the source and the first microphone was set to $D = 3$ m and the reverberation time was set to $T_{60} = 0.35$ s. As shown in Subsection 9.6.1 there is a larger tradeoff between speech dereverberation and noise reduction when a limited amount of microphones is used. In order to emphasize the tradeoff we used four microphones. The AIRs were generated using the source-image method and are 1024 coefficients long. For this scenario long filters are required to estimate the direct response of the desired speech signal. The total length of the non-causal filter was 8192, of which 4096 coefficients correspond to the causal part of the filter. To avoid pre-echoes (i.e., echoes that arrive before the arrival of the direct sound), the non-causal part of the filter was properly truncated to a length of 1024 coefficients (128 ms). The filters and the second-order statistics of the signals are computed on a frame-by-frame basis in the discrete Fourier transform domain. The filter process is performed using the overlap-save technique [29].

In Fig. 9.8(a) the spectrogram and waveform of the noisy and reverberant microphone signal $y_1(k)$ are depicted. In Fig. 9.8(b) the processed signal is shown with $\alpha = 1$, i.e., when desiring noise reduction only. Finally, in Fig. 9.8(c) the processed signal is shown with $\alpha = 0$, i.e. when desiring dereverberation and noise reduction. By comparing the spectrograms one can see that the processed signal shown in Fig. 9.8(c) contains less reverberation

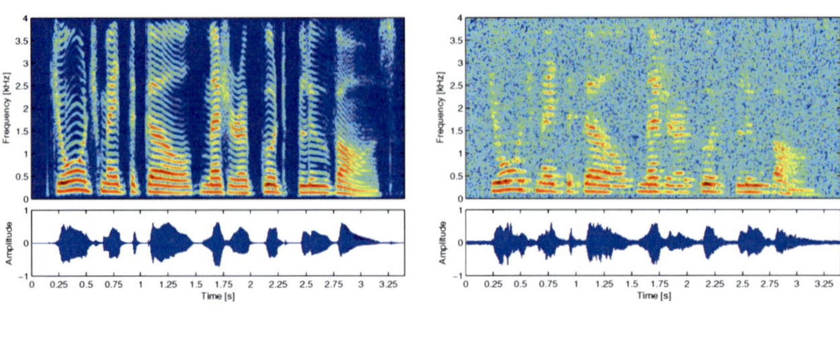

(a) The direct signal received by the first microphone.

(b) The microphone signal $y_1(k)$.

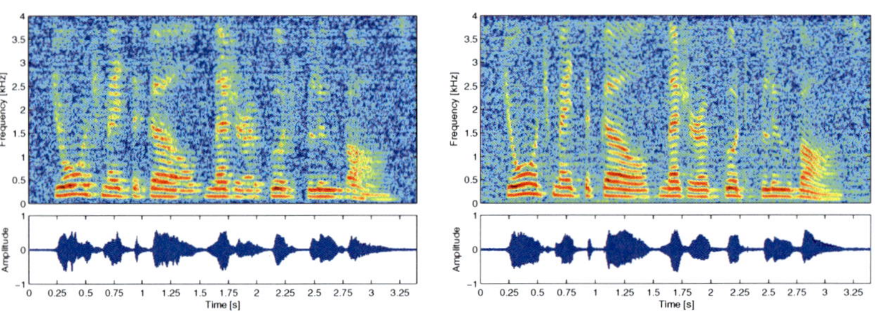

(c) The processed signal with $\alpha = 1$ (noise reduction only).

(d) The processed signal using $\alpha = 0$ (dereverberation and noise reduction).

Fig. 9.8 Spectrograms and waveforms of the unprocessed and processed signals ($\text{iSNR}_{\text{nc}} = 10$ dB, $T_{60} = 0.35$ s, $N = 4$, $D = 3$ m).

compared to the signals shown in Fig. 9.8(a) and Fig. 9.8(b). Specifically, the smearing in time is reduced and the harmonic structure of the speech are restored. In addition, we observe that there is a tradeoff between speech dereverberation and noise reduction as before. As expected, the processed signal in Fig. 9.8(c) contains more noise compared to the processed signal in Fig. 9.8(b).

9.7 Conclusions

In this chapter we studied the MVDR beamformer in room acoustics. The tradeoff between speech dereverberation and noise reduction was analyzed. The results of the theoretical performance analysis are supported by the performance evaluation. The results indicate that there is a tradeoff between the achievable noise reduction and speech dereverberation. The amount of noise reduction that is sacrificed when complete dereverberation is required depends on the direct-to-reverberation ratio of the acoustic impulse response between the source and the reference microphone. The performance evaluation supports the theoretical analysis and demonstrates the tradeoff between speech dereverberation and noise reduction. When desiring both speech dereverberation and noise reduction the results also demonstrate that the amount of noise reduction that is sacrificed decreases when the number of microphones increases.

Appendix

Proof (Property 9.1).

Before we proceed we define the magnitude squared coherence function (MSCF), which is the frequency-domain counterpart of the squared Pearson correlation coefficient (SPCC), which was used in [30] to analyze the noise reduction performance of the single-channel Wiener filter. Let $A(\omega)$ and $B(\omega)$ be the DTFTs of the two zero-mean real-valued random sequences a and b. Then the MSCF between $A(\omega)$ and $B(\omega)$ at frequency ω is defined as

$$\left| \rho\left[A(\omega), B(\omega) \right] \right|^2 = \frac{\left| E\left[A(\omega)B^*(\omega) \right] \right|^2}{E\left[|A(\omega)|^2 \right] E\left[|B(\omega)|^2 \right]} \tag{9.62}$$

$$= \frac{\left| \phi_{ab}(\omega) \right|^2}{\phi_{aa}(\omega)\phi_{bb}(\omega)}.$$

It is clear that the MSCF always takes its values between zero and one.

Let us first evaluate the MSCF $\left| \rho\left[X_1(\omega), Y_1(\omega) \right] \right|^2$ [using (9.2) and (9.35)] and $\left| \rho\left[\mathbf{h}_{\mathrm{MVDR}}^H(\omega)\mathbf{x}(\omega), \mathbf{h}_{\mathrm{MVDR}}^H(\omega)\mathbf{y}(\omega) \right] \right|^2$ [using (9.5) and (9.37)]:

$$\left| \rho\left[X_1(\omega), Y_1(\omega) \right] \right|^2 = \frac{|G_1(\omega)|^2\, \phi_{ss}(\omega)}{|G_1(\omega)|^2\, \phi_{ss}(\omega) + \phi_{v_1 v_1}(\omega)}$$

$$= \frac{\mathrm{iSNR}\left[Q(\omega) \right]}{\frac{|Q(\omega)|^2}{|G_1(\omega)|^2} + \mathrm{iSNR}\left[Q(\omega) \right]}, \tag{9.63}$$

$$\left|\rho\left[\mathbf{h}_{\mathrm{MVDR}}^{H}(\omega)\mathbf{x}(\omega),\mathbf{h}_{\mathrm{MVDR}}^{H}(\omega)\mathbf{y}(\omega)\right]\right|^{2}=\frac{\mathrm{oSNR}\left[\mathbf{h}_{\mathrm{MVDR}}(\omega)\right]}{1+\mathrm{oSNR}\left[\mathbf{h}_{\mathrm{MVDR}}(\omega)\right]}. \tag{9.64}$$

In addition, we evaluate the MSCF between $Y_{1}(\omega)$ and $\mathbf{h}_{\mathrm{MVDR}}^{H}(\omega)\mathbf{y}(\omega)$

$$\begin{aligned}
\left|\rho\left[Y_{1}(\omega),\mathbf{h}_{\mathrm{MVDR}}^{H}(\omega)\mathbf{y}(\omega)\right]\right|^{2}&=\frac{\left|\mathbf{u}^{T}\boldsymbol{\Phi}_{yy}(\omega)\mathbf{h}_{\mathrm{MVDR}}(\omega)\right|^{2}}{\phi_{y_{1}y_{1}}(\omega)\cdot\mathbf{h}_{\mathrm{MVDR}}^{H}(\omega)\boldsymbol{\Phi}_{yy}(\omega)\mathbf{h}_{\mathrm{MVDR}}(\omega)}\\
&=\frac{\phi_{x_{1}x_{1}}(\omega)}{\phi_{y_{1}y_{1}}(\omega)}\cdot\frac{C(\omega)\phi_{x_{1}x_{1}}(\omega)}{\mathbf{u}^{T}\boldsymbol{\Phi}_{xx}(\omega)\mathbf{h}_{\mathrm{MVDR}}(\omega)}\\
&=\frac{\left|\rho\left[X_{1}(\omega),Y_{1}(\omega)\right]\right|^{2}}{\left|\rho\left[X_{1}(\omega),\mathbf{h}_{\mathrm{MVDR}}^{H}(\omega)\mathbf{y}(\omega)\right]\right|^{2}}.
\end{aligned} \tag{9.65}$$

From (9.65) and the fact that $|\rho[A(\omega),B(\omega)]|^{2}\leq 1$, we have

$$\begin{aligned}
\left|\rho\left[X_{1}(\omega),Y_{1}(\omega)\right]\right|^{2}&=\left|\rho\left[Y_{1}(\omega),\mathbf{h}_{\mathrm{MVDR}}^{H}(\omega)\mathbf{y}(\omega)\right]\right|^{2}\times\\
&\quad\left|\rho\left[X_{1}(\omega),\mathbf{h}_{\mathrm{MVDR}}^{H}(\omega)\mathbf{y}(\omega)\right]\right|^{2}\\
&\leq\left|\rho\left[X_{1}(\omega),\mathbf{h}_{\mathrm{MVDR}}^{H}(\omega)\mathbf{y}(\omega)\right]\right|^{2}.
\end{aligned} \tag{9.66}$$

In addition, it can be shown that

$$\begin{aligned}
\left|\rho\left[X_{1}(\omega),\mathbf{h}_{\mathrm{MVDR}}^{H}(\omega)\mathbf{y}(\omega)\right]\right|^{2}&=\left|\rho\left[X_{1}(\omega),\mathbf{h}_{\mathrm{MVDR}}^{H}(\omega)\mathbf{x}(\omega)\right]\right|^{2}\times\\
&\quad\left|\rho\left[\mathbf{h}_{\mathrm{MVDR}}^{H}(\omega)\mathbf{x}(\omega),\mathbf{h}_{\mathrm{MVDR}}^{H}(\omega)\mathbf{y}(\omega)\right]\right|^{2}\\
&\leq\left|\rho\left[\mathbf{h}_{\mathrm{MVDR}}^{H}(\omega)\mathbf{x}(\omega),\mathbf{h}_{\mathrm{MVDR}}^{H}(\omega)\mathbf{y}(\omega)\right]\right|^{2}.
\end{aligned} \tag{9.67}$$

From (9.66) and (9.67), we know that

$$\left|\rho\left[X_{1}(\omega),Y_{1}(\omega)\right]\right|^{2}\leq\left|\rho\left[\mathbf{h}_{\mathrm{MVDR}}^{H}(\omega)\mathbf{x}(\omega),\mathbf{h}_{\mathrm{MVDR}}^{H}(\omega)\mathbf{y}(\omega)\right]\right|^{2}. \tag{9.68}$$

Hence, by substituting (9.63) and (9.64) in (9.68), we obtain

$$\frac{\mathrm{iSNR}\left[Q(\omega)\right]}{\frac{|Q(\omega)|^{2}}{|G_{1}(\omega)|^{2}}+\mathrm{iSNR}\left[Q(\omega)\right]}\leq\frac{\mathrm{oSNR}\left[\mathbf{h}_{\mathrm{MVDR}}(\omega)\right]}{1+\mathrm{oSNR}\left[\mathbf{h}_{\mathrm{MVDR}}(\omega)\right]}. \tag{9.69}$$

As a result

$$|Q(\omega)|^{2}\cdot\mathrm{oSNR}\left[\mathbf{h}_{\mathrm{MVDR}}(\omega)\right]\geq|G_{1}(\omega)|^{2}\cdot\mathrm{iSNR}\left[Q(\omega)\right],\ \forall\omega, \tag{9.70}$$

which is equal to (9.46).

$$\square$$

References

1. J. Benesty, J. Chen, and Y. Huang, *Microphone Array Signal Processing.* Berlin, Germany: Springer-Verlag, 2008.
2. S. Gannot and I. Cohen, "Adaptive beamforming andpostfiltering," in *Springer Handbook of Speech Processing*, J. Benesty, M. M. Sondhi, and Y. Huang, Eds. Springer-Verlag, 2007, book chapter 48.
3. J. Chen, J. Benesty, Y. Huang, and S. Doclo, "New insights into the noise reduction Wiener filter," *IEEE Trans. Audio, Speech, Lang. Process.*, vol. 14, no. 4, pp. 1218–1234, July 2006.
4. S. Doclo and M. Moonen, "GSVD-based optimal filtering for single and multimicrophone speech enhancement," *IEEE Trans. Signal Process.*, vol. 50, no. 9, pp. 2230–2244, Sep. 2002.
5. A. Spriet, M. Moonen, and J. Wouters, "Spatially pre-processed speech distortion weighted multi-channel Wiener filtering for noise reduction'," *Signal Processing*, vol. 84, no. 12, pp. 2367–2387, Dec. 2004.
6. J. Capon, "High resolution frequency-wavenumber spectrum analysis," *Proc. IEEE*, vol. 57, pp. 1408–1418, Aug. 1969.
7. S. Darlington, "Linear least-squares smoothing and prediction with applications," *Bell Syst. Tech. J.*, vol. 37, pp. 1121–94, 1952.
8. M. Er and A. Cantoni, "Derivative constraints for broad-band element space antenna array processors," *IEEE Trans. Acoust., Speech, Signal Process.*, vol. 31, no. 6, pp. 1378–1393, Dec. 1983.
9. O. Frost, "An algorithm for linearly constrained adaptive array processing," *Proceedings of the IEEE*, vol. 60, no. 8, pp. 926–935, Jan. 1972.
10. Y. Kaneda and J. Ohga, "Adaptive microphone-array system for noise reduction," *IEEE Trans. Acoust., Speech, Signal Process.*, vol. 34, no. 6, pp. 1391–1400, Dec 1986.
11. L. J. Griffiths and C. W. Jim, "An alternative approach to linearly constrained adaptive beamforming," *IEEE Trans. Antennas Propag.*, vol. 30, no. 1, pp. 27–34, Jan. 1982.
12. B. R. Breed and J. Strauss, "A short proof of the equivalence of LCMV and GSC beamforming," *IEEE Signal Process. Lett.*, vol. 9, no. 6, pp. 168–169, June 2002.
13. S. Affes and Y. Grenier, "A source subspace tracking array of microphones for double talk situations," in *IEEE Int. Conf. Acoust. Speech and Sig. Proc. (ICASSP)*, Munich, Germany, Apr. 1997, pp. 269–272.
14. S. Gannot, D. Burshtein, and E. Weinstein, "Signal enhancement using beamforming and nonstationarity with applications to speech," *IEEE Trans. Signal Process.*, vol. 49, no. 8, pp. 1614–1626, Aug. 2001.
15. J. Benesty, J. Chen, Y. Huang, and J. Dmochowski, "On microphone array beamforming from a MIMO acoustic signal processing perspective," *IEEE Trans. Audio, Speech, Lang. Process.*, vol. 15, no. 3, pp. 1053–1065, Mar. 2007.
16. Y. Huang, J. Benesty, and J. Chen, "Adaptive blind multichannel identification," in *Springer Handbook of Speech Processing*, J. Benesty, M. M. Sondhi, and Y. Huang, Eds. Springer-Verlag, 2007, book chapter 13.
17. S. Gannot and M. Moonen, "Subspace methods for multimicrophone speech dereverberation," *EURASIP Journal on Applied Signal Processing*, vol. 2003, no. 11, pp. 1074–1090, Oct. 2003.
18. E. Habets, J. Benesty, I. Cohen, S. Gannot, and J. Dmochowski, "New insights into the MVDR beamformer in room acoustics," *IEEE Trans. Audio, Speech, Language Process.*, 2010.
19. M. Brandstein and D. B. Ward, Eds., *Microphone Arrays: Signal Processing Techniques and Applications.* Berlin, Germany: Springer-Verlag, 2001.
20. A. Oppenheim, A. Willsky, and H. Nawab, *Signals and Systems.* Upper Saddle River, NJ: Prentice Hall, 1996.

21. S. Doclo, A. Spriet, J. Wouters, and M. Moonen, "Frequency-domain criterion for the speech distortion weighted multichannel Wiener filter for robust noise reduction," *Speech Communication*, vol. 49, no. 7–8, pp. 636–656, Jul.–Aug. 2007.
22. A. Antoniou and W.-S. Lu, *Practical Optimization: Algorithms and Engineering Applications*. New York, USA: Springer, 2007.
23. H. Cox, R. Zeskind, and M. Owen, "Robust adaptive beamforming," *IEEE Trans. Acoust., Speech, Signal Process.*, vol. 35, no. 10, pp. 1365–1376, Oct. 1987.
24. J. B. Allen and D. A. Berkley, "Image method for efficiently simulating small room acoustics," *Journal of the Acoustical Society of America*, vol. 65, no. 4, pp. 943–950, 1979.
25. E. E. Jan and J. Flanagan, "Sound capture from spatial volumes: Matched-filter processing of microphone arrays having randomly-distributed sensors," in *IEEE Int. Conf. Acoust. Speech and Sig. Proc. (ICASSP)*, Atlanta, Georgia, USA, May 1996, pp. 917–920.
26. G. Reuven, S. Gannot, and I. Cohen, "Performance analysis of the dual source transfer-function generalized sidelobe canceller," *Speech Communication*, vol. 49, pp. 602–622, Jul.–Aug. 2007.
27. P. M. Peterson, "Simulating the response of multiple microphones to a single acoustic source in a reverberant room," *Journal of the Acoustical Society of America*, vol. 80, no. 5, pp. 1527–1529, Nov. 1986.
28. G. Lindsey, A. Breen, and S. Nevard, "SPAR's archivable actual-word databases," University College London, Tech. Rep., Jun. 1987.
29. A. Oppenheim and R. W. Schafer, *Discrete-Time Signal Processing*, 2nd ed. Upper Saddle River, NJ: Prentice-Hall, Inc., 1999.
30. J. Benesty, J. Chen, and Y. Huang, "On the importance of the Pearson correlation coefficient in noise reduction," *IEEE Trans. Audio, Speech, Lang. Process.*, vol. 16, pp. 757–765, May 2008.

Chapter 10
Extraction of Desired Speech Signals in Multiple-Speaker Reverberant Noisy Environments

Shmulik Markovich, Sharon Gannot, and Israel Cohen

Abstract In many practical environments we wish to extract several desired speech signals, which are contaminated by non-stationary and stationary interfering signals. The desired signals may also be subject to distortion imposed by the acoustic room impulse response (RIR). In this chapter, a linearly constrained minimum variance (LCMV) beamformer is designed for extracting the desired signals from multi-microphone measurements. The beamformer satisfies two sets of linear constraints. One set is dedicated to maintaining the desired signals, while the other set is chosen to mitigate both the stationary and non-stationary interferences. Unlike classical beamformers, which approximate the RIRs as delay-only filters, we take into account the entire RIR [or its respective acoustic transfer function (ATF)]. We show that the relative transfer functions (RTFs), which relate the speech sources and the microphones, and a basis for the interference subspace suffice for constructing the beamformer. Additionally, in the case of one desired speech signal, we compare the proposed LCMV beamformer and the minimum variance distortionless response (MVDR) beamformer. These algorithms differ in their treatment of the interference sources. A comprehensive experimental study in both simulated and real environments demonstrates the performance of the proposed beamformer. Particularly, it is shown that the LCMV beamformer outperforms the MVDR beamformer provided that the acoustic environment is time-invariant.

Shmulik Markovich and Sharon Gannot
Bar-Ilan University, Israel, e-mail: `shmulik.markovich@gmail.com,gannot@eng.biu.ac.il`

Israel Cohen
Technion–Israel Institute of Technology, Israel, e-mail: `icohen@ee.technion.ac.il`

I. Cohen et al. (Eds.): Speech Processing in Modern Communication, STSP 3, pp. 255–279.

10.1 Introduction

Speech enhancement techniques, utilizing microphone arrays, have attracted the attention of many researchers for the last thirty years, especially in hands-free communication tasks. Usually, the received speech signals are contaminated by interfering sources, such as competing speakers and noise sources, and also distorted by the reverberating environment. Whereas single microphone algorithms might show satisfactory results in noise reduction, they are rendered useless in competing speaker mitigation task, as they lack the spatial information, or the statistical diversity used by multi-microphone algorithms. Here we address the problem of extracting several desired sources in a reverberant environment containing both non-stationary (competing speakers) and stationary interferences.

Two families of microphone array algorithms can be defined, namely, the blind source separation (BSS) family and the beamforming family. BSS aims at separating all the involved sources, by exploiting their statistical independence, regardless of their attribution to the desired or interfering sources [1]. On the other hand, the beamforming family of algorithms, concentrate on enhancing the sum of the desired sources while treating all other signals as interfering sources.

We will focus on the beamformers family of algorithms. The term beamforming refers to the design of a spatio-temporal filter. Broadband arrays comprise a set of filters, applied to each received microphone signal, followed by a summation operation. The main objective of the beamformer is to extract a desired signal, impinging on the array from a specific position, out of noisy measurements thereof. The simplest structure is the *delay-and-sum* beamformer, which first compensates for the relative delay between distinct microphone signals and then sums the steered signal to form a single output. This beamformer, which is still widely used, can be very effective in mitigating noncoherent, i.e., spatially white, noise sources, provided that the number of microphones is relatively high. However, if the noise source is coherent, the noise reduction (NR) is strongly dependent on the direction of arrival of the noise signal. Consequently, the performance of the delay-and-sum beamformer in reverberant environments is often insufficient. Jan and Flanagan [2] extended the delay-and-sum concept by introducing the so called *filter-and-sum* beamformer. This structure, designed for multipath environments, namely reverberant enclosures, replaces the simpler delay compensator with a matched filter. The array beam-pattern can generally be designed to have a specified response. This can be done by properly setting the values of the multichannel filters weights. Statistically optimal beamformers are designed based on the statistical properties of the desired and interference signals. In general, they aim at enhancing the desired signals, while rejecting the interfering signals. Several criteria can be applied in the design of the beamformer, e.g., maximum signal-to-noise-ratio (MSNR), minimum mean-squared error (MMSE), MVDR, and LCMV. A summary of several design criteria can be found in [3, 4]. Cox et al. [5] introduced an improved adaptive beamformer that maintains a set of linear constraints as well as a quadratic inequality constraint.

In [6] a multichannel Wiener filter (MWF) technique has been proposed that produces an MMSE estimate of the desired speech component in one of the microphone signals, hence simultaneously performing noise reduction and limiting speech distortion. In addition, the MWF is able to take speech distortion into account in its optimization criterion, resulting in the speech distortion weighted (SDW)-MWF [7]. In an MVDR beamformer [8, 9], the power of the output signal is minimized under the constraint that signals arriving from the assumed direction of the desired speech source are processed without distortion. A widely studied adaptive implementation of this beamformer is the generalized sidelobe canceler (GSC) [10]. Several researchers (e.g. Er and Cantoni [11]) have proposed modifications to the MVDR for dealing with multiple linear constraints, denoted LCMV. Their work was motivated by the desire to apply further control to the array/beamformer beam-pattern, beyond that of a steer-direction gain constraint. Hence, the LCMV can be applied for constructing a beam-pattern satisfying certain constraints for a set of directions, while minimizing the array response in all other directions. Breed and Strauss [12] proved that the LCMV extension has also an equivalent GSC structure, which decouples the constraining and the minimization operations. The GSC structure was reformulated in the frequency domain, and extended to deal with the more complicated general ATFs case by Affes and Grenier [13] and later by Gannot et al. [14]. The latter frequency-domain version, which takes into account the reverberant nature of the enclosure, was nicknamed the transfer function GSC (TF-GSC).

Several beamforming algorithms based on subspace methods were developed. Gazor et al. [15] propose to use a beamformer based on the MVDR criterion and implemented as a GSC to enhance a narrowband signal contaminated by additive noise and received by multiple sensors. Under the assumption that the direction-of-arrival (DOA) entirely determines the transfer function relating the source and the microphones, it is shown that determining the signal subspace suffices for the construction of the algorithm. An efficient DOA tracking system, based on the projection approximation subspace tracking deflation (PASTd) algorithm [16] is derived. An extension to the wide-band case is presented by the same authors [17]. However the demand for a delay-only impulse response is still not relaxed. Affes and Grenier [13] apply the PASTd algorithm to enhance speech signal contaminated by spatially white noise, where arbitrary ATFs relate the speaker and the microphone array. The algorithm proves to be efficient in a simplified trading-room scenario, where the direct to reverberant ratio (DRR) is relatively high and the reverberation time relatively low. Doclo and Moonen [18] extend the structure to deal with the more complicated colored noise case by using the generalized singular value decomposition (GSVD) of the received data matrix. Warsitz et al. [19] propose to replace the blocking matrix (BM) in [14]. They use a new BM based on the generalized eigenvalue decomposition (GEVD) of the received microphone data, providing an indirect estimation of the ATFs relating the desired speaker and the microphones.

Affes et al. [20] extend the structure presented in [15] to deal with the multi-source case. The constructed multi-source GSC, which enables multiple target tracking, is based on the PASTd algorithm and on constraining the estimated steering vector to the array manifold. Asano et al. [21] address the problem of enhancing multiple speech sources in a non-reverberant environment. The multiple signal classification (MUSIC) method, proposed by Schmidt [22], is utilized to estimate the number of sources and their respective steering vectors. The noise components are reduced by manipulating the generalized eigenvalues of the data matrix. Based on the subspace estimator, an LCMV beamformer is constructed. The LCMV constraints set consists of two subsets: one for maintaining the desired sources and the second for mitigating the interference sources. Benesty et al. [23] also address beamforming structures for multiple input signals. In their contribution, derived in the time-domain, the microphone array is treated as a multiple input multiple output (MIMO) system. In their experimental study, it is assumed that the filters relating the sources and the microphones are a priori known, or alternatively, that the sources are not active simultaneously. Reuven et al. [24] deal with the scenario in which one desired source and one competing speech source coexist in noisy and reverberant environment. The resulting algorithm, denoted dual source TF-GSC (DTF-GSC) is tailored to the specific problem of two sources and cannot be easily generalized to the multiple desired and interference sources.

In this chapter, we present a novel beamforming technique, aiming at the extraction of multiple desired speech sources, while attenuating several interfering sources by using an LCMV beamformer (both stationary and non-stationary) in a reverberant environment. We derive a practical method for estimating all components of the eigenspace-based beamformer. We first show that the desired signals' RTFs (defined as the ratio between ATFs which relate the speech sources and the microphones) and a basis of the interference subspace suffice for the construction of the beamformer. The RTFs of the desired signals are estimated by applying the GEVD procedure to the received signals' power spectral density (PSD) matrix and the stationary noise PSD matrix. A basis spanning the interference subspace is estimated by collecting eigenvectors, calculated in segments in which the non-stationary signals are active and the desired signals are inactive. A novel method, based on the orthogonal triangular decomposition (QRD), of reducing the rank of interference subspace is derived. This procedure relaxes the common requirement for non-overlapping activity periods of the interference signals.

The structure of the chapter is as follows. In Section 10.2 the problem of extracting multiple desired sources contaminated by multiple interference in a reverberant environment is introduced. In Section 10.3 the multiple constrained LCMV beamformer is presented. In Section 10.4 we describe a novel method for estimating the interferences' subspace as well as a GEVD based method for estimating the RTFs of the desired sources. The entire algorithm is summarized in Section 10.5. In Section 10.6 we present typical test scenarios, discuss some implementation considerations of the algorithm, and show experimental results for both a simulated room and a real conference room scenarios. We address both the problem of extracting multiple desired sources as well as single desired source. In the later case, we compare the performance of the novel beamformer with the TF-GSC. We draw some conclusions and summarize our work in Section 10.7.

10.2 Problem Formulation

Consider the general problem of extracting K desired sources, contaminated by N_s stationary interfering sources and N_{ns} non-stationary sources. The signals are received by M sensors arranged in an arbitrary array. Each of the involved signals undergo filtering by the RIR before being picked up by the microphones. The reverberation effect can be modeled by a finite impulse response (FIR) filter operating on the sources. The signal received by the mth sensor is given by

$$z_m(n) = \sum_{i=1}^{K} s_i^d(n) * h_{im}^d(n) + \sum_{i=1}^{N_s} s_i^s(n) * h_{im}^s(n) + \sum_{i=1}^{N_{ns}} s_i^{ns}(n) * h_{im}^{ns}(n) + v_m(n),$$

(10.1)

where $s_1^d(n), \ldots, s_K^d(n)$, $s_1^s(n), \ldots, s_{N_s}^s(n)$ and $s_1^{ns}(n), \ldots, s_{N_{ns}}^{ns}(n)$ are the desired sources, the stationary and non-stationary interfering sources in the room, respectively. We define $h_{im}^d(n)$, $h_{im}^s(n)$ and $h_{im}^{ns}(n)$ to be the linear time-invariant (LTI) RIRs relating the desired sources, the interfering sources, and each sensor m, respectively. $v_m(n)$ is the sensor noise. $z_m(n)$ is transformed into the short-time Fourier transform (STFT) domain with a rectangular window of length N_{DFT}, yielding:

$$z_m(\ell, k) = \sum_{i=1}^{K} s_i^d(\ell, k) h_{im}^d(\ell, k) +$$

(10.2)

$$\sum_{i=1}^{N_s} s_i^s(\ell, k) h_{im}^s(\ell, k) + \sum_{i=1}^{N_{ns}} s_i^{ns}(\ell, k) h_{im}^{ns}(\ell, k) + v_m(\ell, k),$$

where ℓ is the frame number and k is the frequency index. The assumption that the window length is much larger then the RIR length ensures the multiplicative transfer function (MTF) approximation [25] validness.

The received signals in (10.2) can be formulated in a vector notation:

$$\mathbf{z}(\ell, k) = H^d(\ell, k)\mathbf{s}^d(\ell, k) + H^s(\ell, k)\mathbf{s}^s(\ell, k) + H^{ns}(\ell, k)\mathbf{s}^{ns}(\ell, k) + \mathbf{v}(\ell, k)$$
$$= H(\ell, k)\mathbf{s}(\ell, k) + \mathbf{v}(\ell, k), \tag{10.3}$$

where

$$\mathbf{z}(\ell, k) \triangleq \left[z_1(\ell, k) \ldots z_M(\ell, k) \right]^T$$
$$\mathbf{v}(\ell, k) \triangleq \left[v_1(\ell, k) \ldots v_M(\ell, k) \right]^T$$
$$\mathbf{h}_i^d(\ell, k) \triangleq \left[h_{i1}^d(\ell, k) \ldots h_{iM}^d(\ell, k) \right]^T \qquad i = 1, \ldots, K$$
$$\mathbf{h}_i^s(\ell, k) \triangleq \left[h_{i1}^s(\ell, k) \ldots h_{iM}^s(\ell, k) \right]^T \qquad i = 1, \ldots, N_s$$
$$\mathbf{h}_i^{ns}(\ell, k) \triangleq \left[h_{i1}^{ns}(\ell, k) \ldots h_{iM}^{ns}(\ell, k) \right]^T \qquad i = 1, \ldots, N_{ns}$$

$$H^d(\ell, k) \triangleq \left[\mathbf{h}_1^d(\ell, k) \ldots \mathbf{h}_K^d(\ell, k) \right]$$
$$H^s(\ell, k) \triangleq \left[\mathbf{h}_1^s(\ell, k) \ldots \mathbf{h}_{N_s}^s(\ell, k) \right]$$
$$H^{ns}(\ell, k) \triangleq \left[\mathbf{h}_1^{ns}(\ell, k) \ldots \mathbf{h}_{N_{ns}}^{ns}(\ell, k) \right]$$
$$H^i(\ell, k) \triangleq \left[H^s(\ell, k) \; H^{ns}(\ell, k) \right]$$
$$H(\ell, k) \triangleq \left[H^d(\ell, k) \; H^s(\ell, k) \; H^{ns}(\ell, k) \right]$$

$$\mathbf{s}^d(\ell, k) \triangleq \left[s_1^d(\ell, k) \ldots s_K^d(\ell, k) \right]^T$$
$$\mathbf{s}^s(\ell, k) \triangleq \left[s_1^s(\ell, k) \ldots s_{N_s}^s(\ell, k) \right]^T$$
$$\mathbf{s}^{ns}(\ell, k) \triangleq \left[s_1^{ns}(\ell, k) \ldots s_{N_{ns}}^{ns}(\ell, k) \right]^T$$
$$\mathbf{s}(\ell, k) \triangleq \left[(\mathbf{s}^d(\ell, k))^T \; (\mathbf{s}^s(\ell, k))^T \; (\mathbf{s}^{ns}(\ell, k))^T \right]^T .$$

Assuming the desired speech signals, the interference and the noise signals to be uncorrelated, the received signals' correlation matrix is given by

$$\Phi_{zz}(\ell, k) = H^d(\ell, k)\Lambda^d(\ell, k)\big(H^d(\ell, k)\big)^\dagger + \tag{10.4}$$
$$H^s(\ell, k)\Lambda^s(\ell, k)\big(H^s(\ell, k)\big)^\dagger + H^{ns}(\ell, k)\Lambda^{ns}(\ell, k)\big(H^{ns}(\ell, k)\big)^\dagger + \Phi_{vv}(\ell, k),$$

where

$$\Lambda^d(\ell, k) \triangleq \mathrm{diag}\left(\left[(\sigma_1^d(\ell, k))^2 \ldots (\sigma_K^d(\ell, k))^2 \right] \right),$$
$$\Lambda^s(\ell, k) \triangleq \mathrm{diag}\left(\left[(\sigma_1^s(\ell, k))^2 \ldots (\sigma_{N_s}^s(\ell, k))^2 \right] \right),$$
$$\Lambda^{ns}(\ell, k) \triangleq \mathrm{diag}\left(\left[(\sigma_1^{ns}(\ell, k))^2 \ldots (\sigma_{N_{ns}}^{ns}(\ell, k))^2 \right] \right).$$

$(\bullet)^\dagger$ is the conjugate-transpose operation, and $\mathrm{diag}\,(\bullet)$ is a square matrix with the vector in brackets on its main diagonal. $\Phi_{vv}(\ell, k)$ is the sensor noise correlation matrix assumed to be spatially-white, i.e. $\Phi_{vv}(\ell, k) = \sigma_v^2 I_{M \times M}$ where $I_{M \times M}$ is the identity matrix.

In the special case of a single desired source, i.e. $K = 1$, the following definition applies: $H(\ell, k) \triangleq \left[\, \mathbf{h}_1^d(\ell, k)\ H^s(\ell, k)\ H^{ns}(\ell, k) \,\right]$ and $\mathbf{s}(\ell, k) \triangleq \left[\, s_1^d(\ell, k)\ (\mathbf{s}^s(\ell, k))^T\ (\mathbf{s}^{ns}(\ell, k))^T \,\right]^T$.

10.3 Proposed Method

In this section the proposed algorithm is derived. In the following subsections we adopt the LCMV structure and define a set of constraints used for extracting the desired sources and mitigating the interference sources. Then we replace the constraints set by an equivalent set which can be more easily estimated. Finally, we relax our constraint for extracting the exact input signals, as transmitted by the sources, and replace it by the extraction of the desired speech components at an arbitrarily chosen microphone. The outcome of the latter, a modified constraints set, will constitute a feasible system. In the case of single desired source and multiple interference signals, the MVDR strategy can be adopted instead of the derived LCMV strategy. Hence, in this case, both beamformers are presented.

10.3.1 The LCMV and MVDR Beamformers

A beamformer is a system realized by processing each of the sensor signals $z_m(k, \ell)$ by the filters $w_m^*(\ell, k)$ and summing the outputs. The beamformer output $y(\ell, k)$ is given by

$$y(\ell, k) = \mathbf{w}^\dagger(\ell, k)\mathbf{z}(\ell, k), \tag{10.5}$$

where

$$\mathbf{w}(\ell, k) = \left[\, w_1(\ell, k), \ldots, w_M(\ell, k) \,\right]^T. \tag{10.6}$$

The filters are set to satisfy the LCMV criterion with multiple constraints:

$$\mathbf{w}(\ell, k) = \underset{\mathbf{w}}{\mathrm{argmin}}\{\mathbf{w}^\dagger(\ell, k)\Phi_{zz}(\ell, k)\mathbf{w}(\ell, k)\}$$

$$\text{subject to } C^\dagger(\ell, k)\mathbf{w}(\ell, k) = \mathbf{g}(\ell, k), \tag{10.7}$$

where

$$C^\dagger(\ell, k)\mathbf{w}(\ell, k) = \mathbf{g}(\ell, k) \tag{10.8}$$

is the constraints set. The well-known solution to (10.7) is given by [3]:

$$\mathbf{w}(\ell,k) = \Phi_{zz}^{-1}(\ell,k)C(\ell,k)\left(C^\dagger(\ell,k)\Phi_{zz}^{-1}(\ell,k)C(\ell,k)\right)^{-1}\mathbf{g}(\ell,k). \qquad (10.9)$$

Projecting (10.9) to the column space of the constraints matrix yields a beamformer which satisfies the constraint set but not necessarily minimizes the noise variance at the output. This beamformer is given by [3]

$$\mathbf{w}_0(\ell,k) = C(\ell,k)\left(C^\dagger(\ell,k)C(\ell,k)\right)^{-1}\mathbf{g}(\ell,k). \qquad (10.10)$$

It is shown in [26] that in the case of spatially-white sensor noise, i.e. $\Phi_{vv}(\ell,k) = \sigma_v^2 I_{M \times M}$, and when the constraint set is accurately known, both beamformers defined by (10.9) and (10.10) are equivalent.

Two paradigms can be adopted in the design of a beamformer which is aimed at enhancing a single desired signal contaminated by both noise and interference. These paradigms differ in their treatment of the interference (competing speech and/or directional noise), which is manifested by the definition of the constraints set, namely $C(\ell,k)$ and $\mathbf{g}(\ell,k)$.

The straightforward alternative is to apply a single constraint beamformer, usually referred to as MVDR beamformer, which was efficiently implemented by the TF-GSC [14], for the reverberant case. Another alternative suggests defining constraints for both the desired and the interference sources. Two recent contributions [24] and [26] adopt this alternative. It is shown in [27] that in static scenarios, well-designed nulls towards all interfering signals (as proposed by the LCMV structure) result in an improved undesired signal cancelation compared with the MVDR structure [14]. Naturally, while considering time-varying environments this advantage cannot be guaranteed.

10.3.2 The Constraints Set

We start with the straightforward approach, in which the beam-pattern is constrained to cancel out all interfering sources while maintaining all desired sources (for each frequency bin). Note, that unlike the DTF-GSC approach [24], the stationary noise sources are treated similarly to the interference (non-stationary) sources. We therefore define the following constraints. For each desired source $\{s_i^d\}_{i=1}^{K}$ we apply the constraint

$$\left(\mathbf{h}_i^d(\ell,k)\right)^\dagger \mathbf{w}(\ell,k) = 1, \ i = 1,\dots,K. \qquad (10.11)$$

For each interfering source, both stationary and non-stationary, $\{s_i^s\}_{i=1}^{N_s}$ and $\{s_j^{ns}\}_{j=1}^{N_{ns}}$, we apply

$$\left(\mathbf{h}_i^s(\ell,k)\right)^\dagger \mathbf{w}(\ell,k) = 0, \qquad (10.12)$$

and

$$\left(\mathbf{h}_j^{ns}(\ell,k)\right)^\dagger \mathbf{w}(\ell,k) = 0. \qquad (10.13)$$

Define $N \triangleq K + N_s + N_{ns}$ the total number of signals in the environment (including the desired sources, stationary interference signals, and the non-stationary interference signals). Assuming the column-space of $H(\ell, k)$ is linearly independent (i.e. the ATFs are independent), it is obvious that for the solution in (10.10) to exist we require that the number of microphones will be greater or equal the number of constraints, namely $M \geq N$. It is also understood that whenever the constraints contradict each other, the desired signal constraints will be preferred.

Summarizing, we have a constraint matrix

$$C(\ell, k) \triangleq H(\ell, k), \tag{10.14}$$

and a desired response vector

$$\mathbf{g} \triangleq \left[\underbrace{1 \ \ldots \ 1}_{K} \ \underbrace{0 \ \ldots \ 0}_{N-K} \right]^T. \tag{10.15}$$

Evaluating the beamformer (10.10) output for the input (10.3) and constraints set (10.8) gives:

$$y(\ell, k) = \mathbf{w}_0^{\dagger}(\ell, k)\mathbf{z}(\ell, k) =$$
$$\sum_{i=1}^{K} s_i^d(\ell, k) + \mathbf{g}^{\dagger}\left(H^{\dagger}(\ell, k)H(\ell, k)\right)^{-1}H^{\dagger}(\ell, k)\mathbf{v}(\ell, k). \tag{10.16}$$

The output comprises a sum of two terms: the first is the sum of all the desired sources and the second is the response of the array to the sensor noise.

For the single desired sources scenario we get:

$$y(\ell, k) = s_1^d(\ell, k) + \mathbf{g}^{\dagger}\left(H^{\dagger}(\ell, k)H(\ell, k)\right)^{-1}H^{\dagger}(\ell, k)\mathbf{v}(\ell, k). \tag{10.17}$$

10.3.3 Equivalent Constraints Set

The matrix $C(\ell, k)$ in (10.14) comprises the ATFs relating the sources and the microphones $\mathbf{h}_i^d(\ell, k)$, $\mathbf{h}_i^s(\ell, k)$ and $\mathbf{h}_i^{ns}(\ell, k)$. Hence, the solution given in (10.10) requires an estimate of the various filters. Obtaining such estimates might be a cumbersome task in practical scenarios, where it is usually required that the sources are not active simultaneously (see e.g. [23]). We will show now that the actual ATFs of the interfering sources can be replaced by the basis vectors spanning the same interference subspace, without sacrificing the accuracy of the solution.

Let

$$N_i \triangleq N_s + N_{ns} \tag{10.18}$$

be the number of interferences, both stationary and non-stationary, in the environment. For conciseness we assume that the ATFs of the interfering sources are linearly independent at each frequency bin, and define $E \triangleq [\mathbf{e}_1 \ \ldots \ \mathbf{e}_{N_i}]$ to be any basis[1] that spans the column space of the interfering sources $H^i(\ell, k) = [H^s(\ell, k) \ H^{ns}(\ell, k)]$. Hence, the following identity holds:

$$H^i(\ell, k) = E(\ell, k)\Theta(\ell, k), \tag{10.19}$$

where $\Theta_{N_i \times N_i}(\ell, k)$ is comprised of the projection coefficients of the original ATFs on the basis vectors. When the ATFs associated with the interference signals are linearly independent, $\Theta_{N_i \times N_i}(\ell, k)$ is an invertible matrix.

Define

$$\tilde{\Theta}(\ell, k) \triangleq \begin{bmatrix} I_{K \times K} & \mathcal{O}_{K \times N_i} \\ \mathcal{O}_{N_i \times K} & \Theta(\ell, k) \end{bmatrix}_{N \times N}, \tag{10.20}$$

where $I_{K \times K}$ is a $K \times K$ identity matrix. Multiplication by $(\tilde{\Theta}^\dagger(\ell, k))^{-1}$ of both sides of the original constraints set in (10.8), with the definitions (10.14)–(10.15) and using the equality $\tilde{\Theta}^\dagger(\ell, k))^{-1}\mathbf{g} = \mathbf{g}$, yields an equivalent constraint set:

$$\dot{C}^\dagger(\ell, k)\mathbf{w}(\ell, k) = \mathbf{g}, \tag{10.21}$$

where the equivalent constraint matrix is

$$\dot{C}(\ell, k) = (\tilde{\Theta}^\dagger(\ell, k))^{-1}C^\dagger(\ell, k) = \begin{bmatrix} H^d(\ell, k) \ E(\ell, k) \end{bmatrix}. \tag{10.22}$$

10.3.4 Modified Constraints Set

Both the original and equivalent constraints sets in (10.14) and (10.22) respectively, require estimates of the desired sources ATFs $H^d(\ell, k)$. Estimating these ATFs might be a cumbersome task, due to the large order of the respective RIRs. In the current section we relax our demand for a distortionless beamformer [as depicted in the definition of $\mathbf{g}$ in (10.15)] and replace it by constraining the output signal to be comprised of the desired speech components at an arbitrarily chosen microphone.

Define a modified vector of desired responses:

$$\tilde{\mathbf{g}}(\ell, k) = \Big[\underbrace{(h_{11}^d(\ell, k))^* \ \ldots \ (h_{K1}^d(\ell, k))^*}_{K} \ \underbrace{0 \ \ldots \ 0}_{N-K} \Big]^T,$$

where microphone #1 was arbitrarily chosen as the reference microphone. The modified beamformer satisfying the modified response $\dot{C}^\dagger(\ell, k)\tilde{\mathbf{w}}(\ell, k) =$

---

[1] If this linear independency assumption does not hold, the rank of the basis can be smaller than N_i in several frequency bins. In this contribution we assume the interference subspace to be full rank.

$\tilde{\mathbf{g}}(\ell, k)$ is then given by

$$\tilde{\mathbf{w}}_0(\ell, k) \triangleq \dot{C}(\ell, k)\big(\dot{C}^\dagger(\ell, k)\dot{C}(\ell, k)\big)^{-1}\tilde{\mathbf{g}}(\ell, k). \tag{10.23}$$

Indeed, using the equivalence between the column subspaces of $\dot{C}(\ell, k)$ and $H(\ell, k)$, the beamformer output is now given by

$$y(\ell, k) = \tilde{\mathbf{w}}_0^\dagger(\ell, k)\mathbf{z}(\ell, k) =$$
$$\sum_{i=1}^{K} h_{i1}^d(\ell, k)s_i^d(\ell, k) + \tilde{\mathbf{g}}^\dagger(\ell, k)\big(\dot{C}^\dagger(\ell, k)\dot{C}(\ell, k)\big)^{-1}\dot{C}^\dagger(\ell, k)\mathbf{v}(\ell, k),$$
$$\tag{10.24}$$

as expected from the modified constraint response.

For the single desired sources scenario the modified constraints set yields the following output:

$$y(\ell, k) = h_{i1}^d(\ell, k)s_1^d(\ell, k) + \tilde{\mathbf{g}}^\dagger(\ell, k)\big(\dot{C}^\dagger(\ell, k)\dot{C}(\ell, k)\big)^{-1}\dot{C}^\dagger(\ell, k)\mathbf{v}(\ell, k). \tag{10.25}$$

As mentioned before, estimating the desired signal ATFs is a cumbersome task. Nevertheless, in Section 10.4 we will show that a practical method for estimating the RTF can be derived. We will therefore reformulate in the sequel the constraints set in terms of the RTFs.

It is easily verified that the modified desired response is related to the original desired response (10.15) by

$$\tilde{\mathbf{g}}(\ell, k) = \tilde{\Psi}^\dagger(\ell, k)\mathbf{g},$$

where

$$\Psi(\ell, k) = \mathrm{diag}\left(\begin{bmatrix} h_{11}^d(\ell, k) \ \ldots \ h_{K1}^d(\ell, k) \end{bmatrix}\right),$$

and

$$\tilde{\Psi}(\ell, k) = \begin{bmatrix} \Psi(\ell, k) & \mathcal{O}_{K \times N_i} \\ \mathcal{O}_{N_i \times K} & I_{N_i \times N_i} \end{bmatrix}.$$

Now, a beamformer having the modified beam-pattern should satisfy the modified constraints set:

$$\dot{\mathbf{C}}^\dagger(\ell, k)\tilde{\mathbf{w}}(\ell, k) = \tilde{\mathbf{g}}(\ell, k) = \tilde{\Psi}^\dagger(\ell, k)\mathbf{g}.$$

Hence,

$$(\tilde{\Psi}^{-1}(\ell, k))^\dagger\dot{C}^\dagger(\ell, k)\tilde{\mathbf{w}}(\ell, k) = \mathbf{g}.$$

Define

$$\tilde{C}(\ell, k) \triangleq \dot{C}(\ell, k)\tilde{Psi}^{-1}(\ell, k) = \begin{bmatrix} \tilde{H}^d(\ell, k) \ E(\ell, k) \end{bmatrix}, \tag{10.26}$$

where

$$\tilde{H}^d(\ell, k) \triangleq \left[\, \tilde{\mathbf{h}}_1^d(\ell, k) \ldots \tilde{\mathbf{h}}_K^d(\ell, k) \,\right], \tag{10.27}$$

with

$$\tilde{\mathbf{h}}_i^d(\ell, k) \triangleq \frac{\mathbf{h}_i^d(\ell, k)}{h_{i1}^d(\ell, k)} \tag{10.28}$$

defined as the RTF with respect to microphone #1.

Finally, the modified beamformer is given by

$$\tilde{\mathbf{w}}_0(\ell, k) \triangleq \tilde{C}(\ell, k)\big(\tilde{C}(\ell, k)^\dagger \tilde{C}(\ell, k)\big)^{-1}\mathbf{g} \tag{10.29}$$

and its corresponding output is indeed given by

$$y(\ell, k) = \tilde{\mathbf{w}}_0^\dagger(\ell, k)\mathbf{z}(\ell, k) =$$
$$\sum_{i=1}^{K} s_i^d(\ell, k)h_{i1}^d(\ell, k) + \mathbf{g}^\dagger\big(\tilde{C}^\dagger(\ell, k)\tilde{C}(\ell, k)\big)^{-1}\tilde{C}^\dagger(\ell, k)\mathbf{v}(\ell, k).$$
$$\tag{10.30}$$

Therefore, the modified beamformer output comprises the sum of the desired sources as measured at the reference microphone (arbitrarily chosen as microphone #1) and the sensor noise contribution.

For the single desired sources scenario the modified beamformer output is reduced to

$$y(\ell, k) = s_1^d(\ell, k)h_{11}^d(\ell, k) + \mathbf{g}^\dagger\big(H^\dagger(\ell, k)H(\ell, k)\big)^{-1}H^\dagger(\ell, k)\mathbf{v}(\ell, k). \tag{10.31}$$

10.4 Estimation of the Constraints Matrix

In the previous sections we have shown that knowledge of the RTFs related to the desired sources and a basis that spans the subspace of the interfering sources suffice for implementing the beamforming algorithm. This section is dedicated to the estimation procedure necessary to acquire this knowledge. We start by making some restrictive assumptions regarding the activity of the sources. First, we assume that there are time segments for which none of the non-stationary sources is active. These segments are used for estimating the stationary noise PSD. Second, we assume that there are time segments in which all the desired sources are inactive. These segments are used for estimating the interfering sources subspace (with arbitrary activity pattern). Third, we assume that for every desired source, there is at least one time segment when it is the only non-stationary source active. These segments are used for estimating the RTFs of the desired sources. These assumptions, although restrictive, can be met in realistic scenarios, for which double talk only rarely occurs. A possible way to extract the activity information can be a video signal acquired in parallel to the sound acquisition. In this contribution it is however assumed that the number of desired sources and their activity pattern is available.

In the rest of this section we discuss the subspace estimation procedure. The RTF estimation procedure can be regarded, in this respect, as a multi-source, colored-noise, extension of the single source subspace estimation method proposed by Affes and Grenier [13]. We further assume that the various filters are slowly time-varying filters, i.e $H(\ell, k) \approx H(k)$. Due to inevitable estimation errors, the constraints set is not exactly satisfied, resulting in leakage of residual interference signals to the beamformer output, as well as desired signal distortion. This leakage reflects on the spatially white sensors noise assumption, and is dealt with in [26].

10.4.1 Interferences Subspace Estimation

Let $\ell = \ell_1, \ldots, \ell_{N_{seg}}$, be a set of N_{seg} frames for which all desired sources are inactive. For every segment we estimate the subspace spanned by the active interferences (both stationary and non-stationary). Let $\hat{\Phi}_{zz}(\ell_i, k)$ be a PSD estimate at the interference-only frame ℓ_i. Using the EVD we have $\hat{\Phi}_{zz}(\ell_i, k) = E_i(k)\Lambda_i(k)E_i^{\dagger}(k)$. Interference-only segments consist of both directional interference and noise components and spatially-white sensor noise. Hence, the larger eigenvalues can be attributed to the coherent signals while the lower eigenvalues to the spatially-white signals.

Define two values $\Delta\mathrm{EV}_{\mathrm{TH}}(k)$ and $\mathrm{MEV}_{\mathrm{TH}}$. All eigenvectors corresponding to eigenvalues that are more than $\Delta\mathrm{EV}_{\mathrm{TH}}$ below the largest eigenvalue or not higher than $\mathrm{MEV}_{\mathrm{TH}}$ above the lowest eigenvalue, are regarded as sensor noise eigenvectors and are therefore discarded from the interference signal subspace. Assuming that the number of sensors is larger than the number of directional sources, the lowest eigenvalue level will correspond to the sensor noise variance σ_v^2. The procedure is demonstrated in Fig. 10.1 for the 11 microphone test scenario presented in Section 10.6. A segment which comprises three directional sources (one stationary and two non-stationary interferences) is analyzed using the EVD by 11 microphone array (i.e. the dimensions of the multi-sensor correlation matrix is 11×11). The eigenvalue level as a function of the frequency bin is depicted in the figure. The blue line depicts $\mathrm{MEV}_{\mathrm{TH}}$ threshold and the dark green frequency-dependent line depicts the threshold $\mathrm{EV}_{\mathrm{TH}}(k)$. All eigenvalues that do not meet the thresholds, depicted as gray lines in the figure, are discarded from the interference signal subspace. It can be seen from the figure that in most frequency bins the algorithm correctly identified the three directional sources. Most of the erroneous reading are found in the lower frequency band, where the directivity of the array is low, and in the upper frequency band, where the signals'

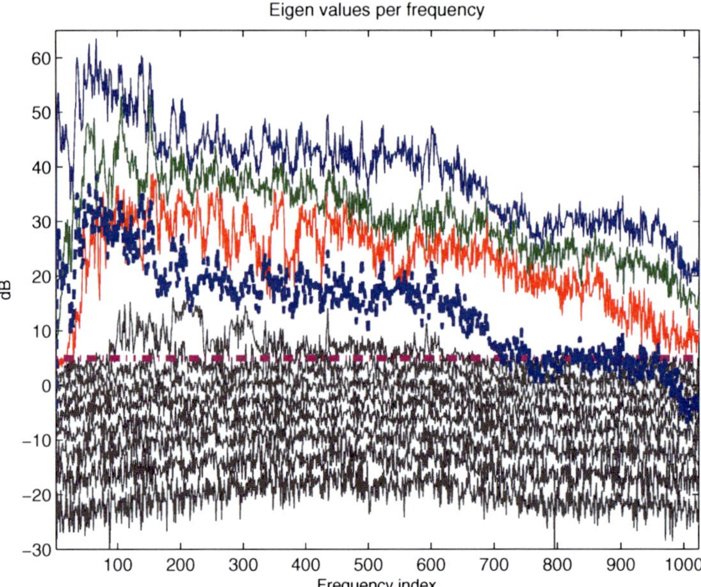

Fig. 10.1 Eigenvalues of an interference-only segment as a function of the frequency bin (solid thin lines). Eigenvalues that do not meet the thresholds $\mathrm{MEV_{TH}}$ (thick black horizontal line) and $\mathrm{EV_{TH}}(k)$ (thick black curve) are depicted in grey and discarded from the interference signal subspace.

power is low. The use of two thresholds is shown to increase the robustness of the procedure.

We denote the eigenvectors that passed the thresholds as $\hat{E}_i(k)$, and their corresponding eigenvalues as $\hat{\Lambda}_i(k)$. This procedure is repeated for each segment ℓ_i; $i = 1, 2, \ldots, N_{seg}$. These vectors should span the basis of the entire interference subspace:

$$H^i(\ell, k) = E(\ell, k)\Theta(\ell, k)$$

defined in (10.19). To guarantee that the eigenvectors $i = 1, 2, \ldots, N_{seg}$ that are common to more than one segment are not counted more than once they should be collected by the union operator:

$$\hat{E}(k) \triangleq \bigcup_{i=1}^{N_{seg}} \hat{E}_i(k), \tag{10.32}$$

where $\hat{E}(k)$ is an estimate for the interference subspace basis $E(\ell, k)$ assumed to be time-invariant in the observation period. Unfortunately, due to arbitrary activity of sources and estimation errors, eigenvectors that correspond to the same source can be manifested as a different eigenvector in each segment. These differences can unnecessarily inflate the number of estimated interference sources. Erroneous rank estimation is one of causes to the well-known desired signal cancellation phenomenon in beamformer structures, since desired signal components may be included in the null subspace. The union operator can be implemented in many ways. Here we chose to use the QRD.

Consider the following QRD of the subspace spanned by the major eigenvectors (weighted in respect to their eigenvalues) obtained by the previous procedure:

$$\left[\hat{E}_1(k)\hat{\Lambda}_1^{\frac{1}{2}}(k) \, \ldots \, \hat{E}_{N_{seg}}(k)\hat{\Lambda}_{N_{seg}}^{\frac{1}{2}}(k) \right] P(k) = Q(k)R(k), \tag{10.33}$$

where $Q(k)$ is a unitary matrix, $R(k)$ is an upper triangular matrix with decreasing diagonal absolute values, $P(k)$ is a permutation matrix and $(\cdot)^{\frac{1}{2}}$ is a square root operation performed on each of the diagonal elements.

All vectors in $Q(k)$ that correspond to values on the diagonal of $R(k)$ that are lower than ΔU_{TH} below their largest value, or less then MU_{TH} above their lowest value are not counted as basis vectors of the directional interference subspace. The collection of all vectors passing the designated thresholds, constitutes $\hat{E}(k)$, the estimate of the interference subspace basis. The novel procedure relaxes the widely-used requirement for non-overlapping activity periods of the distinct interference sources. Moreover, since several segments are collected, the procedure tends to be more robust than methods that rely on PSD estimates obtained by only one segment.

10.4.2 Desired Sources RTF Estimation

Consider time frames for which only the stationary sources are active and estimate the corresponding PSD matrix

$$\hat{\Phi}_{zz}^s(\ell, k) \approx H^s(\ell, k)\Lambda^s(\ell, k)\big(H^s(\ell, k)\big)^{\dagger} + \sigma_v^2 I_{M \times M}. \tag{10.34}$$

Assume that there exists a segment ℓ_i during which the only active non-stationary signal is the ith desired source $i = 1, 2, \ldots, K$. The corresponding PSD matrix will then satisfy

$$\hat{\Phi}_{zz}^{d,i}(\ell_i, k) \approx (\sigma_i^d(\ell_i, k))^2 \mathbf{h}_i^d(\ell_i, k)\big(\mathbf{h}_i^d(\ell_i, k)\big)^{\dagger} + \hat{\Phi}_{zz}^s(\ell, k). \tag{10.35}$$

Now, applying the GEVD to $\hat{\Phi}_{zz}^{d,i}(\ell_i, k)$ and the stationary-noise PSD matrix $\hat{\Phi}_{zz}^s(\ell, k)$ we have:

$$\hat{\Phi}_{zz}^{d,i}(\ell_i, k)\mathbf{f}_i(k) = \lambda_i(k)\hat{\Phi}_{zz}^s(\ell, k)\mathbf{f}_i(k). \tag{10.36}$$

The generalized eigenvectors corresponding to the generalized eigenvalues with values other than 1 span the desired sources subspace. Since we assumed that only source i is active in segment ℓ_i, this eigenvector corresponds to a scaled version of the source ATF. To prove this relation for the single eigenvector case, let $\lambda_i(k)$ correspond the largest eigenvalue at segment ℓ_i and $\mathbf{f}_i(k)$ its corresponding eigenvector. Substituting $\hat{\Phi}_{zz}^{d,i}(\ell_i, k)$ as defined in (10.35) in the left-hand side of (10.36) yields

$$(\sigma_i^d(\ell_i, k))^2 \mathbf{h}_i^d(\ell_i, k)\big(\mathbf{h}_i^d(\ell_i, k)\big)^\dagger \mathbf{f}_i(k) + \hat{\Phi}_{zz}^s(\ell, k)\mathbf{f}_i(k) = \lambda_i(k)\hat{\Phi}_{zz}^s(\ell, k)\mathbf{f}_i(k),$$

therefore

$$(\sigma_i^d(\ell_i, k))^2 \mathbf{h}_i^d(\ell_i, k)\underbrace{\big(\mathbf{h}_i^d(\ell_i, k)\big)^\dagger \mathbf{f}_i(k)}_{\text{scalar}} = \big(\lambda_i(k) - 1\big)\hat{\Phi}_{zz}^s(\ell, k)\mathbf{f}_i(k),$$

and finally,

$$\mathbf{h}_i^d(\ell_i, k) = \underbrace{\frac{\lambda_i(k) - 1}{(\sigma_i^d(\ell_i, k))^2 \big(\mathbf{h}_i^d(\ell_i, k)\big)^\dagger \mathbf{f}_i(k)}}_{\text{scalar}} \hat{\Phi}_{zz}^s(\ell, k)\mathbf{f}_i(k) \;\therefore$$

Hence, the desired signal ATF $\mathbf{h}_i^d(\ell_i, k)$ is a scaled and rotated version of the eigenvector $\mathbf{f}_i(k)$ (with eigenvalue other than 1). As we are interested in the RTFs rather than the entire ATFs the scaling ambiguity can be resolved by the following normalization:

$$\hat{\tilde{\mathbf{h}}}_i^d(\ell, k) \triangleq \frac{\Phi_{zz}^s(\ell, k)\mathbf{f}_i(k)}{\big(\Phi_{zz}^s(\ell, k)\mathbf{f}_i(k)\big)_1}, \tag{10.37}$$

where $(\cdot)_1$ is the first component of the vector corresponding to the reference microphone (arbitrarily chosen to be the first microphone). We repeat this estimation procedure for each desired source $i = 1, 2, \ldots, K$. The value of K is a design parameter of the algorithm. An alternative method for estimating the RTFs based on the non-stationarity of the speech is developed for single source scenario in [14], but can be used as well for the general scenario with multiple desired sources, provided that time frames for each the desired sources are not simultaneously active exist.

10.5 Algorithm Summary

The entire algorithm is summarized in Alg. 1. The algorithm is implemented almost entirely in the STFT domain, using a rectangular analysis window of length N_{DFT}, and a shorter rectangular synthesis window, resulting in the

---

Algorithm 1 Summary of the proposed LCMV beamformer.

---

1) beamformer with modified constraints set:

$y(\ell, k) \triangleq \tilde{\mathbf{w}}_0^\dagger(\ell, k)\mathbf{z}(\ell, k)$

where

$\tilde{\mathbf{w}}_0(\ell, k) \triangleq \tilde{C}(\ell, k)\big(\tilde{C}(\ell, k)^\dagger \tilde{C}(\ell, k)\big)^{-1}\mathbf{g}$

$\tilde{C}(\ell, k) \triangleq \big[\, \tilde{H}^d(\ell, k) \ E(\ell, k)\,\big]$

$$\mathbf{g} \triangleq \left[\, \underbrace{1 \ldots 1}_{K} \ \underbrace{0 \ldots 0}_{N-K}\,\right]^T.$$

$\tilde{\mathbf{H}}^d(\ell, k)$ are the RTFs in respect to microphone #1.

2) Estimation:

   a) Estimate the stationary noise PSD using Welch method: $\hat{\Phi}_{zz}^s(\ell, k)$

   b) Estimate time-invariant desired sources RTFs $\tilde{H}^d(k) \triangleq \big[\, \tilde{\mathbf{h}}_1^d(k) \ldots \tilde{\mathbf{h}}_K^d(k)\,\big]$

      Using GEVD and normalization:

        i) $\hat{\Phi}_{zz}^{d,i}(\ell_i, k)\mathbf{f}_i(k) = \lambda_i \hat{\Phi}_{zz}^s(\ell, k)\mathbf{f}_i(k) \Rightarrow \mathbf{f}_i(k)$

        ii) $\hat{\tilde{\mathbf{h}}}_i^d(\ell, k) \triangleq \dfrac{\hat{\Phi}_{zz}^s(\ell, k)\mathbf{f}_i(k)}{\big(\hat{\Phi}_{zz}^s(\ell, k)\mathbf{f}_i(k)\big)_1}.$

   c) Interferences subspace:

      QRD factorization of eigen-spaces $\left[\, E_1(k)\Lambda_1^{\frac{1}{2}}(k) \ \ldots \ E_{N_{seg}}(k)\Lambda_{N_{seg}}^{\frac{1}{2}}(k)\,\right]$

      Where $\hat{\Phi}_{zz}(\ell_i, k) = E_i(k)\Lambda_i(k)E_i^\dagger(k)$ for time segment ℓ_i.

---

overlap & save procedure [28], avoiding any cyclic convolution effects. The PSD of the stationary interferences and the desired sources are estimated using the Welch method, with a Hamming window of length $D \times N_{\mathrm{DFT}}$ applied to each segment, and $(D-1) \times N_{\mathrm{DFT}}$ overlap between segments. However, since only lower frequency resolution is required, we wrapped each segment to length N_{DFT} before the application of the discrete Fourier transform operation. The interference subspace is estimated from a $L_{seg} \times N_{\mathrm{DFT}}$ length segment. The overlap between segments is denoted OVRLP. The resulting beamformer estimate is tapered by a Hamming window resulting in a smooth filter in the coefficient range $[-FL_l, FL_r]$. The parameters used for the simulation are given in Table 10.1. In cases where the sensor noise is not spatially-white or when the estimation of the constraint matrix is not accurate, the entire LCMV procedure (10.9) should be implemented. In these cases, the presented algorithm will be accompanied by an adaptive noise canceler (ANC) branch constituting a GSC structure, as presented in [26].

10.6 Experimental Study

In this section, we evaluate the performance of the proposed subspace beamformer. In case of one desired source we compare the presented algorithm with the TF-GSC algorithm [14].

Table 10.1 Parameters used by the subspace beamformer algorithm.

Parameter	Description	Value
	General Parameters	
f_s	Sampling frequency	8KHz
	Desired signal to sensor noise ratio (determines σ_v^2)	41dB
	PSD Estimation using Welch Method	
N_{DFT}	DFT length	2048
D	Frequency decimation factor	6
JF	Time offset between segments	2048
	Interferences' subspace Estimation	
L_{seg}	Number of DFT segments used for estimating a single interference subspace	24
OVRLP	The overlap between time segments that are used for interferences subspace estimation	50%
$\Delta\mathrm{EV}_{\mathrm{TH}}$	Eigenvectors corresponding to eigenvalues that are more than $\mathrm{EV}_{\mathrm{TH}}$ lower below the largest eigenvalue are discarded from the signal subspace	40dB
$\mathrm{MEV}_{\mathrm{TH}}$	Eigenvectors corresponding to eigenvalues not higher than $\mathrm{MEV}_{\mathrm{TH}}$ above the sensor noise are discarded from the signal subspace	5dB
$\Delta\mathrm{U}_{\mathrm{TH}}$	vectors of $\mathbf{Q}(k)$ corresponding to values of $\mathbf{R}(k)$ that are more than U_{TH} below the largest value on the diagonal of $\mathbf{R}(k)$	40dB
$\mathrm{MU}_{\mathrm{TH}}$	vectors of $\mathbf{Q}(k)$ corresponding to values of $\mathbf{R}(k)$ not higher than $\mathrm{MU}_{\mathrm{TH}}$ above the lowest value on the diagonal of $\mathbf{R}(k)$	5dB
	Filters Lengths	
FL_r	Causal part of the beamformer filters	1000 taps
FL_l	Noncausal part of the beamformer filters	1000 taps

10.6.1 The Test Scenario

The proposed algorithm was tested both in simulated and real room environments in several test scenarios. In test scenario #1 five directional signals, namely two (male and female) desired speech sources, two (other male and female) speakers as competing speech signals, and a stationary speech-like noise drawn from NOISEX-92 [29] database were mixed.

In test scenarios #2-#4 the performance of the multi-constraints algorithm was compared to the TF-GSC algorithm [14] in a simulated room environment, using one desired speech source, one stationary speech-like noise drawn from NOISEX-92 [29] database, and various number of competing speakers (ranging from zero to two). For the simulated room scenario the image method [30] was used to generate the RIR. The implementation is described in [31]. All the signals $i = 1, 2, \ldots, N$ were then convolved with the corresponding time-invariant RIRs. The microphone signals $z_m(\ell, k)$; $m = 1, 2, \ldots, M$ were finally obtained by summing up the contributions of all directional sources with an additional uncorrelated sensor noise.

The level of all desired sources is equal. The desired signal to sensor noise ratio was set to 41dB (this ratio determines σ_v^2). The relative power between the desired sources and all interference sources are depicted in Table 10.2 and Table 10.3 for scenario #1 and scenarios #2-#4, respectively.

In the real room scenario each of the signals was played by a loudspeaker located in a reverberant room (each signal was played by a different loudspeaker) and captured by an array of M microphones. The signals $\mathbf{z}(\ell, k)$ were finally constructed by summing up all recorded microphone signals with a gain related to the desired input signal to interference ratio (SIR).

For evaluating the performance of the proposed algorithm, we applied the algorithms in two phases. During the first phase, the algorithm was applied to an input signal, comprised of the sum of the desired speakers, the competing speakers, and the stationary noise (with gains in accordance with the respective SIR. In this phase, the algorithm performed the various estimations yielding $y(\ell, k)$, the actual algorithm output. In the second phase, the beamformer was *not* recalculated. Instead, the beamformer obtained in the first phase was applied to each of the unmixed sources.

Denote by $y_i^d(\ell, k)$; $i = 1, \ldots, K$, the desired signals components at the beamformer output, $y_i^{ns}(\ell, k)$; $i = 1, \ldots, N_{ns}$ the corresponding non-stationary interference components, $y_i^s(\ell, k)$; $i = 1, \ldots, N_s$ the stationary interference components, and $y^v(\ell, k)$ the sensor noise component at the beamformer output respectively.

One quality measure used for evaluating the performance of the proposed algorithm is the improvement in the SIR level. Since, generally, there are several desired sources and interference sources we will use all pairs of SIR for quantifying the performance. The SIR of desired signal i relative to the non-stationary signal j as measured on microphone m_0 is defined as follows:

$$\mathrm{SIR}_{\mathrm{in},ij}^{ns}[\mathrm{dB}] = 10 \log_{10} \frac{\sum_\ell \sum_{k=0}^{N_{\mathrm{DFT}}-1} \left(s_i^d(\ell, k) h_{im_0}^d(\ell, k) \right)^2}{\sum_\ell \sum_{k=0}^{N_{\mathrm{DFT}}-1} \left(s_j^{ns}(\ell, k) h_{jm_0}^{ns}(\ell, k) \right)^2}$$

$$1 \le i \le K, 1 \le j \le N_{ns}.$$

Similarly, the input SIR of the desired signal i relative to the stationary signal j:

$$\mathrm{SIR}_{\mathrm{in},ij}^{s}[\mathrm{dB}] = 10 \log_{10} \frac{\sum_\ell \sum_{k=0}^{N_{\mathrm{DFT}}-1} \left(s_i^d(\ell, k) h_{im_0}^d(\ell, k) \right)^2}{\sum_\ell \sum_{k=0}^{N_{\mathrm{DFT}}-1} \left(s_j^s(\ell, k) h_{jm_0}^s(\ell, k) \right)^2}$$

$$1 \le i \le K, 1 \le j \le N_s.$$

These quantities are compared with the corresponding beamformer outputs SIR:

$$\mathrm{SIR}^{ns}_{\mathrm{out},ij}[\mathrm{dB}] = 10\log_{10}\frac{\sum_\ell \sum_{k=0}^{N_{\mathrm{DFT}}-1}\left(y_i^d(\ell,k)\right)^2}{\sum_\ell \sum_{k=0}^{N_{\mathrm{DFT}}-1}\left(y_j^{ns}(\ell,k)\right)^2}$$

$$1 \leq i \leq K, 1 \leq j \leq N_{ns},$$

$$\mathrm{SIR}^{s}_{\mathrm{out},ij}[\mathrm{dB}] = 10\log_{10}\frac{\sum_\ell \sum_{k=0}^{N_{\mathrm{DFT}}-1}\left(y_i^d(\ell,k)\right)^2}{\sum_\ell \sum_{k=0}^{N_{\mathrm{DFT}}-1}\left(y_j^{s}(\ell,k)\right)^2}$$

$$1 \leq i \leq K, 1 \leq j \leq N_{s}.$$

For evaluating the distortion imposed on the desired sources we also calculated the squared error distortion (SED) and log spectral distance (LSD) distortion measures relating each desired source component $1 \leq i \leq K$ at the output, namely $y_i^d(\ell,k)$ and its corresponding component received by microphone #1, namely $s_i^d(\ell,k)h_{i1}^d$. Define the SED and the LSD distortion for each desired source $1 \leq i \leq K$:

$$\mathrm{SED}_{out,i}[\mathrm{dB}] = \tag{10.38}$$

$$10\log_{10}\frac{\sum_\ell \sum_{k=0}^{N_{\mathrm{DFT}}-1}\left(s_i^d(\ell,k)h_{i1}^d(\ell,k)\right)^2}{\sum_\ell \sum_{k=0}^{N_{\mathrm{DFT}}-1}\left(s_i^d(\ell,k)h_{i1}^d(\ell,k) - y_i^d(\ell,k)\right)^2},$$

$$\mathrm{LSD}_{out,i} = \tag{10.39}$$

$$\frac{1}{L'}\sum_\ell \sqrt{\frac{1}{N_{\mathrm{DFT}}}\sum_{k=0}^{N_{\mathrm{DFT}}-1}\left[20\log_{10}|s_i^d(\ell,k)h_{i1}^d(\ell,k)| - 20\log_{10}|y_i^d(\ell,k)|\right]^2},$$

where L' is the number of speech active frames and $\{\ell \in \text{Speech Active}\}$. These figures-of-merit are also depicted in the Tables.

10.6.2 Simulated Environment

The RIRs were simulated with a modified version [31] of Allen and Berkley's *image method* [30] with various reverberation levels ranging between 150–300mSec. The simulated environment was a $4m \times 3m \times 2.7m$ room. A nonuniform linear array consisting of 11 microphones with inter-microphone distances ranging from $5cm$ to $10cm$. The microphone array and the various sources positions are depicted in Fig. 2(a). A typical RIR relating a source and one of the microphones is depicted in Fig. 2(c). The SIR improvements, as a function of the reverberation time T_{60}, obtained by the LCMV beamformer for scenario 1 are depicted in Table 10.2. The SED and the LSD distortion measures are also depicted for each source. Since the desired sources RTFs are estimated when the competing speech signals are inactive, their relative

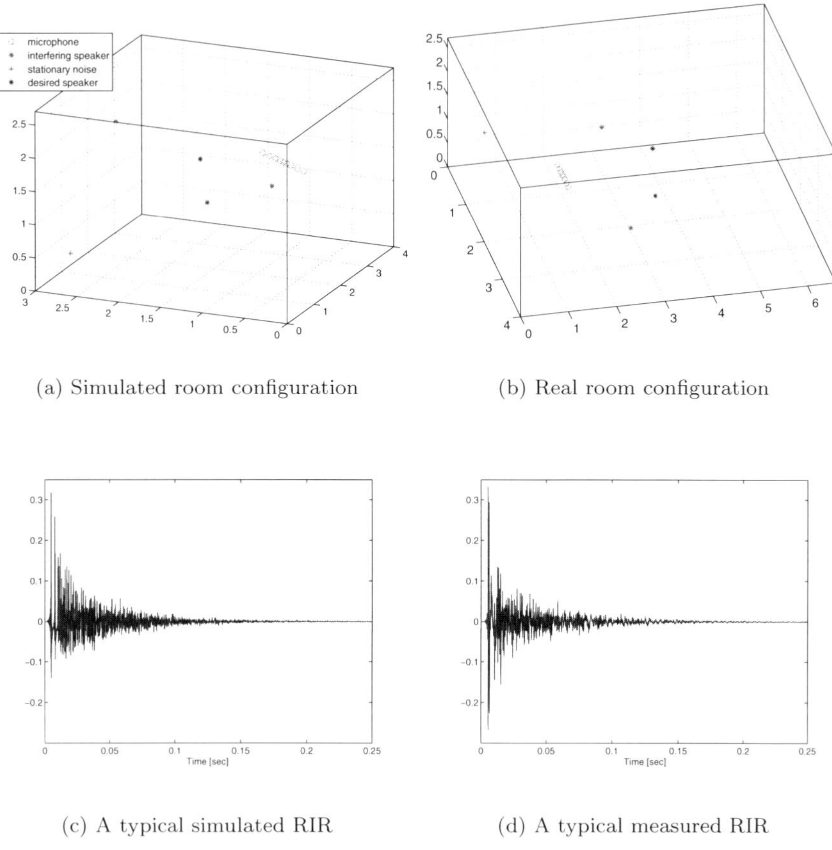

(a) Simulated room configuration   (b) Real room configuration

(c) A typical simulated RIR   (d) A typical measured RIR

Fig. 10.2 Room configuration and the corresponding typical RIR for simulated and real scenarios.

power has no influence on the obtained performance, and is therefore kept fixed during the simulations.

In Table 10.3 the multi-constraints algorithm and the TF-GSC are compared in terms of the objective quality measures, as explained above, for various number of interference sources. Since the TF-GSC contains an ANC branch, we compared it to a multi-constraint beamformer that also incorporates an ANC [27]. It is evident from Table 10.3 that the multi-constraint beamformer outperforms the TF-GSC algorithm in terms of SIR improvement, as well as distortion level measured by the SED and LSD values. The lower distortion of the multi-constraint beamformer can be attributed to the stable nature of the nulls in the beam-pattern as compared with the adaptive nulls of the TF-GSC structure. The results in the Tables were obtained using

Table 10.2 Test scenario #1: 11 microphone array, 2 desired speakers, 2 interfering speakers at $6dB$ SIR, and one stationary noise at $13dB$ SIR with various reverberation levels. SIR improvement in dB for the LCMV output and speech distortion measures (SED and LSD in dB) between the desired source component received by microphone #1 and respective component at the LCMV output.

T_{60}	Source	BF SIR imp.			SED	LSD
		s_1^{ns}	s_2^{ns}	s_1^{s}		
150ms	s_1^{d}	12.53	14.79	13.07	11.33	1.12
	s_2^{d}	12.39	14.98	12.93	13.41	1.13
200ms	s_1^{d}	10.97	12.91	11.20	9.51	1.39
	s_2^{d}	12.13	13.07	11.36	10.02	1.81
250ms	s_1^{d}	10.86	12.57	11.07	8.49	1.56
	s_2^{d}	11.19	12.90	11.40	8.04	1.83
300ms	s_1^{d}	11.53	11.79	11.21	7.78	1.86
	s_2^{d}	11.49	11.75	11.17	7.19	1.74

Table 10.3 Test scenario #2-#4: Simulated room environment with reverberation time $T_{60} = 300\text{mS}$, 11 microphones, one desired speaker, one stationary noise at $13dB$ SIR, and various number of interfering speakers at $6dB$ SIR. SIR improvement, SED and LSD in dB relative to microphone #1 as obtained by the TF-GSC [14] and the multi-constraint [26] beamformers.

N_{ns}	TF-GSC					Multi-Constraint				
	SIR imp.			SED	LSD	SIR imp.			SED	LSD
	s_1^{ns}	s_2^{ns}	s_1^{s}			s_1^{ns}	s_2^{ns}	s_1^{s}		
0	–	–	15.62	6.97	2.73	–	–	27.77	14.66	1.31
1	9.54	–	13.77	6.31	2.75	21.01	–	23.95	12.72	1.35
2	7.86	10.01	10.13	7.06	2.77	17.58	20.70	17.70	11.75	1.39

the second phase of the test procedure. It is shown that for test scenario #1 the multi-constraint beamformer can gain an average value of 12.1dB SIR improvement for both stationary and non-stationary interferences.

The multi-constraints algorithm and the TF-GSC were also subjectively compared by informal listening tests and by the assessment of waveforms and sonograms. The outputs of the TF-GSC [14] the multi-constraint algorithm [26] algorithm for test scenario #3 (namely, one competing speaker) are depicted in Fig. 3(a) and Fig. 3(b) respectively. It is evident that the multi-constraint beamformer outperforms the TF-GSC beamformer especially in terms of the competing speaker cancellation. Speech samples demonstrating the performance of the proposed algorithm can be downloaded from [32].

10.6.3 Real Environment

In the real room environment we used as the directional signals four speakers drawn from the TIMIT [33] database and the speech-like noise described above. The performance was evaluated using real medium-size conference

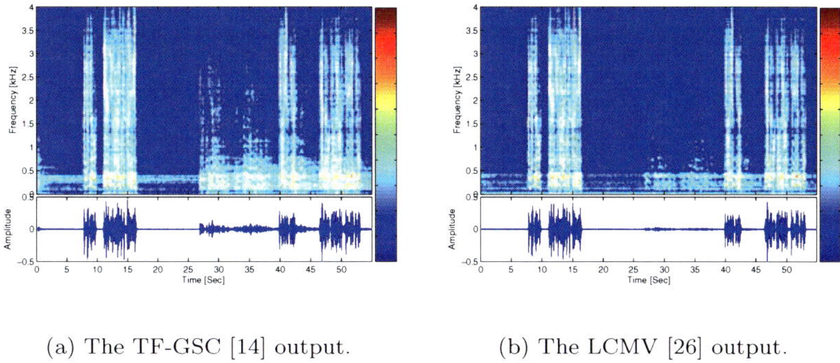

(a) The TF-GSC [14] output.   (b) The LCMV [26] output.

Fig. 10.3 Test scenario #3: Sonograms depicting the difference between TF-GSC and LCMV.

room equipped with furniture, book shelves, a large meeting table, chairs and other standard items. The room dimensions are $6.6m \times 4m \times 2.7m$. A linear nonuniform array consisting of 8 omni-directional microphones (AKG CK32) was used to pick up the various sources that were played separately from point loudspeakers (FOSTEX 6301BX). The algorithm's input was constructed by summing up all non-stationary components contributions with a 6dB SIR, the stationary noise with 13dB SIR and additional, spatially white, computer-generated sensor noise signals. The source-microphone constellation is depicted in Fig. 2(b). The RIR and the respective reverberation time were estimated using the WinMLS2004 software (a product of Morset Sound Development). A typical RIR, having $T_{60} = 250$mSec, is depicted in Fig. 2(d). A total SIR improvement of $15.28dB$ was obtained for the interfering speakers and $16.23dB$ for the stationary noise.

10.7 Conclusions

We have addressed the problem of extracting several desired sources in a reverberant environment contaminated by both non-stationary (competing speakers) and stationary interferences. The LCMV beamformer was designed to satisfy a set of constraints for the desired and interference sources. A novel and practical method for estimating the interference subspace was presented. A two phase off-line procedure was applied. First, the test scene (comprising the desired and interference sources) was analyzed using few seconds of data for each source. We therefore note, that this version of the algorithm can be applied for time-invariant scenarios. Recursive estimation methods for time-varying environments is a topic of ongoing research. Experimental results for both simulated and real environments have demonstrated that the proposed method can be applied for extracting several desired sources from a combination of multiple sources in a complicated acoustic environment. In the case of one desired source, two alternative beamforming strategies for interference cancellation in noisy and reverberant environment were compared. The TF-GSC, which belongs to the MVDR family, applies a single constraint towards the desired signal, leaving the interference mitigation adaptive. Alternatively, the multi-constraint beamformer implicitly applies carefully designed nulls towards all interference signals. It is shown that for the time-invariant scenario the later design shows a significant advantage over the former beamformer design. It remains an open question what is the preferred strategy in slowly time-varying scenarios.

References

1. J. Cardoso, "Blind signal separation: Statistical principles," *Proc. of the IEEE*, vol. 86, no. 10, pp. 2009–2025, Oct. 1998.
2. E. Jan and J. Flanagan, "Microphone arrays for speech processing," *Int. Symposium on Signals, Systems, and Electronics (ISSSE)*, pp. 373–376, Oct. 1995.
3. B. D. Van Veen and K. M. Buckley, "Beamforming: a versatile approach to spatial filtering," *IEEE Trans. Acoust., Speech, Signal Processing*, vol. 5, no. 2, pp. 4–24, Apr. 1988.
4. S. Gannot and I. Cohen, *Springer Handbook of Speech Processing*. Springer, 2007, ch. Adaptive Beamforming and Postfitering, pp. 199–228.
5. H. Cox, R. Zeskind, and M. Owen, "Robust adaptive beamforming," *IEEE Trans. Acoust., Speech, Signal Processing*, vol. 35, no. 10, pp. 1365–1376, Oct. 1987.
6. S. Doclo and M. Moonen, "Combined frequency-domain dereverberation and noise reduction technique for multi-microphone speech enhancement," in *Proc. Int. Workshop on Acoustic Echo and Noise Control (IWAENC)*, Darmstadt, Germany, Sep. 2001, pp. 31–34.
7. A. Spriet, M. Moonen, and J. Wouters, "Spatially pre-processed speech distortion weighted multi-channel Wiener filtering for noise reduction," *Signal Processing*, vol. 84, no. 12, pp. 2367–2387, 2004.
8. J. Capon, "High-resolution frequency-wavenumber spectrum analysis," *Proc. IEEE*, vol. 57, no. 8, pp. 1408–1418, Aug. 1969.
9. O. Frost, "An algorithm for linearly constrained adaptive array processing," *Proc. IEEE*, vol. 60, no. 8, pp. 926–935, Aug. 1972.
10. L. J. Griffiths and C. W. Jim, "An alternative approach to linearly constrained adaptive beamforming," *IEEE Trans. Antennas Propagate.*, vol. 30, no. 1, pp. 27–34, Jan. 1982.
11. M. Er and A. Cantoni, "Derivative constraints for broad-band element space antenna array processors," *IEEE Trans. Acoust., Speech, Signal Processing*, vol. 31, no. 6, pp. 1378–1393, Dec. 1983.
12. B. R. Breed and J. Strauss, "A short proof of the equivalence of LCMV and GSC beamforming," *IEEE Signal Processing Lett.*, vol. 9, no. 6, pp. 168–169, Jun. 2002.
13. S. Affes and Y. Grenier, "A signal subspace tracking algorithm for microphone array processing of speech," *IEEE Trans. Speech and Audio Processing*, vol. 5, no. 5, pp. 425–437, Sep. 1997.

14. S. Gannot, D. Burshtein, and E. Weinstein, "Signal enhancement using beamforming and nonstationarity with applications to speech," *Signal Processing*, vol. 49, no. 8, pp. 1614–1626, Aug. 2001.
15. S. Gazor, S. Affes, and Y. Grenier, "Robust adaptive beamforming via target tracking," *IEEE Trans. Signal Processing*, vol. 44, no. 6, pp. 1589–1593, Jun. 1996.
16. B. Yang, "Projection approximation subspace tracking," *IEEE Trans. Signal Processing*, vol. 43, no. 1, pp. 95–107, Jan. 1995.
17. S. Gazor, S. Affes, and Y. Grenier, "Wideband multi-source beamforming with adaptive array location calibration and direction finding," *Proc. IEEE Int. Conf. Acoustics, Speech, and Signal Processing (ICASSP)*, vol. 3, pp. 1904–1907, May 1995.
18. S. Doclo and M. Moonen, "GSVD-based optimal filtering for single and multimicrophone speech enhancement," *IEEE Trans. Signal Processing*, vol. 50, no. 9, pp. 2230–2244, 2002.
19. E. Warsitz, A. Krueger, and R. Haeb-Umbach, "Speech enhancement with a new generalized eigenvector blocking matrix for application in generalized sidelobe canceler," *Proc. IEEE Int. Conf. Acoustics, Speech, and Signal Processing (ICASSP)*, pp. 73–76, Apr. 2008.
20. S. Affes, S. Gazor, and Y. Grenier, "An algorithm for multi-source beamforming and multi-target tracking," *IEEE Trans. Signal Processing*, vol. 44, no. 6, pp. 1512–1522, Jun. 1996.
21. F. Asano, S. Hayamizu, T. Yamada, and S. Nakamura, "Speech enhancement based on the subspace method," *IEEE Trans. Speech and Audio Processing*, vol. 8, no. 5, pp. 497–507, Sep. 2000.
22. R. Schmidt, "Multiple emitter location and signal parameter estimation," *IEEE Trans. Antennas Propagate.*, vol. 34, no. 3, pp. 276–280, Mar. 1986.
23. J. Benesty, J. Chen, Y. Huang, and J. Dmochowski, "On microphone-array beamforming from a MIMO acoustic signal processing perspective," *IEEE Trans. Audio, Speech and Language Processing*, vol. 15, no. 3, pp. 1053–1065, Mar. 2007.
24. G. Reuven, S. Gannot, and I. Cohen, "Dual-source transfer-function generalized sidelobe canceler," *IEEE Trans. Audio, Speech and Language Processing*, vol. 16, no. 4, pp. 711–727, May 2008.
25. Y. Avargel and I. Cohen, "System identification in the short-time Fourier transform domain with crossband filtering," *IEEE Trans. Audio, Speech and Language Processing*, vol. 15, no. 4, pp. 1305–1319, May 2007.
26. S. Markovich, S. Gannot, and I. Cohen, "Multichannel eigenspace beamforming in a reverberant noisy environment with multiple interfering speech signals," *submitted to IEEE Transactions on Audio, Speech and Language Processing*, Jul. 2008.
27. ——, "A comparison between alternative beamforming strategies for interference cancelation in noisy and reverberant environment," in *the 25th convention of the Israeli Chapter of IEEE*, Eilat, Israel, Dec. 2008.
28. J. J. Shynk, "Frequency-domain and multirate adaptive filtering," *IEEE Signal Processing Magazine*, vol. 9, no. 1, pp. 14–37, Jan. 1992.
29. A. Varga and H. J. M. Steeneken, "Assessment for automatic speech recognition: II. NOISEX-92: A database and an experiment to study the effect of additive noise on speech recognition systems," *Speech Communication*, vol. 12, no. 3, pp. 247–251, Jul. 1993.
30. J. Allen and D. Berkley, "Image method for efficiently simulating small-room acoustics," *Journal of the Acoustical Society of America*, vol. 65, no. 4, pp. 943–950, Apr. 1979.
31. E. Habets, "Room impulse response (RIR) generator," http://home.tiscali.nl/ehabets/rir_generator.html, Jul. 2006.
32. S. Gannot, "Audio sample files," http://www.biu.ac.il/~gannot, Sep. 2008.
33. J. S. Garofolo, "Getting started with the DARPA TIMIT CD-ROM: An acoustic phonetic continuous speech database," National Institute of Standards and Technology (NIST), Gaithersburg, Maryland, Tech. Rep., 1988, (prototype as of December 1988).

Chapter 11
Spherical Microphone Array Beamforming

Boaz Rafaely, Yotam Peled, Morag Agmon, Dima Khaykin, and Etan Fisher

Abstract Spherical microphone arrays have been recently studied for spatial sound recording, speech communication, and sound field analysis for room acoustics and noise control. Complementary studies presented progress in beamforming methods. This chapter reviews beamforming methods recently developed for spherical arrays, from the widely used delay-and-sum and Dolph-Chebyshev, to the more advanced optimal methods, typically performed in the spherical harmonics domain.

11.1 Introduction

The growing interest in spherical microphone arrays can probably be attributed to the ability of such arrays to measure and analyze three-dimensional sound fields in an effective manner. The other strong point of these arrays is the ease of array processing performed in the spherical harmonics domain. The papers by Meyer and Elko [1] and Abhayapala and Ward [2] presented the use of spherical harmonics in spherical array processing and inspired a growing research activity in this field. The studies that followed provided further insight into spherical arrays from both theoretical and experimental view points.

This chapter presents an overview of some recent results in one important aspect of spherical microphone arrays, namely beamforming. The wide range of beamforming methods available in the standard array literature have been recently adapted for spherical arrays, typically formulated in the spherical harmonics domain, facilitating spatial filtering in three-dimensional sound fields. The chapter starts with an overview of spherical array processing,

Boaz Rafaely, Yotam Peled, Morag Agmon, Dima Khaykin, and Etan Fisher
Ben-Gurion University of the Negev, Israel, e-mail: `{br,yotamp,moraga,khaykin,fisher}@ee.bgu.ac.il`

I. Cohen et al. (Eds.): Speech Processing in Modern Communication, STSP 3, pp. 281–305.
springerlink.com

presenting the spherical Fourier transform and array processing in the space and transform domains. Then, standard beam pattern design methods such as regular, delay-and-sum, and Dolph-Chebyshev are developed for spherical arrays. Optimal and source-dependent methods such as minimum variance and null-steering based methods are presented next. The chapter concludes with more specific beamforming techniques, such as steering beam patterns of arbitrary shapes, beamforming for sources in the near-field of the array, and direction-of-arrival estimation.

11.2 Spherical Array Processing

The theory of spherical array processing is briefly outlined in this section. Consider a sound field with pressure denoted by $p(k, r, \Omega)$, where k is the wave number, and $(r, \Omega) \equiv (r, \theta, \phi)$ is the spatial location in spherical coordinates [3]. The spherical Fourier transform of the pressure is given by [3]

$$p_{nm}(k, r) = \int_{\Omega \in S^2} p(k, r, \Omega) Y_n^{m^*}(\Omega)\, d\Omega, \tag{11.1}$$

with the inverse transform relation:

$$p(k, r, \Omega) = \sum_{n=0}^{\infty} \sum_{m=-n}^{n} p_{nm}(k, r) Y_n^m(\Omega), \tag{11.2}$$

where $\int_{\Omega \in S^2} d\Omega \equiv \int_0^{2\pi} \int_0^{\pi} \sin \theta d\theta d\phi$ and $Y_n^m(\Omega)$ is the spherical harmonics of order n and degree m [3]:

$$Y_n^m(\Omega) \equiv \sqrt{\frac{(2n+1)}{4\pi} \frac{(n-m)!}{(n+m)!}} P_n^m(\cos \theta) e^{im\phi}, \tag{11.3}$$

where $i = \sqrt{-1}$ and $P_n^m(\cdot)$ is the associated Legendre function. A sound field composed of a single plane wave is of great importance for beamforming because beam patterns are typically measured as the array response to a single plane wave [4]. Therefore, we consider a sound field composed of a single plane wave of amplitude $a(k)$, with an arrival direction Ω_0, in which case p_{nm} can be written as [5]

$$p_{nm}(k, r) = a(k) b_n(kr) Y_n^{m^*}(\Omega_0), \tag{11.4}$$

where $b_n(kr)$ depends on the sphere boundary and has been presented for rigid sphere, open sphere:

$$b_n(kr) = \begin{cases} 4\pi i^n \left[j_n(kr) - \frac{j_n'(kr_a)}{h_n'(kr_a)} h_n(kr) \right] & \text{rigid sphere} \\ 4\pi i^n j_n(kr) & \text{open sphere} \end{cases}, \tag{11.5}$$

and other array configurations [6], where j_n and h_n are the spherical Bessel and Hankel functions, j_n' and h_n' are their derivatives, and r_a is the radius of the rigid sphere.

Spherical array processing typically includes a first stage of approximating p_{nm}, followed by a second beamforming stage performed in the spherical harmonics domain [1, 2, 7]. For the first stage, we can write

$$p_{nm}(k) \approx \sum_{j=1}^{M} g_{nm}^{j}(k) p(k, r_j, \Omega_j). \tag{11.6}$$

Array input is measured by M pressure microphones located at (r_j, Ω_j), and is denoted by $p(k, r_j, \Omega_j)$. Coefficients g_{nm}^{j}, which may be frequency-dependent, are selected to ensure accurate approximation of the integral in (11.1) by the summation in (11.6). The accuracy of the approximation in (11.6) is affected by several factors including the choice of sampling points and the coefficients g_{nm}^{j}; the operating frequency range; array radius; and highest array order N, typically satisfying $(N+1)^2 \leq M$ [7]. Several sampling methods are available which provide good approximation by (11.6) for limited frequency ranges and array orders [7]. Analysis of sampling and the effect of aliasing have been previously presented [8], while in this chapter it is assumed for simplicity that p_{nm} can be measured accurately.

Once p_{nm} has been measured, array output y can be computed by [7, 9]

$$y(k) = \sum_{n=0}^{N} \sum_{m=-n}^{n} w_{nm}^{*} \, p_{nm}(k), \tag{11.7}$$

where w_{nm} are the beamforming weights represented in the spherical harmonics domain. Equation (11.7) can also be written in the space domain [7]:

$$y(k) = \int_{\Omega \in S^2} w^*(\Omega) p(k, r, \Omega) d\Omega, \tag{11.8}$$

where $w(\Omega)$, calculated as the inverse spherical Fourier transform of w_{nm}, represents spatial weighting of the sound pressure over the surface of a sphere. Although array processing directly in the space domain is common, recently, spherical array processing has been developed in the spherical harmonics domain. The latter presents several advantages, such as ease of beam pattern steering and a unified formulation for a wide range of array configurations.

In many applications a beam pattern which is rotationally symmetric around the array look direction is sufficient, in which case array weights can be written as [1]

$$w_{nm}^*(k) = \frac{d_n}{b_n(kr)} Y_n^{m^*}(\Omega_l),\qquad(11.9)$$

where d_n controls the beam pattern and Ω_l is the array look direction. Substituting (11.4) and (11.9) into (11.7), we get

$$y = \sum_{n=0}^{N}\sum_{m=-n}^{n} d_n Y_n^{m^*}(\Omega_0) Y_n^{m}(\Omega_l).\qquad(11.10)$$

Using the spherical harmonics addition theorem [10], this can be further simplified to

$$y = \sum_{n=0}^{N} d_n \frac{2n+1}{4\pi} P_n(\cos\Theta),\qquad(11.11)$$

where $P_n(\cdot)$ is the Legendre polynomial, and Θ is the angle between the look direction Ω_l and the plane wave arrival direction Ω_0. Note that in this case the beam pattern is a function of a one-dimensional parameter Θ (and n in the transform domain), although it operates in a three-dimensional sound field.

Methods for the design of d_n are presented in this chapter. Measures for arrays performance based on the choice of d_n are an important tool to assess array performance and compare array designs. Two such measures, namely the directivity (Q) and the white-noise gain (WNG) are presented here. Derivation of these measures can be found elsewhere [7, 11]. First, the directivity, which measures the array output due to a plane wave arriving from the look direction, relative to the output of an omni-directional sensor, is given by

$$Q = \frac{\left|\sum_{n=0}^{N} d_n(2n+1)\right|^2}{\sum_{n=0}^{N}|d_n|^2(2n+1)},\qquad(11.12)$$

with the directivity index (DI) calculated as $10\log_{10} Q$. The WNG, which is a measure of array robustness against sensor noise and other uncertainties is given by

$$WNG = \frac{M}{(4\pi)^2}\frac{\left|\sum_{n=0}^{N} d_n(2n+1)\right|^2}{\sum_{n=0}^{N}\frac{|d_n|^2}{|b_n|^2}(2n+1)}.\qquad(11.13)$$

11.3 Regular Beam Pattern

A common choice for d_n is simply $d_n = 1$. The beam pattern achieved is known as a regular beam pattern [12]. In this case, (11.11) becomes [5]

$$y = \frac{N+1}{4\pi(\cos\Theta - 1)}[P_{N+1}(\cos\Theta) - P_N(\cos\Theta)]. \tag{11.14}$$

This can lead to plane-wave decomposition for $N \to \infty$ [5], such that $y(\Omega_l) = \delta(\Omega_0 - \Omega_l)$, showing that the array output for a single plane wave is a delta function pointing to the arrival direction of the plane wave.

Another interesting result for the regular beam pattern is that this choice of d_n leads to a beam pattern with a maximum directivity, which is equivalent to maximum signal-to-noise ratio (SNR) for spatially-white noise [13]. The directivity given by (11.12) can be written in a matrix notation [13]:

$$Q = \frac{\mathbf{d}^H \mathbf{B} \mathbf{d}}{\mathbf{d}^H \mathbf{H} \mathbf{d}}, \tag{11.15}$$

where

$$\mathbf{d} = [d_0, d_1, \cdots, d_N]^T, \tag{11.16}$$

is the $(N+1)\times 1$ vector of beamforming weights, superscript "H" representing Hermitian conjugate, and $(N+1) \times (N+1)$ matrix $\mathbf{B}$, given by

$$\mathbf{B} = \mathbf{b}\mathbf{b}^T, \tag{11.17}$$

is composed of $(N+1) \times 1$ vector $\mathbf{b}$ given by

$$\mathbf{b} = [1, 3, \ldots, 2N+1]^T. \tag{11.18}$$

Finally, $(N+1) \times (N+1)$ matrix $\mathbf{H}$ is given by

$$\mathbf{H} = \texttt{diag}(\mathbf{b}). \tag{11.19}$$

Equation (11.15) represents a generalized Rayleigh quotient of two Hermitian forms. The maximum value of the generalized Rayleigh quotient is the largest generalized eigenvalue of the equivalent generalized eigenvalue problem:

$$\mathbf{B}\mathbf{x} = \lambda \mathbf{H}\mathbf{x}, \tag{11.20}$$

where λ is the generalized eigenvalue and $\mathbf{x}$ is the corresponding generalized eigenvector. The eigenvector corresponding to the largest eigenvalue will hold the coefficients vector $\mathbf{d}$ which maximize the directivity. Since $\mathbf{B}$ equals a dyadic product it has only one eigenvector $\mathbf{x} = \mathbf{H}^{-1}\mathbf{b}$ with an eigenvalue $\lambda = \mathbf{b}^T \mathbf{H}^{-1} \mathbf{b}$. The coefficients vector that maximizes the directivity is therefore given by

$$\mathbf{d} = \arg\max_{\mathbf{d}} Q = \mathbf{H}^{-1}\mathbf{b} = [1, 1, \ldots, 1]^T, \tag{11.21}$$

with a corresponding maximum directivity of

Table 11.1 Polynomial coefficients for regular or hyper-cardioid beam patterns of orders $N = 1, ..., 4$.

Order N	a_0	a_1	a_2	a_3	a_4
1	1/4	3/4	0	0	0
2	−3/8	3/4	15/8	0	0
3	−3/8	−15/8	15/8	35/8	0
4	15/32	−15/8	−105/16	35/8	315/32

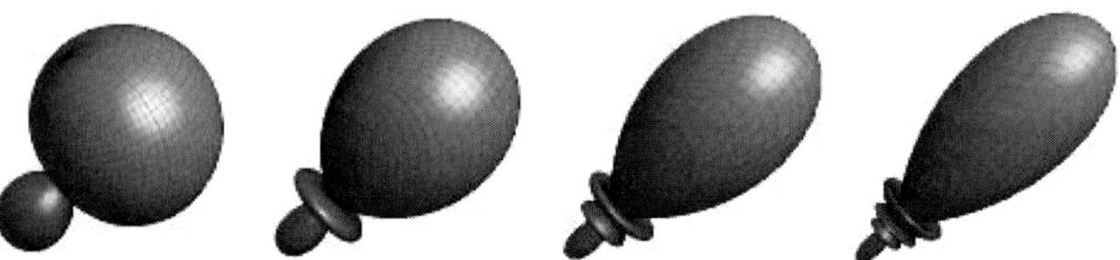

Fig. 11.1 Regular beam pattern for a spherical array of orders (left to right) $N = 1, 2, 3$, and 4.

$$\max_{\mathbf{d}} Q = \mathbf{b}^T \mathbf{H}^{-1} \mathbf{b} = \sum_{n=0}^{N} (2n + 1) = (N + 1)^2. \tag{11.22}$$

Equation (11.22) shows that the maximum value of the directivity is dependent on the array order and is given by $(N + 1)^2$, i.e. proportional to the number of microphones in the array. The directivity of a regular beam pattern is identical to a hypercardioid beam pattern [14]. Equation (11.14) represents trigonometric polynomials in $\cos(\Theta)$, which may be written as

$$y = \frac{1}{\pi} \left[a_0 + a_1 \cos(\Theta) + \cdots + a_N \cos^N(\Theta) \right]. \tag{11.23}$$

The coefficients for array orders $N = 1, ..., 4$ are presented in Table 11.1, see also [1]. Figure 11.1 illustrates the beam pattern of these arrays.

11.4 Delay-and-Sum Beam Pattern

Delay-and-sum is one of the most common beamformers due to the simplicity of its realization and its robustness against noise and uncertainties [13]. In order to implement a delay-and-sum beamformer, the signal from each sensor is aligned by adding a time delay corresponding to the arrival time of the signal, or by introducing phase shifts [4]. Plane waves in a Cartesian coordinate system can be written as $e^{-i\mathbf{k}_0 \cdot \mathbf{r}}$, where $\mathbf{k}_0 = (k, \theta_0, \phi_0)$ in spherical coordinates is the wave vector, (θ_0, ϕ_0) is the propagation direction of the plane wave, and $\mathbf{r} = (r, \theta, \phi)$ is the spatial location. The phase shifts aligning sensor outputs in this case are therefore generally given by $e^{i\mathbf{k}_l \cdot \mathbf{r}}$, with perfect alignment achieved when $\mathbf{k}_l = \mathbf{k}_0$, i.e. when the look direction, given by (θ_l, ϕ_l) with $\mathbf{k}_l = (k, \theta_l, \phi_l)$, is equal to the wave arrival direction. Substituting the plane wave and the phase shifts expressions into (11.8), replacing $p(k, r, \Omega)$ and $w^*(\Omega)$ respectively, gives

$$y(k) = \int_{\Omega \in S^2} e^{i\mathbf{k}_l \cdot \mathbf{r}} e^{-i\mathbf{k}_0 \cdot \mathbf{r}} d\Omega. \tag{11.24}$$

Substituting the spherical harmonics expansion for the exponential functions, or plane waves, as given in (11.4), and using the orthogonality of the spherical harmonics gives [11]

$$y(k) = \sum_{n=0}^{\infty} |b_n(kr)|^2 \frac{2n+1}{4\pi} P_n(\cos \Theta). \tag{11.25}$$

This summation is equivalent to array output, (11.11), with $d_n = |b_n(kr)|^2$. Note that in this case b_n is due to an open sphere denoting plane waves in free field.

A known result for delay-and-sum beamformer is that it has an optimal WNG with a value equal to the number of microphones in the array [13]. The WNG given by (11.13) can be written in a matrix form as

$$WNG = M \frac{\mathbf{d}^H \mathbf{B} \mathbf{d}}{\mathbf{d}^H \mathbf{H} \mathbf{d}}, \tag{11.26}$$

where $\mathbf{d}$ is given by (11.16), $\mathbf{B}$ is given by (11.17), with $\mathbf{b}$ in this case defined as

$$\mathbf{b} = \frac{1}{4\pi} [1, 3, \ldots, 2N+1]^T, \tag{11.27}$$

and matrix $\mathbf{H}$ is defined as

$$\mathbf{H} = \texttt{diag}(1/|b_0|^2, 3/|b_1|^2 \ldots, (2N+1)/|b_N|^2). \tag{11.28}$$

According to (11.11) and given $P_n(1) = 1$, the array output at the look direction is simply $\mathbf{b}^T \mathbf{d}$. Maximizing the WNG with a distortionless constraint at the look direction, $\mathbf{b}^T \mathbf{d} = 1$, is therefore equivalent to solving the following minimization problem:

$$\min_{\mathbf{d}} \mathbf{d}^H \mathbf{H} \mathbf{d} \quad \text{subject to} \quad \mathbf{d}^H \mathbf{B} \mathbf{d} = 1. \tag{11.29}$$

The solution for this minimization problem is [4]

$$\mathbf{d} = \frac{\mathbf{H}^{-1}\mathbf{b}}{\mathbf{b}^{T}\mathbf{H}^{-1}\mathbf{b}}. \tag{11.30}$$

Substituting $\mathbf{b}$ and $\mathbf{H}$ gives:

$$d_n = \frac{4\pi|b_n|^2}{\sum_{n=0}^{N}|b_n|^2(2n+1)}. \tag{11.31}$$

Note that b_n depends on the array configuration. In the case of an open-sphere array, this result tends to the delay-and-sum beamformer as $N \to \infty$, since the summation in the denominator of (11.31) tend to $(4\pi)^2$ in this case. Note that d_n in this case is $|b_n|^2/4\pi$ and not $|b_n|^2$ in order to maintain the constraint. Substituting (11.30) in (11.26) gives the maximum value of the WNG, which is

$$WNG_{max} = M \cdot \mathbf{b}^{T}\mathbf{H}^{-1}\mathbf{b} = \frac{M}{(4\pi)^2}\sum_{n=0}^{N}|b_n|^2(2n+1). \tag{11.32}$$

The maximum WNG approaches M as $N \to \infty$.

11.5 Dolph-Chebyshev Beam Pattern

Dolph-Chebychev beam pattern, widely used in array processing due to the direct control over main-lobe width and maximum side-lobe level [4], has been recently developed for spherical arrays [15]. The development is based on the similarity between the Legendre polynomials that define the spherical array beam pattern and the Chebyshev polynomials that define the desired Dolph-Chebyshev beam pattern, producing array weights that can be computed directly and accurately given desired main-lobe width or side-lobe level.

The Dolph-Chebyshev beampattern is of the form [4]:

$$B(\alpha) = \frac{1}{R}T_M\left(x_0 \cos\left(\alpha/2\right)\right), \tag{11.33}$$

where $T_M(\cdot)$ is the Chebyshev polynomial of order M [10], R is the ratio between the maximum main-lobe level and the maximum side-lobes level, and x_0 is a parameter controlling the null-to-null beamwidth $2\alpha_0$, given by $x_0 = \cos\left(\frac{\pi}{2M}\right)/\cos(\alpha_0/2)$. R and x_0 are also related through $R = \cosh(M\cosh^{-1}(x_0))$.

Equating the Dolph-Chebyshev beampattern (11.33) with the spherical array beampattern given by (11.11), a relation between the spherical array weights d_n and the Dolph-Chebyshev polynomial coefficients can be derived. Details of the derivation have been recently presented [15], its result is presented here:

$$d_n = \frac{2\pi}{R} \sum_{l=0}^{n} \sum_{j=0}^{N} \sum_{m=0}^{j} \frac{1-(-1)^{m+l+1}}{m+l+1} \frac{j!}{m!(j-m)!} \left(\frac{1}{2^j}\right) t_{2j}^{2N} p_l^n x_0^{2j}, \qquad (11.34)$$

where p_l^n denote the coefficients of the l-th power of the n-th order Legendre polynomials, and t_{2j}^{2N} denote the coefficients of the $2j$-th power of the $2N$-th order Chebyshev polynomials. The procedure for designing a Dolph-Chebyshev beam pattern for a spherical array of order N starts with selecting either a desired sidelobe level $1/R$ or a desired main-lobe width $2\alpha_0$, making both x_0 and R available through the relations presented above, and finally calculating the beam pattern coefficients using (11.34).

Equation (11.34) can be written in a compact matrix form as

$$\mathbf{d} = \frac{2\pi}{R} \mathbf{PACTx}_0, \qquad (11.35)$$

where $\mathbf{d}$ is an $(N+1) \times 1$ vector of the beam pattern coefficients given by (11.16), and

$$\mathbf{x}_0 = [1, x_0^2, x_0^4, ..., x_0^{2N}]^T \qquad (11.36)$$

is an $(N+1) \times 1$ vector of the beamwidth parameter x_0. Matrix $\mathbf{P}$ is a lower triangular matrix with each row l consisting of the coefficients of the Legendre polynomial of order l:

$$\mathbf{P} = \begin{bmatrix} p_0^0 & 0 & \cdots & 0 \\ p_0^1 & p_1^1 & \cdots & 0 \\ \vdots & \vdots & \ddots & \vdots \\ p_0^N & p_1^N & \cdots & p_N^N \end{bmatrix}. \qquad (11.37)$$

Matrix $\mathbf{A}$ has elements (l,m) given by $\frac{1-(-1)^{m+l+1}}{m+l+1}$, and matrix $\mathbf{C}$ is an upper triangular matrix, with non-zero (m,l) elements given by $\frac{j!}{m!(j-m)!2^j}$:

$$\mathbf{A} = \begin{bmatrix} 2 & 0 & \cdots & \frac{1-(-1)^{N+1}}{N+1} \\ 0 & \frac{2}{3} & \cdots & \frac{1-(-1)^{N+2}}{N+2} \\ \vdots & \vdots & \ddots & \vdots \\ \frac{1-(-1)^{N+1}}{N+1} & \frac{1-(-1)^{N+2}}{N+2} & \cdots & \frac{1-(-1)^{2N+1}}{2N+1} \end{bmatrix}, \qquad (11.38)$$

$$\mathbf{C} = \begin{bmatrix} 1 & \frac{1}{2} & \cdots & \frac{1}{2^N} \\ 0 & \frac{1}{2} & \cdots & \frac{N}{2^N} \\ \vdots & \vdots & \ddots & \vdots \\ 0 & 0 & \cdots & \frac{1}{2^N} \end{bmatrix}. \qquad (11.39)$$

Finally, matrix $\mathbf{T}$ is a diagonal matrix with diagonal elements j consisting of the Chebyshev polynomial coefficients t_{2j}^{2N}:

$$\mathbf{T} = \mathrm{diag}(t_0^{2N}, t_2^{2N}, ..., t_{2N}^{2N}). \qquad (11.40)$$

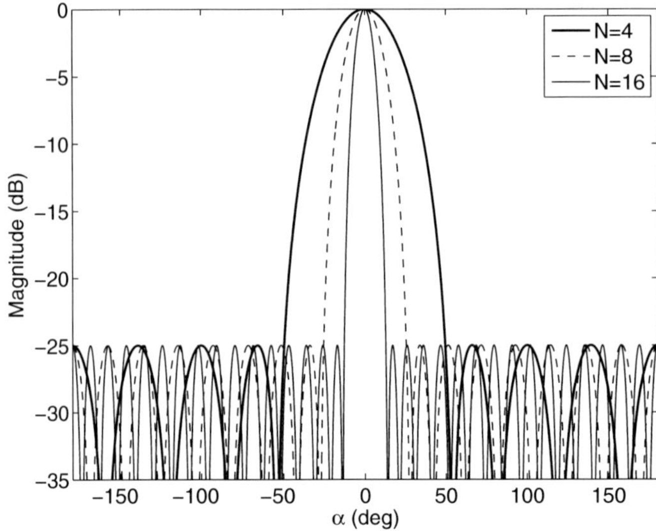

Fig. 11.2 Dolph-Chebyshev beampattern for a spherical array of orders $N = 4, 8, 16$, designed with a side-lobe level constraint of $-25\,$dB.

All four matrices are of size $(N+1) \times (N+1)$. Figure 11.2 presents Dolph-Chebyshev beam patterns for a spherical array of various orders N designed using the method presented in the section. The figure shows that equal side-lobe level is achieved, and that the width of the main lobe is decreased for increasing array order.

11.6 Optimal Beamforming

A wide range of algorithms for optimal beamforming have been previously proposed, with minimum-variance distortionless response (MVDR) an example of a widely used method [4]. In this section the formulation of optimal beamforming is presented, both in the more common space domain, but also in the spherical harmonics domain [12, 16, 17]. The aim of presenting both is twofold. First, to show that the matrix formulation of the algorithms is the same in both domains, so standard methods such as MVDR, presented in this section, but also other methods such as generalized side-lobe canceler (GSC) and linearly constrained minimum variance (LCMV), usually presented in the space domain [4], can be directly applied to designs in the spherical harmonics domain. Second, some advantages of designs in the spherical harmonics over the space domain will be presented.

Spherical microphone array output can be calculated in the space domain using the spatially sampled version of (11.8), or in the spherical harmonics domain, (11.7), as [7]

$$y = \sum_{j=1}^{M} w^* \left(k, \Omega_j\right) p \left(k, r, \Omega_j\right) = \sum_{n=0}^{N} \sum_{m=-n}^{n} w_{nm}^* \left(k\right) p_{nm} \left(k\right), \quad (11.41)$$

where M is the number of spatial samples or microphones. Equation (11.41) can be written in a matrix form by defining the following vectors:

$$\mathbf{p} = [p(k, r_1, \Omega_1), p(k, r_2, \Omega_2), \ldots, p(k, r_M, \Omega_M)]^T \quad (11.42)$$

is the $M \times 1$ vector of sound pressure at the microphones, and

$$\mathbf{w} = [w(k, \Omega_1), w(k, \Omega_2), \ldots, w(k, \Omega_M)]^T \quad (11.43)$$

is the corresponding $M \times 1$ vector of beamforming weights. Similar vectors can be defined in the spherical harmonics domain:

$$\mathbf{p}_{nm} = [p_{00}(k), p_{1(-1)}(k), p_{10}(k), p_{11}(k), \ldots, p_{NN}(k)]^T \quad (11.44)$$

is the $(N+1)^2 \times 1$ vector of the spherical harmonics coefficients of the pressure, and

$$\mathbf{w}_{nm} = [w_{00}(k), w_{1(-1)}(k), w_{10}(k), w_{11}(k), \ldots, w_{NN}(k)]^T \quad (11.45)$$

is the $(N+1)^2 \times 1$ vector of the spherical harmonics coefficients of the weights. Array output can now be written in both domains as

$$y = \mathbf{w}^H \mathbf{p} = \mathbf{w}_{nm}^H \mathbf{p}_{nm}. \quad (11.46)$$

Array manifold vectors, which in this case represent the microphones output due to a unit amplitude plane wave arriving from direction Ω_0, are defined as follows:

$$\mathbf{v} = [v_1, v_2, \ldots, v_M]^T, \quad (11.47)$$

where each element can be calculated from (11.4) and (11.2):

$$v_j = \sum_{n=0}^{\infty} \sum_{m=-n}^{n} b_n(kr) Y_n^{m^*}(\Omega_0) Y_n^m(\Omega_j), \quad (11.48)$$

and

$$\mathbf{v}_{nm} = [v_{00}, v_{1(-1)}, v_{10}, v_{11} \ldots, v_{NN}]^T, \quad (11.49)$$

with

$$v_{nm} = b_n(kr)Y_n^{m^*}(\Omega_0).$$

(11.50)

Array response to a unit amplitude plane wave arriving from Ω_0 can now be written as

$$y = \mathbf{w}^H\mathbf{v} = \mathbf{w}_{nm}^H\mathbf{v}_{nm}.$$

(11.51)

Due to the similarity in formulation in both the space domain and the spherical harmonics domain, the optimal beamformer can be derived in both domain using the same method. Presented here is a derivation for an MVDR beamformer, although a similar derivation can be formulated for other optimal beamformers as detailed in [4], for example.

An MVDR beamformer is designed to minimize the variance of array output, $E[|y|^2]$, with a constraint that forces undistorted response at the array look direction, therefore preserving the source signal while attenuating all other signals to the minimum. The minimization problem for an MVDR beamformer can be formulated in the space domain as [4]

$$\min_{\mathbf{w}} \; \mathbf{w}^H\mathbf{S_p}\mathbf{w} \quad \text{subject to} \quad \mathbf{w}^H\mathbf{v} = 1,$$

(11.52)

where $\mathbf{S_p} = E[\mathbf{pp}^H]$ is the cross-spectrum matrix. The solution to this problem is [4]

$$\mathbf{w} = \frac{\mathbf{S_p}^{-1}\mathbf{v}}{\mathbf{v}^H\mathbf{S_p}^{-1}\mathbf{v}}.$$

(11.53)

In the spherical harmonics domain, the problem formulation is similar:

$$\min_{\mathbf{w}_{nm}} \; \mathbf{w}_{nm}^H\mathbf{S}_{\mathbf{p}_{nm}}\mathbf{w}_{nm} \quad \text{subject to} \quad \mathbf{w}_{nm}^H\mathbf{v}_{nm} = 1,$$

(11.54)

where $\mathbf{S}_{\mathbf{p}_{nm}} = E[\mathbf{p}_{nm}\mathbf{p}_{nm}^H]$, with a solution:

$$\mathbf{w}_{nm} = \frac{\mathbf{S}_{\mathbf{p}_{nm}}^{-1}\mathbf{v}_{nm}}{\mathbf{v}_{nm}^H\mathbf{S}_{\mathbf{p}_{nm}}^{-1}\mathbf{v}_{nm}}.$$

(11.55)

It is clear that the formulation in the space and spherical harmonics domains are similar. However, the formulation in the spherical harmonics domain may have advantages. First, in practice, typical arrays employ spatial over-sampling, so vector $\mathbf{w}_{nm}$ and matrix $\mathbf{S}_{\mathbf{p}_{nm}}$ will be of lower dimensions compared to $\mathbf{w}$ and $\mathbf{S_p}$, and so the spherical harmonics formulation may be more efficient. Also, the components of the steering vector, $\mathbf{v}_{nm}$, have a simpler form compared to $\mathbf{v}$, the latter involving a double summation. Furthermore, the same formulation can be used for various array configurations, such as open sphere, rigid sphere, and an array with cardioid microphones.

11.7 Beam Pattern with Desired Multiple Nulls

A multiple-null beamformer for spherical arrays, useful when the sources arrive from known directions, is presented in this section. Recently, such multiple-null beamformer has been employed in the analysis of directional room impulse responses [18]. Improved performance can be achieved by a multiple-null beamformer compared to a regular beamformer due to the ability of the former to significantly suppress undesired interfering sources. Formulation of the multiple-null beamformer is presented both in the space domain and in the spherical harmonics domain.

Consider a sound field composed of L plane waves arriving from directions denoted by Ω_l, $l = 1, \ldots, L$, and a spherical array of order N. Following the notation used in section 11.6, $\mathbf{w}$ and $\mathbf{w}_{nm}$ are array weights in the space and spherical harmonics domains, as in (11.43) and (11.45) respectively, and $\mathbf{v}$ and $\mathbf{v}_{nm}$ are the steering vectors as in (11.47) and (11.49). In this section, steering vectors due to a plane wave arriving from Ω_l are denoted by $\mathbf{v}^l$ and $\mathbf{v}^l_{nm}$. Now, if the plane wave from direction Ω_l is a desired source, or signal, a constraint of $\mathbf{w}^H \mathbf{v}^l = 1$ will ensure receiving this source without distortion. If, on the other hand, the plane wave is an interfering source, or noise, a constraint of $\mathbf{w}^H \mathbf{v}^l = 0$ will ensure cancellation of this source. Assuming L_s out of the L plane waves are desired sources, and $L - L_s$ are interfering sources, we get L constraints that can be written in a matrix form as

$$\mathbf{w}^H \mathbf{V} = \mathbf{c}, \tag{11.56}$$

where matrix $\mathbf{V}$ of dimensions $M \times L$ is a steering matrix composed of L columns, each column containing a steering vector for the l-th plane wave:

$$\mathbf{V} = [\mathbf{v}^1, \mathbf{v}^2, \ldots, \mathbf{v}^L]. \tag{11.57}$$

Vector $\mathbf{c}$ of dimensions $1 \times L$ contains the constraints values, i.e. L_s unit values and $L - L_s$ zero values:

$$\mathbf{c} = [1, 1, \ldots, 1, 0, 0, \ldots, 0]. \tag{11.58}$$

A similar formulation can be derived for the array in the spherical harmonics domain, replacing $\mathbf{w}$, $\mathbf{v}$, and $\mathbf{V}$ with $\mathbf{w}_{nm}$, $\mathbf{v}_{nm}$, and $\mathbf{V}_{nm}$, respectively:

$$\mathbf{V}_{nm} = [\mathbf{v}^1_{nm}, \mathbf{v}^2_{nm}, \ldots, \mathbf{v}^L_{nm}], \tag{11.59}$$

and

$$\mathbf{w}^H_{nm} \mathbf{V}_{nm} = \mathbf{c}. \tag{11.60}$$

A least-squares solution can be applied to (11.56) and (11.60) to solve for the coefficients, i.e.,

$$\mathbf{w} = \mathbf{V}^\dagger \mathbf{c}^T, \tag{11.61}$$

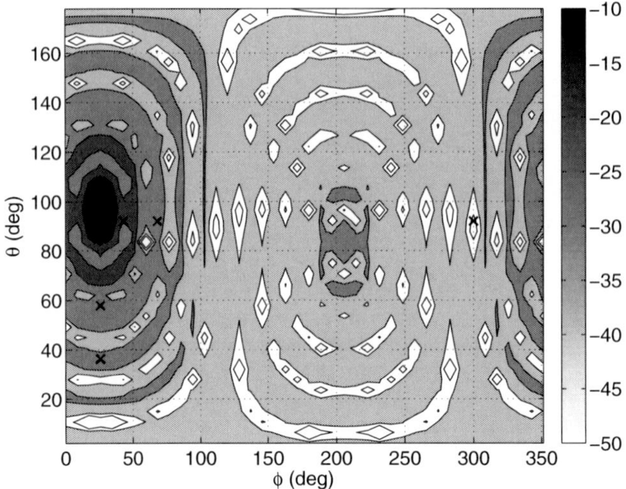

Fig. 11.3 Regular beampattern.

where $\mathbf{V}^{\dagger}$ is the pseudo-inverse of $\mathbf{V}$. Similarly,

$$\mathbf{w}_{nm} = \mathbf{V}_{nm}^{\dagger}\mathbf{c}^{T}, \qquad (11.62)$$

is the solution in the spherical harmonics domain. As detailed in Section 11.6, array processing in the spherical harmonics domain may have several advantages over array processing in the space domain.

In the following example multiple-null beamforming is compared to regular beamforming. A spherical array composed of a rotating microphone was used to measure room impulse responses in an auditorium, see experiment details in [19]. A loudspeaker was placed on the stage at the Sonnenfeldt auditorium, Ben-Gurion University, and the scanning dual-sphere microphone array [6] was placed at the seating area. The array provided sound pressure data in the frequency range 500-2800 Hz. Direct sound and early room reflections have been identified [19]. In this example, multiple-null beamforming is realized by setting a distortionless constraint at the the direction of the direct sound, and setting nulls at the directions of the five early reflections, such that $L = 6$ and $L_s = 1$. Array order of $N = 10$ has been used. Figures 11.3 and 11.4 present the regular and multiple-null beampatterns, respectively. Although the beampatterns look similar, the multiple-null beampattern has zeros, or low response, at the directions of the five room reflections, denoted by "$\times$" marks on the figures. The multiple-null beampattern was shown to reduce the effect of the early reflections, relative to the regular beampattern, when analyzing the directional impulse response in the direction of the direct sound [18].

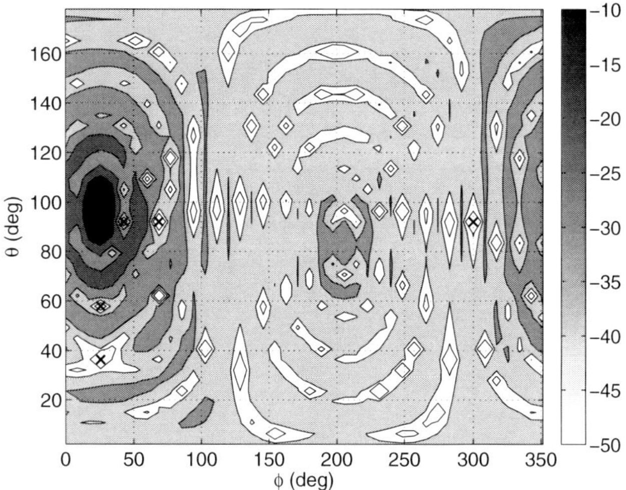

Fig. 11.4 Multiple-null beampattern.

11.8 2D Beam Pattern and its Steering

The arrays presented above employed beam patterns that are rotationally symmetric about the look direction. The advantages of these beam patterns are simplicity of the 1D beam-pattern design through d_n and the ease of beam steering through Ω_l, see (11.9). However, in some cases we may be interested in beam patterns that are not rotationally-symmetric about the look direction. A situation may arise where the sound sources occupy a wide region in directional space, such as a stage in an auditorium, or few speakers positioned in proximity. In this case the main lobe should be wide over the azimuth and narrow over the elevation, and so beam patterns that are rotationally-symmetric about the look direction may not be suitable. In this case we may want to select more general array coefficients given by [7]

$$w_{nm}^*(k) = \frac{c_{nm}(k)}{b_n(kr)}. \tag{11.63}$$

Array directivity can now be calculated similar to (11.10) as

$$y(\Omega) = \sum_{n=0}^{N} \sum_{m=-n}^{n} c_{nm} Y_n^{m^*}(\Omega). \tag{11.64}$$

Beam pattern y and coefficients c_{nm} are related simply through the spherical Fourier transform and complex-conjugate operations. This provides a simple framework for the design of c_{nm} once the desired beam pattern is available. However, the steering of such beam pattern is more complicated, and has been recently presented [9], making use of the Wigner-D function [20]:

$$D_{mm'}^n(\alpha, \beta, \gamma) = e^{-im\alpha} d_{mm'}^n(\beta) e^{-im'\gamma}, \tag{11.65}$$

where α, β, γ represent rotation angles, and d_{nm}^n is the Wigner-d function, which is real and can be written in terms of the Jacobi polynomial [20]. The Wigner-D functions form a basis for the rotational Fourier transform, applied to functions defined over the rotation group SO(3) [20, 21]. They are useful for beam pattern steering due to the property that a rotated spherical harmonics can be represented using the following summation [21]:

$$\Lambda(\alpha, \beta, \gamma) Y_n^{m'}(\theta, \phi) = \sum_{m=-n}^{n} D_{mm'}^n(\alpha, \beta, \gamma) Y_n^m(\theta, \phi), \tag{11.66}$$

where $\Lambda(\alpha, \beta, \gamma)$ denotes the rotation operation, which can be represented by Euler angles [10]. In this case an initial counter-clockwise rotation of angle γ is performed about the z-axis, followed by a counter-clockwise rotation by angle β about the y-axis, and completed by a counter-clockwise rotation of angle α about the z-axis. See, for example, [10, 20] for more details on Euler angles and rotations.

Using (11.66) and (11.64) a general expression for a rotated beam pattern can be derived:

$$
\begin{aligned}
y^r(\Omega) &= \Lambda(\alpha, \beta, \gamma) y(\Omega) \\
&= \sum_{n=0}^{N} \sum_{m=-n}^{n} c_{nm} \sum_{m'=-n}^{n} D_{m'm}^{n*}(\alpha, \beta, \gamma) Y_n^{m'*}(\Omega) \\
&= \sum_{n=0}^{N} \sum_{m'=-n}^{n} \left[\sum_{m=-n}^{n} c_{nm} D_{m'm}^{n*}(\alpha, \beta, \gamma) \right] Y_n^{m'*}(\Omega) \\
&= \sum_{n=0}^{N} \sum_{m'=-n}^{n} c_{nm'}^r Y_n^{m'*}(\Omega).
\end{aligned}
\tag{11.67}
$$

The result is similar to (11.64) with y^r replacing y and c_{nm}^r replacing c_{nm}. Therefore, rotation of the beam pattern can simply be achieved by weighting the original coefficients with the Wigner-D function at the rotation angles, such that

$$c_{nm'}^r = \sum_{m=-n}^{n} c_{nm} D_{m'm}^{n*}(\alpha, \beta, \gamma). \tag{11.68}$$

The rotated coefficients w_{nm}^r can be written in terms of the original coefficients w_{nm} as

$$w_{nm'}^r = \sum_{m=-n}^{n} w_{nm} D_{m'm}^n(\alpha, \beta, \gamma). \tag{11.69}$$

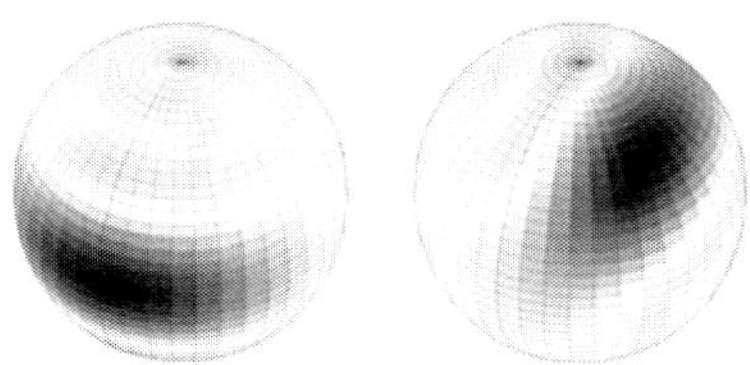

Fig. 11.5 Magnitude of the beam pattern $y(\Omega)$, plotted as gray level on the surface of a unit sphere, left: initial look direction, right: rotated beam pattern.

Equation (11.69) can be written in a matrix form as

$$\mathbf{w}^r = \mathbf{D}\mathbf{w}, \tag{11.70}$$

where $\mathbf{w}^r$ is the $(N+1)^2 \times 1$ vector of coefficients of the rotated beam pattern, $\mathbf{w}$ is the $(N+1)^2 \times 1$ vector of the original coefficients, given by

$$\mathbf{w}^r = [w_{00}^r, w_{1(-1)}^r, w_{10}^r, w_{11}^r, \cdots, w_{NN}^r]^T, \tag{11.71}$$

$$\mathbf{w} = [w_{00}, w_{1(-1)}, w_{10}, w_{11}, \cdots, w_{NN}]^T, \tag{11.72}$$

and $\mathbf{D}$ is an $(N+1)^2 \times (N+1)^2$ block diagonal matrix, having block elements of $\mathbf{D}^0, \mathbf{D}^1, ..., \mathbf{D}^N$. Matrices $\mathbf{D}^n$ are of dimension $(2n+1) \times (2n+1)$ with elements $D_{m'm}^n(\alpha, \beta, \gamma)$. For example, $\mathbf{D}^0 = D_{00}^0$,

$$\mathbf{D}^1 = \begin{pmatrix} D_{(-1)(-1)}^1 & D_{(-1)0}^1 & D_{(-1)1}^1 \\ D_{0(-1)}^1 & D_{00}^1 & D_{01}^1 \\ D_{1(-1)}^1 & D_{10}^1 & D_{11}^1 \end{pmatrix}, \tag{11.73}$$

and so on. As the addition of any two rotations produces another rotation [20], successive rotations $\mathbf{D}_1$ and $\mathbf{D}_2$ can be implemented by multiplying the two rotation matrices, i.e. $\mathbf{D}_2\mathbf{D}_1$, to produce an equivalent rotation.

Figure 11.5 (left) shows an example beam pattern for an array of order $N = 6$. The look direction in this case is $(\theta, \phi) = (90°, 180°)$. Figure 11.5 (right) shows the beam pattern after it has been rotated to a new look direction $(\theta, \phi) = (45°, 240°)$, while another rotation of $\psi_l = 30°$ has been applied about the new look direction.

11.9 Near-Field Beamforming

The beamformers presented in previous sections all assume that the sound sources are in the far-field, producing plane wave sound field around the array. This assumption approximately holds, as long as the sources are far enough from the array. However, for applications in which the sources are close to the array, such as close talk microphones, video conferencing and music recording, the far-field assumption may not hold, and could lead to design or analysis errors. Furthermore, the spherical wavefront of near-field sources includes important spatial information which may be utilized for improved spatial separation and design [22, 23].

The sound field due to a unit amplitude point source located at $\mathbf{r}_s = (r_s, \theta_s, \phi_s)$ is [3]

$$p(k, r, \Omega) = \frac{e^{ik|\mathbf{r}-\mathbf{r}_s|}}{|\mathbf{r} - \mathbf{r}_s|} = \sum_{n=0}^{\infty} \sum_{m=-n}^{n} b_n^s(kr, kr_s) Y_n^{m*}(\Omega_s) Y_n^m(\Omega), \qquad (11.74)$$

where $b_n^s(kr, kr_s)$ depends on the sphere boundary and is related to $b_n(kr)$ through [24]

$$b_n^s(kr, kr_s) = ki^{1-n} b_n(kr) h_n(kr_s). \qquad (11.75)$$

The spherical Fourier transform of (11.74) is [23]

$$p_{nm}(k, r) = b_n^s(kr, kr_s) Y_n^{m*}(\Omega_s). \qquad (11.76)$$

Traditionally, the Fraunhofer or Fresnel distances are used to determine the transition between near-field and far-field of sensor arrays [25]. Although these parameters guarantee a limit on phase error under the far-field assumption, they do not necessarily indicate the extent of the near-field of spherical microphone arrays in the context of capabilities such as radial discrimination. A criterion for defining the near-field of the spherical microphone array was suggested in [23], based on comparison of the magnitude of the far-field and near-field radial components, $b_n(kr)$ and $b_n^s(kr, kr_s)$, respectively, where it has been shown that for $kr_s > n$ the two functions are similar. Therefore, given an array of order N, a point source is considered in the near field if [23]

$$kr_s \leq N. \qquad (11.77)$$

Now, due to the physical constraint that the source is external to the array, $r_s \geq r_a$, the maximum wave number that allows near-field conditions is

$$k_{max} \approx \frac{N}{r_a}. \qquad (11.78)$$

Note that this is also typically the highest wavenumber for which there is no significant spatial aliasing [8]. Using (11.77) and (11.78), the criterion for which a source at distance r_s from the array is considered near-field is given by

$$r_{NF} = r_a \frac{k_{max}}{k}, \qquad (11.79)$$

such that $r_s \leq r_{NF}$ is in the near field. Equation (11.79) suggests that a spherical array with a large near-field extent ($r_{NF} >> r_a$) can be achieved either at low frequencies ($k << k_{max}$) or by over-sampling the array, therefore increasing N and k_{max}.

Equation (11.7) can be used for the computation of array output, only in this case near-field versions of the weights w^*_{nm} and the pressure coefficients p_{nm} are employed. The latter has been presented in (11.76). When the source distance r_s is known, the weights can be selected to eliminate term b^s_n using [26]:

$$w^*_{nm} = \frac{d_n}{b^s_n(kr, kr_s)} Y^m_n(\Omega_l).$$
(11.80)

Given a unit amplitude point source as in (11.76), the array output is the same as in (11.11). However, r_s is not usually known, and so a more general weighting function has been suggested [23]:

$$w^*_{nm} = \frac{d_n}{b^s_n(kr, kr_0)} Y^m_n(\Omega_l),$$
(11.81)

where r_0 is a controllable parameter. It was shown in [23] that the "look distance," r_0, can be adjusted to enhance the gain of sources in the near-field relative to sources in the far field, and vice-versa. The array output due to a unit amplitude point source has a different form in this case, given by

$$y = \sum_{n=0}^{N} d_n \frac{h_n(kr_s)}{h_n(kr_0)} \frac{2n+1}{4\pi} P_n(\cos\Theta).$$
(11.82)

As an example for radial filtering with the near-field beamformer, consider an array of order $N = 4$ and radius $r_a = 0.1$ meters. k_{max} in this case is 40, and so two cases are considered. First, with $k = 4 \ll k_{max}$, and a near-field radius of $r_{NF} = 1$ meter. Second, with $k = 40 = k_{max}$, and a near-field radius of $r_{NF} = r_a = 0.1$ meters. Array weighting $d_n = 1$ was selected in both examples, with array look-distance of $r_0 = 0.2$ meters.

Figures 11.6 and 11.7 show the magnitude of array output as a function of r_s, the distance of the point source from the array center, and the angle of the point source away from the look direction, Θ. Note that the array output has been normalized with the natural point source decay term, $1/r_s$. Figure 11.6 shows that when the near-field criterion is satisfied, the array output decreases as r_s increases, therefore achieving enhancement of near sources relative to far sources. However, when the near-field criterion is not satisfied as in Fig. 11.7, the normalized array output does not allow for the discrimination of sources based on radial distance. Also, both figures show that when the source is not in the look direction, it will be attenuated relative to a source in the look direction.

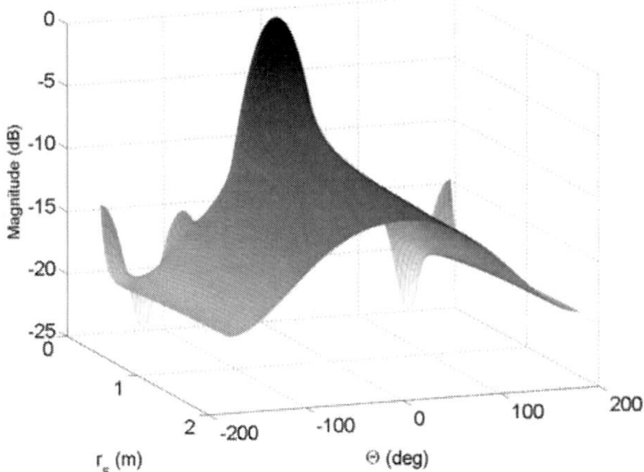

Fig. 11.6 Spatial beam pattern for $k = \frac{k_{max}}{10}$.

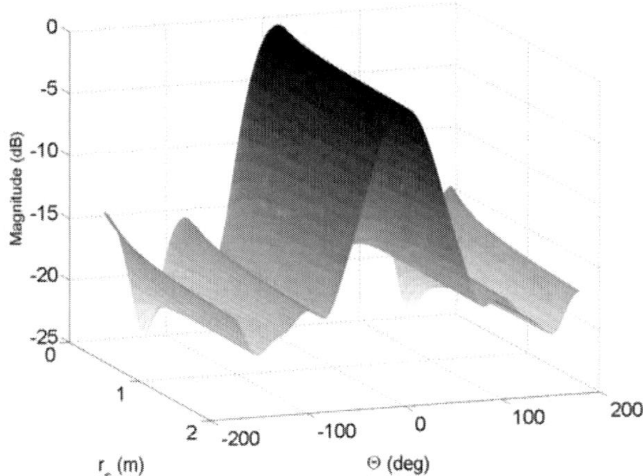

Fig. 11.7 Spatial beam pattern for $k = k_{max}$.

11.10 Direction-of-Arrival Estimation

The task of estimating the direction of arrival (DOA) of sources has been of interest for a wide range of applications [4]. Various DOA estimation methods have been previously studied, based on beamforming, Maximum Likelihood (ML) [27] and subspace techniques such as MUSIC [28] and ESPRIT [29]. Most of these methods can be implemented with arbitrary array configurations, although some methods have been designed for specific configurations

[4]. This section presents DOA estimation methods recently developed for spherical arrays.

MUSIC is a popular DOA estimation algorithm due to its simplicity and high spatial resolution. A MUSIC method for spherical arrays formulated in the spherical harmonics domain is presented next. Consider L plane-waves with amplitudes $a_l(k)$ incident on a spherical array from directions Ω_l, $l = 1, \ldots, L$. Following (11.4), the spherical Fourier coefficients of the pressure on the array can be written as

$$p_{nm}(k, r) = b_n(kr) \sum_{l=1}^{L} a_l(k) Y_n^{m^*}(\Omega_l). \tag{11.83}$$

Defining $s_{nm}(k) = p_{nm}(k, r)/b_n(kr)$, we can write (11.83) as

$$s_{nm}(k) = \sum_{l=1}^{L} a_l(k) Y_n^{m^*}(\Omega_l), \tag{11.84}$$

or in a more compact matrix form:

$$\mathbf{s}_{nm} = \mathbf{Y}^H \mathbf{a}, \tag{11.85}$$

where $\mathbf{a}$ is the $L \times 1$ plane-waves amplitudes vector:

$$\mathbf{a} = [a_1(k), a_2(k), \ldots, a_L(k)]^T, \tag{11.86}$$

and $\mathbf{Y}$ is an $L \times (N+1)^2$ beamspace manifold matrix with the l-th row given by

$$\mathbf{y}(\Omega_l) = \left[Y_0^0(\Omega_l), Y_1^{-1}(\Omega_l), Y_1^0(\Omega_l), Y_1^1(\Omega_l), \ldots, Y_N^N(\Omega_l) \right]. \tag{11.87}$$

Following [4], given J snapshots, an estimate of the cross-spectrum matrix can be calculated as

$$\hat{\mathbf{S}} = \frac{1}{J} \sum_{j=1}^{J} \mathbf{s}_{nm}^j \mathbf{s}_{nm}^{j^H}. \tag{11.88}$$

MUSIC spectrum can then be computed as

$$P_{MU}(\Omega) = \frac{1}{\mathbf{y}(\Omega) \mathbf{E}_N \mathbf{E}_N^H \mathbf{y}^H(\Omega)}, \tag{11.89}$$

where columns of matrix $\mathbf{E}_N$ are the eigenvectors of matrix $\hat{\mathbf{S}}$ associated with the $(N+1)^2 - L$ smallest eigenvalues, forming the noise subspace. These eigenvectors can be found by singular value decomposition of $\hat{\mathbf{S}}$ [4]. Peaks in the MUSIC spectrum indicate DOA of the plane waves. The advantages of using MUSIC in the spherical harmonics domain compared to the space domain are similar to those presented in Section 11.6. In particular, the steering

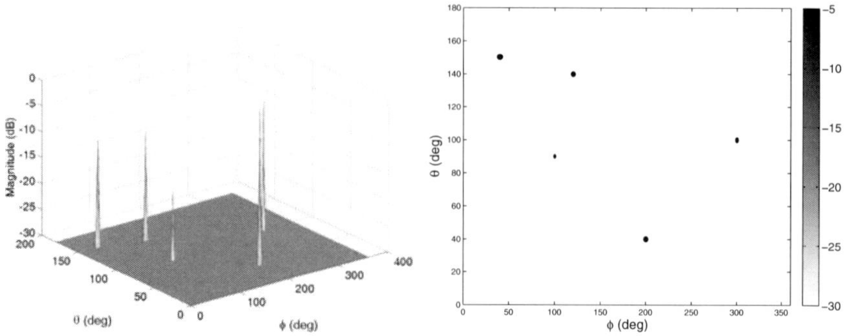

Fig. 11.8 MUSIC spectrum showing peaks at the plane waves arrival directions, left: 3Dplot, right: 2D plot.

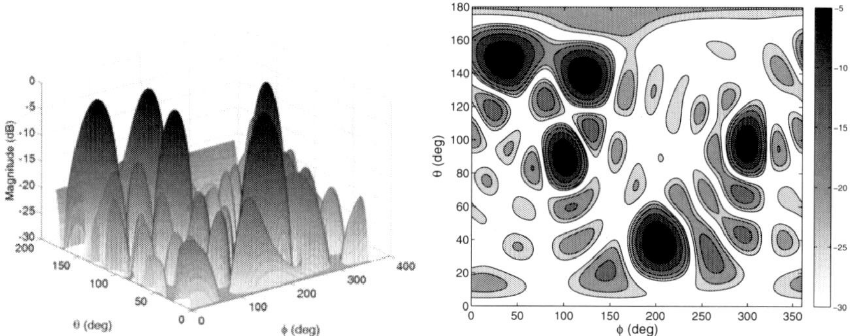

Fig. 11.9 Magnitude of array output with regular beamforming, left: 3Dplot, right: 2D plot.

matrix $\mathbf{Y}^H$ has a simplified form and typically lower size, and is frequency independent and independent on array configuration.

A simulation example is presented next, comparing DOA estimation by MUSIC realized in the spherical harmonics domain and beamforming by a regular beam pattern [5], as presented in Section 11.3. A spherical array of order $N = 8$ is positioned in a sound field composed of five uncorrelated and equal-power plane waves. Figures 11.8 and 11.9 show MUSIC spectrum and the output of the regular beamforming, respectively, illustrating the peaks corresponding to the DOA of the plane waves. The figures show that MUSIC provides sharper and well separated peaks compared to regular beamforming, due to the improved spatial resolution it exhibits.

However, the drawback of MUSIC is the requirement for a two-dimensional search to find both azimuth and elevation angles of the DOA, making it computationally inefficient. Recently, methods for DOA estimation for spherical microphone arrays that avoid the search process were presented [30, 31]. These algorithms are realizations of the well known subspace-based technique ESPRIT in the spherical harmonics domain, exploiting the recurrence relation of spherical harmonics [20].

Nevertheless, MUSIC and even ESPRIT have an additional drawback. They do not work for coherent sources, for example in a multipath environment. A possible remedy comes in the form of spatial smoothing (SS) [32], by decorrelating coherent signals. SS is performed by constructing spatially-shifted sub-arrays, and averaging the cross-spectrum matrices of these sub-arrays to form a higher-rank smoothed sources cross-spectrum matrix. SS was originally developed for uniform linear arrays (ULAs), but [33] presented sufficient and necessary conditions for successful application of SS to arrays of arbitrary configurations. SS was applied in [34] to spherical arrays, where conditions on the shifted arrays were presented, that assure successful application of SS.

11.11 Conclusions

This chapter presented an overview of recent studies of spherical microphone arrays, concentrating on spatial filtering, or beamforming. The results presented in this chapter can be used to design spherical microphone arrays and implement array processing in a wide range of applications.

Acknowledgements This research was supported in part by THE ISRAEL SCIENCE FOUNDATION (grant No. 155/06).

References

1. J. Meyer and G. W. Elko, "A highly scalable spherical microphone array based on an orthonormal decomposition of the soundfield," in *Proc. IEEE ICASSP*, 2002, vol. II, pp. 1781–1784.
2. T. D. Abhayapala and D. B. Ward, "Theory and design of high order sound field microphones using spherical microphone array," in *Proc. IEEE ICASSP*, 2002, vol. II, pp. 1949–1952.
3. E. G. Williams, *Fourier Acoustics: Sound Radiation and Nearfield Acoustical Holography*. New York: Academic Press, 1999.
4. H. L. V. Trees, *Optimum Array Processing (Detection, Estimation, and Modulation Theory, Part IV)*, 1st ed. Wiley-Interscience, 2002.
5. B. Rafaely, "Plane-wave decomposition of the pressure on a sphere by spherical convolution," *J. Acoust. Soc. Am.*, vol. 116, no. 4, pp. 2149–2157, 2004.
6. I. Balmages and B. Rafaely, "Open-sphere designs for spherical microphone arrays," *IEEE Trans. Audio, Speech and Lang. Proc.*, vol. 15, no. 2, pp. 727–732, Feb. 2007.
7. B. Rafaely, "Analysis and design of spherical microphone arrays," *IEEE Trans. Speech Audio Proc.*, vol. 13, no. 1, pp. 135–143, 2005.
8. B. Rafaely, B. Weiss, and E. Bachmat, "Spatial aliasing in spherical microphone arrays," *IEEE Trans. Sig. Proc.*, vol. 55, no. 3, pp. 1003–1010, Mar. 2007.

9. B. Rafaely, "Spherical microphone array beam steering using Wigner-D weighting," *IEEE Sig. Proc. Let.*, vol. 15, pp. 417–420, 2008.

10. G. Arfken and H. J. Weber, *Mathematical Methods for Physicists*, 5th ed. San Diego: Academic Press, 2001.

11. B. Rafaely, "Phase-mode versus delay-and-sum spherical microphone array processing," *IEEE Sig. Proc. Let.*, vol. 12, no. 10, pp. 713–716, 2005.

12. Z. Li and R. Duraiswami, "Flexible and optimal design of spherical microphone arrays for beamforming," *IEEE Trans. Audio, Speech and Lang. Proc.*, vol. 15, no. 2, pp. 702–714, Feb. 2007.

13. M. Brandstein and D. Ward, *Microphone arrays: Signal Processing Techniques and Applications.* Berlin: Springer, 2001.

14. Y. Huang and J. Benesty, *Audio Signal Processing for Next-Generation Multimedia Communication Systems.* Kluwer Academic Publishers, 2004.

15. A. Koretz and B. Rafaely, "Dolph-chebyshev beampattern design for spherical microphone arrays," *IEEE Trans. Sig. Proc.*, Accepted for publication 2009.

16. Y. Peled and B. Rafaely, "Study of speech intelligibility in noisy enclosures using spherical microphones arrays," in *Hands-free Speech Communication and Microphone Arrays (HSCMA 2008)*, Trento, Italy, May 2008, pp. 160–163.

17. ——, "Study of speech intelligibility in noisy enclosures using optimal spherical beamforming," in *IEEE 25th Convention of Electrical and Electronics Engineers in Israel, IEEEI 2008*, Eilat, Israel, Dec. 2008, pp. 285–289.

18. B. Rafaely, "Spherical microphone array with multiple nulls for analysis of directional room impulse responses," in *Proc. IEEE ICASSP*, 2008, pp. 281–284.

19. B. Rafaely, I. Balmages, and L. Eger, "High-resolution plane-wave decomposition in an auditorium using a dual-radius scanning spherical microphone array," *J. Acoust. Soc. Am.*, vol. 122, no. 5, pp. 2661–2668, Nov. 2007.

20. D. A. Varshalovich, A. N. Moskalev, and V. K. Khersonskii, *Quantum Theory of Angular Momentum*, 1st ed. Singapore: World Scientific Publishing, 1988.

21. P. J. Kostelec and D. N. Rockmore, "Ffts on the rotation group," Santa-Fe Institute, Working Paper series, Paper 03-11-060, 2003.

22. J. G. Ryan and R. A. Goubran, "Array optimization applied in the near field of a microphone array," *IEEE Trans. Speech and Audio Proc.*, vol. 8, no. 2, pp. 173–176, 2000.

23. E. Fisher and B. Rafaely, "The near field spherical microphone array," in *Proc. IEEE ICASSP*, 2008, pp. 5272–5275.

24. J. Meyer and G. W. Elko, "Position independent close-talking microphone," *Signal Processing*, vol. 86, no. 6, pp. 1254–1259, June 2006.

25. C. A. Balanis, *Antenna Theory: Analysis and Design*, 3rd ed. Wiley-Interscience, 2005.

26. J. Meyer and G. W. Elko, "Position independent close-talk microphone," *Signal Processing*, vol. 86, pp. 1254–1259, 2005.

27. J. F. Bohme, "Estimation of source parameters by maximum likelihood and nonlinear regression," in *Proc. IEEE ICASSP*, 1984, pp. 7.3.1–7.3.4.

28. R. O. Shmidt, "Multiple emitter location and signal parameter estimation," *IEEE Trans. Antennas and Propagatation*, vol. 34, pp. 276–280, Mar. 1984.

29. R. Roy and T. Kailath, "Esprit-estimation of signal parameters via rotational invariance techniques," *IEEE Trans. Acoust. Speech Sig. Proc.*, vol. 37, pp. 984–995, July 1989.

30. H. Teutsch and W. Kellermann, "Detection and localization of multiple wideband acoustic sources based on wavefield decomposition using spherical apertures," in *Proc. IEEE ICASSP*, 2008, pp. 5276–5279.

31. R. Goossens and H. Rogier, "Closed-form 2d angle estimation with a spherical array via spherical phase mode excitation and esprit," in *Proc. IEEE ICASSP*, 2008, pp. 2321–2324.

32. T. J. Shan, M. Wax, and T. Iiailath, "On spatial smoothing for direction-of-arrival estimation of coherent signals," *IEEE Trans. Acoust. Speech and Sig. Proc.*, vol. 33, no. 4, pp. 806–811, Aug. 1985.
33. H. Wang and K. J. R. Liu, "2-d spatial smoothing for multipath coherent signal separation," *IEEE Trans. Aerosp. Electron. Syst.*, vol. 34, p. 391405, Apr. 1998.
34. D. Khaykin and B. Rafaely, "Signal decorrelation by spatial smoothing for spherical microphone arrays," in *IEEE 25th Convention of Electrical and Electronics Engineers in Israel, IEEEI 2008*, Eilat, Israel, Dec. 2008, pp. 270–274.

Chapter 12
Steered Beamforming Approaches for Acoustic Source Localization

Jacek P. Dmochowski and Jacob Benesty

Abstract Multiple microphone devices aimed at hands-free capabilities and speech recognition via acoustic beamforming require reliable estimates of the position of the acoustic source. Two-stage approaches based on the time-difference-of-arrival (TDOA) are computationally simple but lack robustness in practical environments. This chapter presents a family of broadband source localization algorithms based on parameterized spatiotemporal correlation, including the popular and robust steered response power (SRP) algorithm. Before forming the conventional spatial correlation matrix, the vector of microphone signals is time-aligned with respect to a hypothesized source location. It is shown that this parametrization easily generalizes classical narrowband techniques to the broadband setting. Methods based on minimum information entropy and temporally constrained minimum variance are developed. In order to ease the high computational demands imposed by a location parameterized scheme, a sparse representation of the parameterized spatial correlation matrix is proposed.

12.1 Introduction

The localization of acoustic sources represents both a classical parameter estimation problem and an important component of more general problems such as hands-free voice communication and speech recognition. The problem is significantly more difficult than that of narrowband source localization [1], [2], as the desired signal is wideband and its propagation to the array is convolutive.

Jacek P. Dmochowski
City College of New York, NY, USA, e-mail: `jdmochowski@ccny.cuny.edu`

Jacob Benesty
INRS-EMT, QC, Canada, e-mail: `benesty@emt.inrs.ca`

I. Cohen et al. (Eds.): Speech Processing in Modern Communication, STSP 3, pp. 307–337.
springerlink.com

Simply put, the localization problem is to estimate the position of the source using only the signals observed at an array of microphones. The basic premise behind localization is that sources at different locations will exhibit different relative delays – the differences in propagation time from one microphone to the next. This physical property forms the basis for virtually all localization methods, and also necessitates the use of multiple microphones. While the theoretical minimum number of microphones is indeed two, it will be shown in the remaining sections of this chapter that it is advantageous to include additional microphones in the localization scheme.

The other classical acoustic localization techniques consist of two steps. In the first step, the relative delays across the various microphone pairs are estimated using the generalized cross-correlation (GCC) method described in [3]. The estimates of the relative delays are then mapped to the source location using one of a handful of techniques ranging from maximum-likelihood estimation to spherical interpolation [4], [5], [6]. In practice, the estimates of the relative delays are quite noisy. The second step typically involves a non-linear transformation of these relative delays, leading to performance limitations in harsh environments.

To that end, the methods presented in this chapter refrain from making a hard decision on a single relative delay between each microphone pair. Instead, a spatial statistic is formed for each potential location; having computed such a statistic for each candidate location, the algorithms designate the estimate of the source by the location which either minimizes or maximizes the statistic, depending on the context. Instead of parameterizing the relative delay, these techniques parameterize the cross-correlation functions with the location of the source.

Acoustic source localization is an active research area. We very briefly mention here some recent trends. Particle filtering methods formulate the localization problem in a state space and model the location probabilistically using a Markov process where the current "state" (i.e., location and velocity) evolve as a function of the previous states and estimates of the source location [7], [8], [9], [10], [11]. The particle filtering approach is useful for tracking purposes. While the majority of localization algorithms have focused on the inter-microphone delays, recently approaches utilizing energy measurements have also been proposed [12], [13], [14]. Localization methods employing a set of distributed microphones have also been studied [15], [16]. In this paradigm, the array consists of a set of networked multimedia devices with embedded microphones. Finally, methods based on blind multiple-input multiple-output (MIMO) identification of the acoustic impulse responses are presented in [17], [18]. This chapter is not meant to serve as a comprehensive treatment of the state-of-the-art. Rather, we present herein a group of algorithms based on parameterized spatial correlation which generalize classical narrowband techniques to the broadband setting.

The signal model used throughout the chapter is described in Section 12.2. Section 12.3 briefly covers the fundamentals of spatial and spatiotemporal filtering. Section 12.4 presents the parameterized spatial correlation matrix as the fundamental structure in acoustic source localization and a family of methods based on this structure are detailed in Section 12.5. A sparse representation of the parameterized spatial correlation matrix is described in Section 12.6. A linearly constrained minimum variance approach based on the parameterized spatiotemporal correlation matrix is presented in Section 12.7. Section 12.8 summarizes the current limitations of the state-of-the-art, and concluding statements are made in Section 12.9.

12.2 Signal Model

Assume that an array of N microphones samples the sound field in both space and time; the output of sensor n at time k is modeled as

$$y_n(k) = \alpha_n(\mathbf{r}_s)s\left[k - \tau - \mathcal{F}_{1n}(\mathbf{r}_s)\right] + v_n(k), \tag{12.1}$$

where $\alpha_n(\mathbf{r}_s), n = 1, 2, \ldots, N$, models the attenuation of the source signal at sensor n as a function of the source location $\mathbf{r}_s = \begin{bmatrix} r_s & \phi_s & \theta_s \end{bmatrix}^T$ (superscript T denotes the transpose of a vector or a matrix), where r_s, ϕ_s, and θ_s denote the range, elevation, and azimuth, respectively, in a spherical coordinate system, s is the source signal, τ is the propagation time (in samples) from the source to sensor 1, $\mathcal{F}_{nm}(\mathbf{r}_s)$ is a function that relates the source position to the relative delay between microphones n and m, and v_n is the additive noise at sensor n. In the free-field case,

$$\alpha_n(\mathbf{r}_s) \propto \frac{1}{\|\mathbf{r}_n - \mathbf{r}_s\|}, \tag{12.2}$$

where $\mathbf{r}_n$ is the position of the nth microphone. In the majority of microphone array applications, the distance from the desired source to the array is large relative to the extent of the spatial aperture; in other words, the source is located in the array's far-field and the propagation of the signal is effectively that of a plane wave. As a result, it may be reasonably assumed that the attenuation coefficients are uniform across the array:

$$\alpha_n(\mathbf{r}_s) = 1, \quad \forall n, \mathbf{r}_s. \tag{12.3}$$

Moreover, the relative delays across the array are also independent of the range:

$$\mathcal{F}_{nm}(\phi_s, \theta_s) = \frac{1}{c}\boldsymbol{\zeta}^T(\phi_s, \theta_s)(\mathbf{x}_m - \mathbf{x}_n), \tag{12.4}$$

where

$$\boldsymbol{\zeta}(\phi_s, \theta_s) = \begin{bmatrix} \sin\phi_s\cos\theta_s & \sin\phi_s\sin\theta_s & \cos\phi_s \end{bmatrix}^T \tag{12.5}$$

is a unit vector which points in the direction of propagation of the source, and $\mathbf{x}_n = \begin{bmatrix} x_n & y_n & z_n \end{bmatrix}^T$ is the position vector of the nth microphone in Cartesian co-ordinates. The dimensionality of the location space is further reduced to one if one assumes that the source and the array lie on a plane; in other words, when $\phi_{\mathrm{s}} = \frac{\pi}{2}$:

$$
\begin{aligned}
\mathcal{F}_{nm}(\phi_{\mathrm{s}}, \theta_{\mathrm{s}}) &= \mathcal{F}_{nm}(\theta_{\mathrm{s}}) \\
&= \frac{1}{c}\boldsymbol{\zeta}^T(\theta_{\mathrm{s}})(\mathbf{x}_m - \mathbf{x}_n) \\
&= \frac{1}{c}\left[\cos\theta_{\mathrm{s}}(x_m - x_n) + \sin\theta_{\mathrm{s}}(y_m - y_n)\right].
\end{aligned}
\tag{12.6}
$$

Throughout the chapter, we assume that the source rests in the array's far-field. Moreover, to ease notation, we further assume that the source is at least approximately on the same plane as the microphones. While the latter is a rather lofty assumption, the argument of the relative delay function may be made multivariate to generalize to the case of an elevated source. The methods presented in the remainder of this chapter apply equally to both the two-dimensional elevation-azimuth space and the one-dimensional azimuth space. If the source is located in the array's near-field, $\mathcal{F}_{nm}$ becomes a function of the range r_{s} and the resulting methods may also determine the source range.

Lastly, it should be mentioned that in all practical acoustic environments, each microphone picks up a convolution of the source with a room impulse response. Since only the direct-path component conveys location information, the reverberant components are not included in the desired signal term and may be lumped into the additive noise terms $v_n(k)$, thus making the additive noise temporally correlated with the desired signal. It will be shown that this temporal correlation is one of the challenges to robust localization in real environments.

12.3 Spatial and Spatiotemporal Filtering

Before delving into the steered-beamforming approach to acoustic source localization, we briefly cover the fundamentals of spatiotemporal filtering.

The principle behind linear filtering is to collect the signal across an aperture (i.e., a discrete set of samples), apply a weight to each collected sample, and then sum the weighted samples to form the filter output. Spatiotemporal filtering, more commonly known as broadband beamforming, is a direct application of this principle. In addition to filtering the signal in the spatial aperture, we store the previous $L - 1$ samples of each microphone to form a spatiotemporal aperture of size NL:

$$\bar{\mathbf{y}}(k) = \left[\, \mathbf{y}^T(k)\ \mathbf{y}^T(k-1)\ \cdots\ \mathbf{y}^T(k-L+1)\,\right]^T, \qquad (12.7)$$

where

$$\mathbf{y}(k) = \left[\, y_1(k)\ y_2(k)\ \cdots\ y_N(k)\,\right]^T \qquad (12.8)$$

is the spatial aperture at time k.

A linear spatiotemporal filter $\mathbf{h}$ is then applied to the aperture to form the beamformer output:

$$z(k) = \mathbf{h}^T \bar{\mathbf{y}}(k), \qquad (12.9)$$

where $z(k)$ is the output of the broadband beamformer and

$$\mathbf{h} = \left[\, \mathbf{h}_0^T\ \mathbf{h}_1^T\ \cdots\ \mathbf{h}_{L-1}^T\,\right]^T \qquad (12.10)$$

is the spatiotemporal filter which consists of L spatial filters $\mathbf{h}_l$, where each spatial filter spatially filters the array signals at time $k - l$:

$$\mathbf{h}_l = \left[\, h_{l,1}\ h_{l,2}\ \cdots\ h_{l,N}\,\right]^T, \qquad (12.11)$$

where $h_{l,n}$ is the coefficient applied to $y_n(k-l)$.

The variance of the beamformer output then follows as

$$E\left[z^2(k)\right] = E\left\{\left[\mathbf{h}^T\bar{\mathbf{y}}(k)\right]^2\right\}$$
$$= \mathbf{h}^T \mathbf{R}_{\bar{y}} \mathbf{h}, \qquad (12.12)$$

where $E\left[\cdot\right]$ denotes mathematical expectation and

$$\mathbf{R}_{\bar{y}} = E\left[\bar{\mathbf{y}}(k)\bar{\mathbf{y}}^T(k)\right] \qquad (12.13)$$

is the spatiotemporal correlation matrix (STCM) of the observed signals.

If the signal of interest is narrowband, the temporal aperture length may be taken to be $L = 1$, and the filter consists of a weight for each spatial sample (i.e., microphone). In this case, the STCM simplifies to the spatial correlation matrix (SCM), which is the fundamental structure in narrowband localization methods:

$$\mathbf{R}_y = E\left[\mathbf{y}(k)\mathbf{y}^T(k)\right]. \qquad (12.14)$$

12.4 Parameterized Spatial Correlation Matrix (PSCM)

In source localization applications, one is interested in how the location of the source affects the observed second orders statistics (SOS) at the array. In other words, one would like to parameterize the SCM and STCM with the source location. One way of achieving this is to time-align the microphone signals *prior* to forming the correlation matrix. This is in contrast to the narrowband approach, where a linear weighting is applied to the SCM in order to properly phase-delay the sensor signals. In the broadband case, the time-aligning is performed directly in the correlation matrix.

To that end, consider forming the *parameterized spatial correlation matrix* (PSCM) according to

$$\mathbf{R}_y(\theta) = E\left[\mathbf{y}(k,\theta)\mathbf{y}^T(k,\theta)\right], \tag{12.15}$$

where

$$\mathbf{y}(k,\theta) = \left[\, y_1(k)\ y_2\left[k + \mathcal{F}_{12}(\theta)\right]\ \cdots\ y_N\left[k + \mathcal{F}_{1N}(\theta)\right]\,\right]^T \tag{12.16}$$

is the spatial aperture time aligned with respect to location θ. Substituting (12.1) into (12.16) yields the nth element of $\mathbf{y}(k,\theta)$:

$$y_n(k,\theta) = s\left[k - \tau - \mathcal{F}_{1n}(\theta_s) + \mathcal{F}_{1n}(\theta)\right] + v_n\left[k + \mathcal{F}_{1n}(\theta)\right]. \tag{12.17}$$

Thus, we may write the time-aligned spatial aperture as

$$\mathbf{y}(k,\theta) = \mathbf{s}(k - \tau, \theta) + \mathbf{v}(k,\theta), \tag{12.18}$$

where

$$\mathbf{s}(k,\theta) = \left[\, s(k)\ s\left[k - \mathcal{F}_{12}(\theta_s) + \mathcal{F}_{12}(\theta)\right]\ \cdots\ s\left[k - \mathcal{F}_{1N}(\theta_s) + \mathcal{F}_{1N}(\theta)\right]\,\right]^T$$

and

$$\mathbf{v}(k,\theta) = \left[\, v_1(k)\ v_2\left[k + \mathcal{F}_{12}(\theta)\right]\ \cdots\ v\left[k + \mathcal{F}_{1N}(\theta)\right]\,\right]^T.$$

Notice that when $\theta = \theta_s$,

$$\mathbf{s}(k,\theta_s) = s(k)\mathbf{1}_N, \tag{12.19}$$

where $\mathbf{1}_N$ is a vector of N ones. Conversely, the parameterized noise vector is given by

$$\mathbf{v}(k,\theta_s) = \left[\, v_1(k)\ v_2\left[k + \mathcal{F}_{12}(\theta_s)\right]\ \cdots\ v\left[k + \mathcal{F}_{1N}(\theta_s)\right]\,\right]^T.$$

Assuming that the signal and noise vectors are uncorrelated:

$$E\left[\mathbf{s}(k,\theta)\mathbf{v}^T(k,\theta)\right] = \mathbf{0}_{N \times N}, \tag{12.20}$$

where $\mathbf{0}_{N \times N}$ is an N-by-N matrix of zeros, the PSCM may be written as

$$\mathbf{R}_y(\theta) = \mathbf{R}_s(\theta) + \mathbf{R}_v(\theta), \tag{12.21}$$

where

$$\mathbf{R}_s(\theta) = E\left[\mathbf{s}(k-\tau,\theta)\mathbf{s}^T(k-\tau,\theta)\right] \qquad (12.22)$$

and

$$\mathbf{R}_v(\theta) = E\left[\mathbf{v}(k,\theta)\mathbf{v}^T(k,\theta)\right]. \qquad (12.23)$$

From (12.23), the parameterized noise correlation matrix $\mathbf{R}_v(\theta)$ will typically be full-rank, regardless of the parameter θ. Conversely, $\mathbf{R}_s(\theta)$ is rank-one if $\theta = \theta_s$. This property allows us to localize the source by observing the nature of the PSCM as θ is varied across the space of locations.

12.5 Source Localization Using Parameterized Spatial Correlation

Informally speaking, the elements of the PSCM are highly correlated (i.e., larger in magnitude) when the hypothesized parameter θ matches the true parameter θ_s. There are thus several ways of processing the various PSCMs to identify the actual source location. The remainder of this chapter focuses on how to choose a function $g(\mathbf{M})$, where $\mathbf{M}$ is a matrix, such that

$$\hat{\theta}_s = \arg\max_\theta g\left[\mathbf{R}_y(\theta)\right]$$

$$\approx \theta_s. \qquad (12.24)$$

A convenient aspect of the parametrization of the SCM is that it allows for the extension of the classical narrowband methods such as minimum variance, subspace, and linear prediction [2] to the broadband signal case. This will be clear in the forthcoming subsections.

12.5.1 Steered Response Power

The steered response power (SRP) method [20], [21] is the simplest method based on the PSCM. Note that the correlated nature of $\mathbf{R}_s(\theta_s)$ will increase the off-diagonal values comprising the correctly-steered PSCM $\mathbf{R}_y(\theta_s)$. Thus, one way of localizing the source is to simply sum the elements of the PSCM as a function of the parameter θ. The location which yields the largest sum of elements is designated as the source:

$$\hat{\theta}_s = \arg\max_\theta \mathbf{1}_N^T \mathbf{R}_y(\theta)\mathbf{1}_N. \qquad (12.25)$$

Note that this is equivalent to applying a fixed filter given by

$$\mathbf{h}(\theta) = \mathbf{h}$$
$$= \mathbf{1}_N, \ \forall \theta, \tag{12.26}$$

to the parameterized aperture $\mathbf{y}(k, \theta)$,

$$z(k, \theta) = \mathbf{h}^T \mathbf{y}(k, \theta), \tag{12.27}$$

and computing the resulting output power at each parameter:

$$E\left[z^2(k, \theta)\right] = \mathbf{1}_N^T \mathbf{R}_y(\theta) \mathbf{1}_N. \tag{12.28}$$

Since the filter $\mathbf{h}$ is independent of both the data and the parameter, the SRP approach is computationally attractive.

Figure 12.1 depicts the SRP spatial spectra averaged over one minute of synthetically convolved speech at four levels of room reverberation; the reflection coefficients of the walls, ceiling, and floor are adjusted to generate 60 dB decay times T_{60} of 0, 100, 200, and 300 ms. The reverberation times are measured using the reverse time integrated method of [22]. The image method for simulating room acoustics is employed in order to generate the synthetic impulse responses [23]. The simulated room has the following properties:

- dimensions given by 304.8cm-by-457.2cm-by-381cm,
- a four-element uniform linear array (ULA) with an inter-microphone spacing of $d = 0.0425$ cm with the center of the array located at $(152.4, 19.05, 101.6)$,
- an isotropic speech source located at $(254, 190.5, 101.6)$ cm, and
- a spatially and temporally white Gaussian noise field with an SNR of 30 dB measured with respect to the convolved speech.

The location estimates are computed every 128 ms frame over a 60 second female speech signal (i.e., English). The sampling rate is 48 kHz. In order to increase spatial resolution, the cross-correlation functions are upsampled by a factor of 20 before forming the PSCM.

In order to ease complexity, the simulations assumed (correctly) that the source and array lie on the same plane. Thus, the parameter space was one-dimensional, and ranged from $0°$ to $179°$ degrees azimuth in increments of one degree, measured with respect to the array axis such that an angle of $90°$ indicates array broadside.

Notice that the absolute levels of the spatial spectra are irrelevant to performance. The key factor is the level of the false direction-of-arrivals (DOAs) relative to that of the true DOA, which in this case is 120 degrees azimuth. The bias of the DOA estimator increases as the room reverberation becomes stronger.

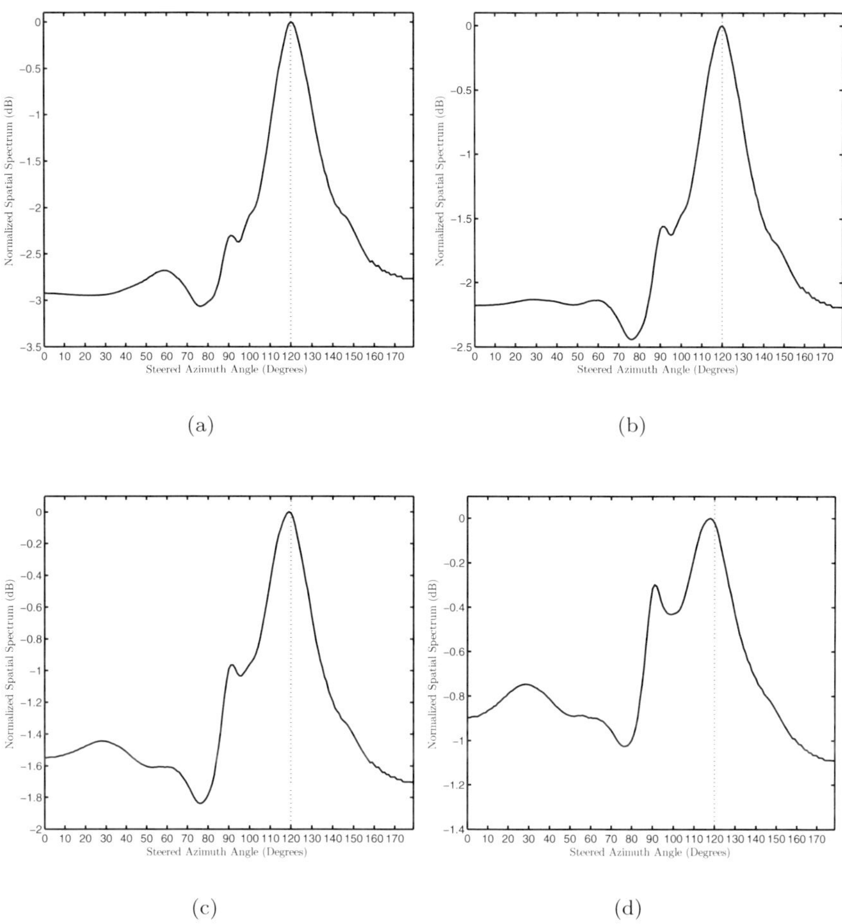

Fig. 12.1 Ensemble averaged SRP spatial spectra: (a) 0 ms, (b) 100 ms, (c) 200 ms, and (d) 300 ms.

12.5.2 Minimum Variance Distortionless Response

The SRP method is commonly employed due to its simplicity – the various PSCMs need to be formed and the elements summed. A more sophisticated method, proposed by Krolik and Swingler [24], attempts to apply a location-dependent filter $\mathbf{h}(\theta)$ to the PSCM in order to minimize the contribution of $\mathbf{R}_v(\theta)$ to $\mathbf{R}_y(\theta)$.

To that end, consider constraining $\mathbf{h}(\theta_\mathrm{s})$ to have unity gain in response to the properly-aligned source vector:

$$\mathbf{h}^T(\theta_s)\mathbf{s}(k, \theta_s) = s(k), \tag{12.29}$$

which is equivalent to

$$\mathbf{h}^T(\theta_s)\mathbf{1}_N = 1, \tag{12.30}$$

while utilizing the remaining degrees of freedom in $\mathbf{h}(\theta_s)$ to minimize the beamformer output power $\mathbf{h}^T(\theta_s)\mathbf{R}_y(\theta_s)\mathbf{h}(\theta_s)$. This brings about the following optimization problem, which needs to be solved for every parameter θ:

$$\mathbf{h}_{\mathrm{mvdr}}(\theta) = \arg\min_{\mathbf{h}(\theta)} \mathbf{h}^T(\theta)\mathbf{R}_y(\theta)\mathbf{h}(\theta) \ \text{ subject to } \ \mathbf{h}^T(\theta)\mathbf{1}_N = 1. \tag{12.31}$$

This is an application of the celebrated minimum variance distortionless response (MVDR) method [25] to the steered-beamforming problem. Using the method of Lagrange multipliers, the optimal weights are found as

$$\mathbf{h}_{\mathrm{mvdr}}(\theta) = \frac{\mathbf{R}_y^{-1}(\theta)\mathbf{1}_N}{\mathbf{1}_N^T\mathbf{R}_y^{-1}(\theta)\mathbf{1}_N}. \tag{12.32}$$

Consequently, the location estimate is given by

$$\hat{\theta}_s = \arg\max_{\theta} \mathbf{h}_{\mathrm{mvdr}}^T(\theta)\mathbf{R}_y(\theta)\mathbf{h}_{\mathrm{mvdr}}(\theta)$$

$$= \arg\max_{\theta} \left[\mathbf{1}_N^T\mathbf{R}_y^{-1}(\theta)\mathbf{1}_N\right]^{-1}. \tag{12.33}$$

It is very interesting to see that while SRP sums the elements of the PSCM, MVDR inverts the sum of the PSCM's inverse's elements.

Figure 12.2 depicts the MVDR spatial spectra for the scenario described in the previous subsection. Notice that the spectra bear a strong resemblance to that of the SRP method.

12.5.3 Maximum Eigenvalue

In addition to selecting a weight vector that minimizes the contribution of the noise, one may also attempt to find a weight vector that simply maximizes the output power of the steered beamformer at each direction in order to identify the location θ at which the steered power $E\left[z^2(k, \theta)\right]$ is the largest [27].

This weight selection may be written as the following constrained optimization problem:

$$\mathbf{h}_{\mathrm{maxeig}}(\theta) = \arg\max_{\mathbf{h}(\theta)} \mathbf{h}^T(\theta)\mathbf{R}_y(\theta)\mathbf{h}(\theta) \ \text{ subject to } \ \mathbf{h}^T(\theta)\mathbf{h}(\theta) = 1.$$

$$\tag{12.34}$$

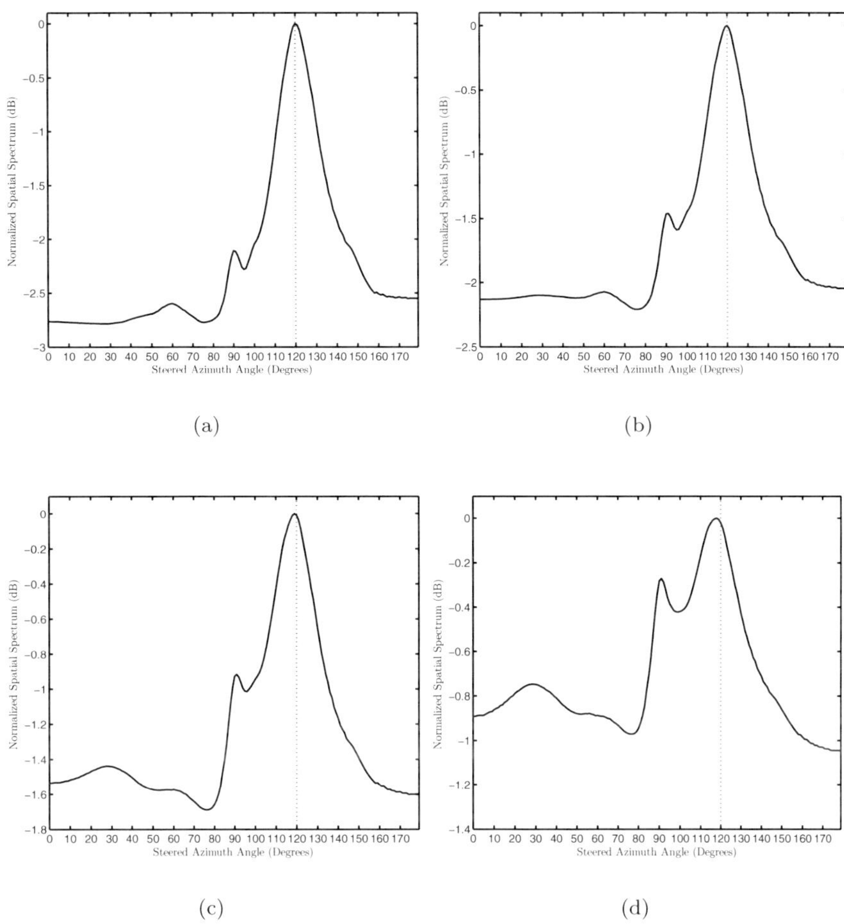

Fig. 12.2 Ensemble averaged MVDR spatial spectra: (a) 0 ms, (b) 100 ms, (c) 200 ms, and (d) 300 ms.

The solution to this optimization problem is the vector that maximizes the Rayleigh quotient

$$\frac{\mathbf{h}^T(\theta)\mathbf{R}_y(\theta)\mathbf{h}(\theta)}{\mathbf{h}^T(\theta)\mathbf{h}(\theta)}.$$

The solution is well known and given by the principal eigenvector of $\mathbf{R}_y(\theta)$:

$$\mathbf{h}_{\mathrm{maxeig}}(\theta) = \mathbf{u}_1(\theta), \tag{12.35}$$

where $\mathbf{u}_1(\theta)$ is the principal eigenvector of $\mathbf{R}_y(\theta)$. Note again that the principal eigenvector must be found for all PSCMs. The location estimate follows as

$$
\begin{aligned}
\hat{\theta}_{\mathrm{s}} &= \arg\max_\theta \mathbf{h}_{\mathrm{maxeig}}^T(\theta)\mathbf{R}_y(\theta)\mathbf{h}_{\mathrm{maxeig}}(\theta) \\
&= \arg\max_\theta \lambda_1(\theta),
\end{aligned}
\tag{12.36}
$$

where $\lambda_1(\theta)$ is the maximum eigenvalue of $\mathbf{R}_y(\theta)$.

Figure 12.3 depicts the maximum eigenvalue spatial spectra. Once again, the PSCM based adaptive weighting has little impact on the resulting spatial spectra.

12.5.4 Broadband MUSIC

The multiple signal classification (MUSIC) method is a classical method for the localization of narrowband signal sources [28], [29]. It is a subspace method which exploits the distinct eigenstructure of the SCM. In this section, it is shown that the parametrization of the SCM leads to a similar eigenstructure in the PSCM, thus allowing for the generalization of the MUSIC method to broadband signals [30].

Recall that the PSCM's signal component $\mathbf{R}_s(\theta)$ is rank-one if $\theta = \theta_{\mathrm{s}}$. Moreover, since the PSCM is a correlation matrix, it is positive-definite and thus has a spectral representation. Consider the case when $\theta = \theta_{\mathrm{s}}$; the PSCM may then be written as

$$
\mathbf{R}_y(\theta_{\mathrm{s}}) = \sigma_s^2 \mathbf{1}_N \mathbf{1}_N^T + \mathbf{R}_v(\theta_{\mathrm{s}}),
\tag{12.37}
$$

where $\sigma_s^2 = E\left[s^2(k)\right]$ is the variance of the source signal. For the time being, assume that the parameterized noise correlation matrix may be written as

$$
\mathbf{R}_v(\theta_{\mathrm{s}}) = \sigma_v^2 \mathbf{I}_{N \times N},
\tag{12.38}
$$

where $\mathbf{I}_{N \times N}$ is the N-by-N identity matrix. It then follows that any vector $\mathbf{u}$ orthogonal to $\mathbf{1}_N$:

$$
\mathbf{1}_N^T \mathbf{u} = 0
\tag{12.39}
$$

is an eigenvector of $\mathbf{R}_y(\theta_{\mathrm{s}})$, since

$$
\mathbf{R}_y(\theta_{\mathrm{s}})\mathbf{u} = \sigma_v^2 \mathbf{u}.
\tag{12.40}
$$

Since the dimensionality of $\mathbf{1}_N$ is N, there are $N-1$ such eigenvectors, which are termed the *noise eigenvectors* as their corresponding eigenvalues are all equal to the noise variance σ_v^2. Moreover, the remaining eigenvector must

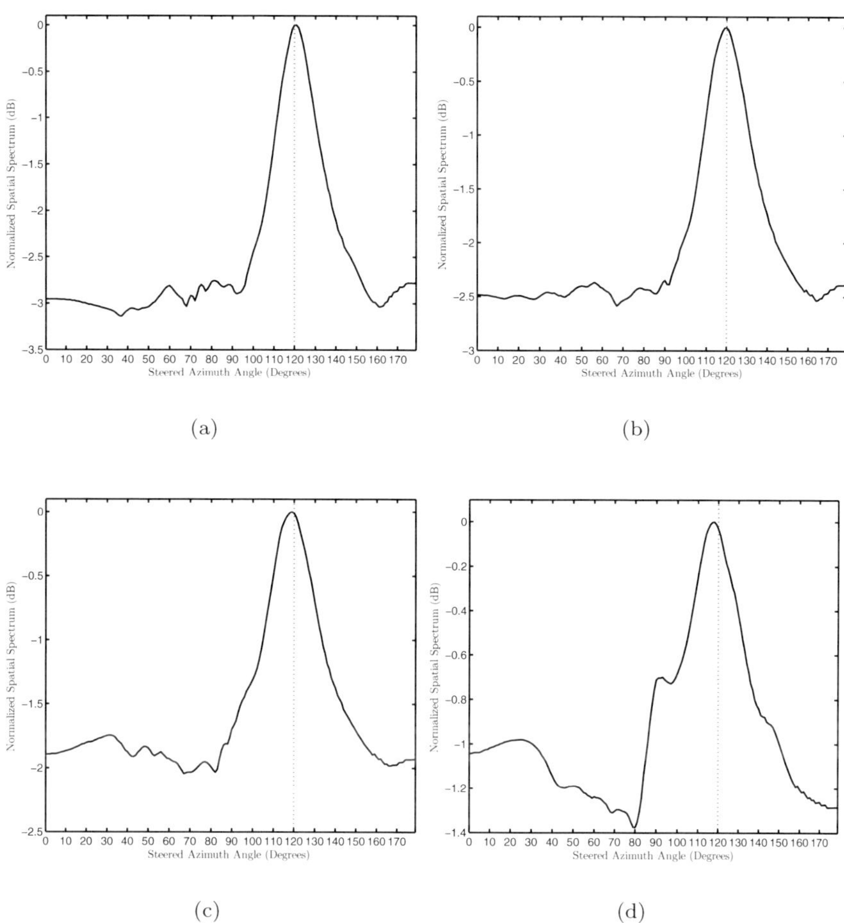

Fig. 12.3 Ensemble averaged maximum eigenvalue spatial spectra: (a) 0 ms, (b) 100 ms, (c) 200 ms, and (d) 300 ms.

necessarily be the principal eigenvector since it is not orthogonal to $\mathbf{1}_N$ – this eigenvector is termed the signal eigenvector, and the spectral representation of the location-matched PSCM is given by

$$\mathbf{R}_y(\theta_{\mathrm{s}}) = \lambda_1 \mathbf{u}_1(\theta_{\mathrm{s}})\mathbf{u}_1^T(\theta_{\mathrm{s}}) + \sigma_v^2 \sum_{i=2}^{N} \mathbf{u}_i(\theta_{\mathrm{s}})\mathbf{u}_i^T(\theta_{\mathrm{s}}), \qquad (12.41)$$

where $\mathbf{u}_1(\theta_{\mathrm{s}})$ is the principal signal eigenvector of $\mathbf{R}_y(\theta_{\mathrm{s}})$ and $\mathbf{u}_i(\theta_{\mathrm{s}})$, $i = 2, \ldots, N$ are the noise eigenvectors.

Notice that when the parameter θ does not match the actual location θ_s, the eigenvalue spectrum of the PSCM will effectively be spread out by the mismatch between θ and θ_s.

Putting all of this together, one may look for the source by examining the $N-1$ lower eigenvalues of the PSCM and test them for orthogonality with $\mathbf{1}_N$. At the actual source location θ_s, these $N-1$ lower eigenvalues theoretically yield

$$\frac{1}{\mathbf{1}_N^T \sum_{i=2}^{N} \mathbf{u}_i(\theta_s)\mathbf{u}_i^T(\theta_s)\mathbf{1}_N} = \infty. \tag{12.42}$$

Thus, the broadband MUSIC location estimate is given by

$$\hat{\theta}_s = \arg\max_{\theta} \left[\mathbf{1}_N^T \left(\sum_{i=2}^{N} \mathbf{u}_i(\theta)\mathbf{u}_i^T(\theta) \right) \mathbf{1}_N \right]^{-1}. \tag{12.43}$$

Figure 12.4 depicts the broadband MUSIC spatial spectra. It is evident that the spectra show increased spatial resolution, analogous to the narrowband MUSIC method. In the presence of reverberation, the resulting spectra reveal false peaks corresponding to both multipath components and the autocorrelation of the speech signal.

12.5.5 Minimum Entropy

Given a random variable x with probability density function (pdf) $p(x)$, the entropy of the random variable is given by [31]

$$\begin{aligned} H(x) &= -E\left[\ln p(x)\right] \\ &= -\int_{-\infty}^{\infty} p(x)\ln p(x)dx, \end{aligned} \tag{12.44}$$

and quantifies the level of uncertainty associated with the random variable x. High probability values of x contribute less in the $-\ln p(x)$ term but are weighted more due to the $p(x)$ term. Conversely, low probability values of x have a large $-\ln p(x)$ contribution but these contributions are weighted less by the low value of $p(x)$. The value of $H(x)$ quantifies the average value of the uncertainty about x. The measure generalizes in a straightforward manner to a random vector $\mathbf{x}$ with joint pdf $p(\mathbf{x})$; the joint entropy of $\mathbf{x}$ is given by

$$\begin{aligned} H(\mathbf{x}) &= -E\left[\ln p(\mathbf{x})\right] \\ &= -\int_{-\infty}^{\infty} p(\mathbf{x})\ln p(\mathbf{x})d\mathbf{x}. \end{aligned} \tag{12.45}$$

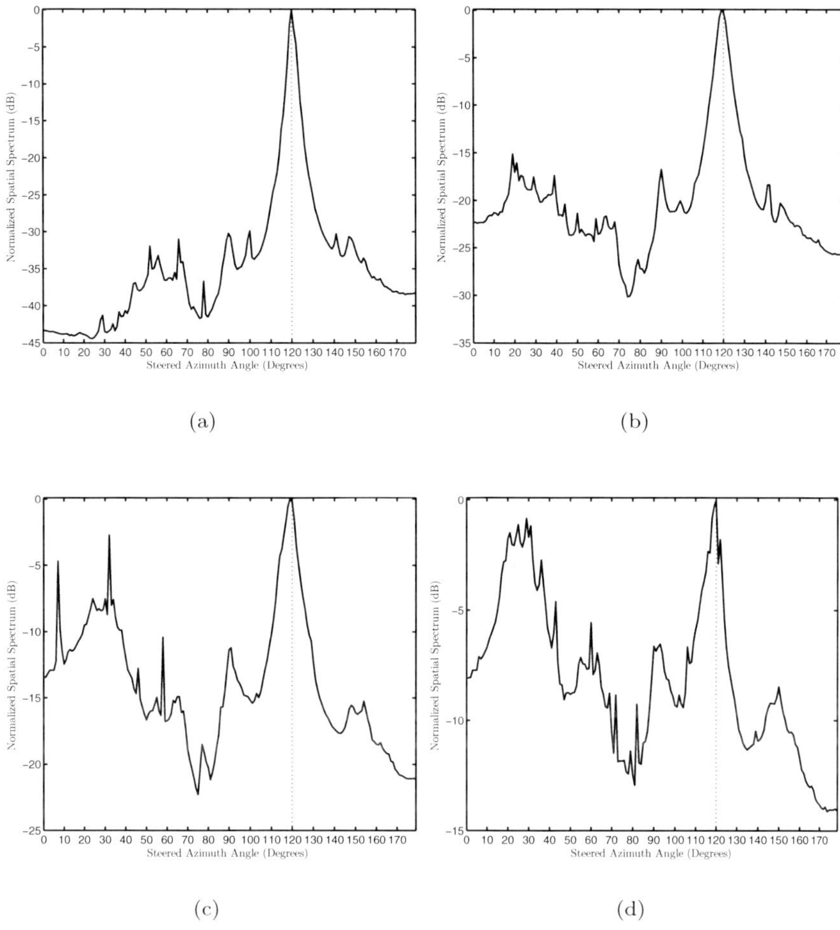

Fig. 12.4 Ensemble averaged MUSIC spatial spectra: (a) 0 ms, (b) 100 ms, (c) 200 ms, and (d) 300 ms.

A measure closely related to entropy is the mutual information, which quantifies the level of dependence between two random variables x and y:

$$I(x;y) = \int_{-\infty}^{\infty} \int_{-\infty}^{\infty} p(x,y) \ln \frac{p(x,y)}{p(x)p(y)} dx dy, \qquad (12.46)$$

where $p(x,y)$ is the joint distribution of x and y. One can show that the mutual information may be written as

$$\begin{aligned}
I(x;y) &= H(x) - H(x|y) \\
&= H(y) - H(y|x) \\
&= H(x) + H(y) - H(x,y),
\end{aligned} \tag{12.47}$$

where $H(x|y)$ is the entropy of x conditioned on y:

$$H(x|y) = -\int_{-\infty}^{\infty} \int_{-\infty}^{\infty} p(x,y) \ln p(x|y) dx dy, \tag{12.48}$$

and $H(x,y)$ is the joint entropy of x and y:

$$H(x,y) = -\int_{-\infty}^{\infty} \int_{-\infty}^{\infty} p(x,y) \ln p(x,y) dx dy. \tag{12.49}$$

Theorem 12.1. *The joint entropy of the random variables x and y attains a minimum when $p(x|y) = \delta(x - y) = p(y|x)$. In other words, joint entropy is minimized when the two random variables are equal at each ensemble.*

Proof. Assuming that $p(x|y) = \delta(x - y)$, the joint entropy of x and y may be written as

$$\begin{aligned}
H(x,y) &= -\int_{-\infty}^{\infty} \int_{-\infty}^{\infty} p(x,y) \ln p(x,y) dx dy \\
&= -\int_{-\infty}^{\infty} \int_{-\infty}^{\infty} p(x)p(y|x) \ln \left[p(x)p(y|x) \right] dx dy \\
&= -\int_{-\infty}^{\infty} \int_{-\infty}^{\infty} p(x)\delta(x - y) \ln \left[p(x)\delta(x - y) \right] dx dy \\
&= -\int_{-\infty}^{\infty} p(y) \ln \left[p(y) \right] dy \\
&= H(y) \\
&= H(x) \\
&= H.
\end{aligned} \tag{12.50}$$

Since the joint entropy is given by

$$H(x,y) = H(x) + H(y) - I(x;y), \tag{12.51}$$

it follows that

$$I(x;y) = H. \tag{12.52}$$

To prove that $H(x,y) \geq H(x)$, suppose that we could find two random variables x and y such that

$$H(x,y) < H(x). \tag{12.53}$$

This would imply that

$$H(x) + H(y) - I(x;y) < H(x), \qquad (12.54)$$

or

$$I(x;y) > H(y). \qquad (12.55)$$

However, since the mutual information is given by

$$I(x;y) = H(y) - H(y|x) \qquad (12.56)$$

and since $H(y|x) = -E\left[\ln p(y|x)\right] \geq 0$, this implies a contradiction. Therefore, $H(x,y)$ is minimized when $p(x|y) = p(y|x) = \delta(x - y)$.

Applying this result to the problem of source localization, recall that $\mathbf{s}(k,\theta) = s(k)\mathbf{1}_N$ when $\theta = \theta_\mathrm{s}$. Thus, when steered to the true location, the elements of the random vector $\mathbf{s}(k,\theta)$ are fully dependent and their joint entropy is minimized. As a result, one can localize the source by scanning the location space for the location that minimizes the joint entropy of the $\mathbf{y}(k,\theta)$. Notice that the noise component is assumed to be incoherent across the array and thus varying the parameter theoretically does not reduce the entropy of $\mathbf{y}(k,\theta)$.

12.5.5.1 Gaussian Signals

In order to compute the minimum entropy estimate of the source location, one must assume a distribution for the random vector $\mathbf{y}(k,\theta)$. An obvious choice is the multivariate Gaussian distribution; the random vector $\mathbf{x}$ follows a multivariate Gaussian distribution with a mean vector of $\mathbf{0}_N$ and a covariance matrix $\mathbf{R}$ if its joint pdf is given by

$$p(\mathbf{x}) = \frac{1}{\left(\sqrt{2\pi}\right)^N \det^{1/2}(\mathbf{R})} e^{-1/2\mathbf{x}^T\mathbf{R}^{-1}\mathbf{x}}, \qquad (12.57)$$

where $\det(\cdot)$ denotes the determinant of a matrix. The joint entropy of a Gaussian random vector is given by [32]

$$H(\mathbf{x}) = \frac{1}{2}\ln\left[(2\pi e)^N \det(\mathbf{R})\right]. \qquad (12.58)$$

Thus, the entropy of a jointly distributed Gaussian vector is proportional to the determinant of the covariance matrix. Applying this to the source localization problem, the minimum entropy estimate of the location θ_s is given by [32]

$$\hat{\theta}_{\mathrm{s}} = \arg\min_{\theta} H\left[\mathbf{y}(k,\theta)\right]$$

$$= \arg\min_{\theta} \det\left[\mathbf{R}_y(\theta)\right]. \tag{12.59}$$

It is interesting to link the minimum entropy approach to the eigenvalue methods presented earlier. To that end, notice that the determinant of a positive definite matrix is given by the product of its eigenvalues; thus, the minimum entropy approach may also be written as

$$\hat{\theta}_{\mathrm{s}} = \arg\min_{\theta} \prod_{n=1}^{N} \lambda_n(\theta). \tag{12.60}$$

Figure 12.5 depicts the minimum entropy spatial spectra. The minimum entropy estimator also shows increased resolution compared to the methods based on steered beamforming (i.e., SRP, MVDR, and maximum eigenvalue). However, as the level of reverberation is increased, spurious peaks are introduced into the spectra.

12.5.5.2 Laplacian Signals

The speech signal is commonly modeled by a Laplacian distribution whose heavier tail models the dynamic nature of speech; the univariate Laplacian distribution is given by

$$p(x) = \frac{\sqrt{2}}{2\sigma_x} e^{-\frac{\sqrt{2}|x|}{\sigma_x}}. \tag{12.61}$$

An N-dimensional zero-mean random vector is said to follow a jointly Laplacian distribution if its joint PDF is given by

$$p(\mathbf{x}) = 2(2\pi)^{-N/2}\det^{-1/2}(\mathbf{R})\left(\mathbf{x}^T\mathbf{R}^{-1}\mathbf{x}\right)^{P/2} K_P\left(\sqrt{2\mathbf{x}^T\mathbf{R}^{-1}\mathbf{x}}\right), \tag{12.62}$$

where $P = \frac{2-N}{2}$ and $K_P(\cdot)$ is the modified Bessel function of the third kind:

$$K_P(a) = \frac{1}{2}\left(\frac{a}{2}\right)^P \int_0^\infty z^{-P-1} e^{-z-\frac{a^2}{4z}} \, dz, \quad a > 0. \tag{12.63}$$

It then follows that the joint entropy of a Laplacian distributed random vector is given by

$$H(\mathbf{x}) = \frac{1}{2}\ln\left[\frac{(2\pi)^N}{4}\det\mathbf{R}\right] - \frac{P}{2}E\left[\ln(\eta/2)\right] - E\left[\ln K_P\left(\sqrt{2\eta}\right)\right], \tag{12.64}$$

where $\eta = \mathbf{x}^T\mathbf{R}^{-1}\mathbf{x}$ and the expectation terms apparently lack closed forms.

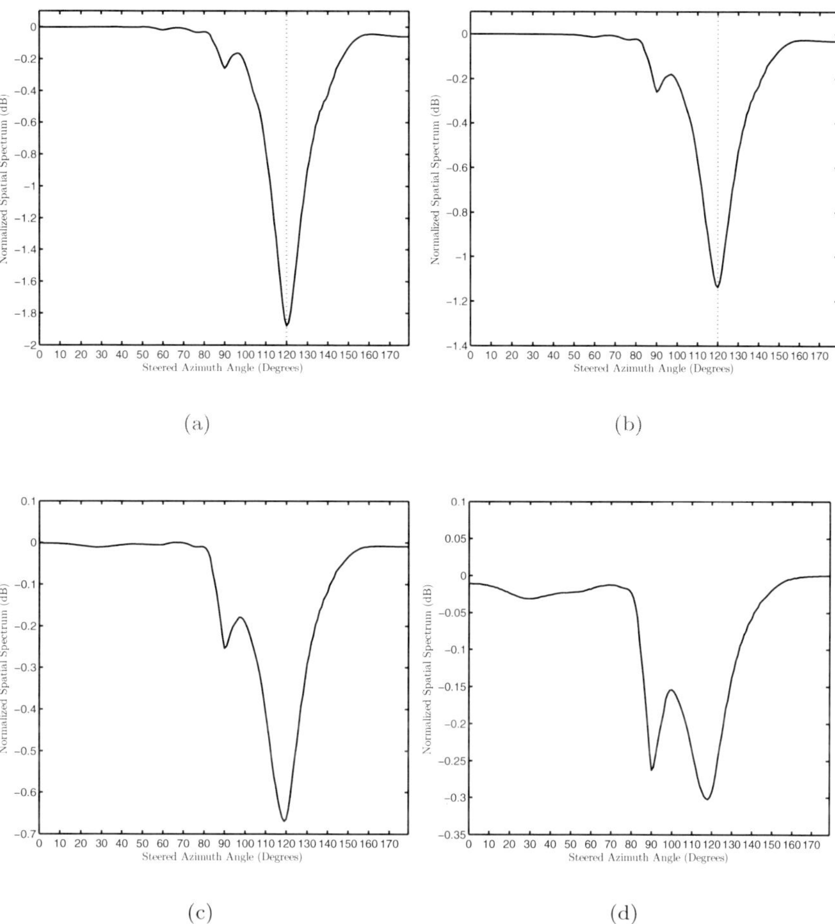

Fig. 12.5 Ensemble averaged minimum entropy spatial spectra (Gaussian assumption): (a) 0 ms, (b) 100 ms, (c) 200 ms, and (d) 300 ms.

A closed-form minimum entropy location estimator for the Laplacian case is thus not available; however, in practice, the assumption of ergodicity for the signals $y_n(k,\theta)$, $n = 1, 2, \ldots, N$ allows us to form an empirical minimum entropy estimator. To that end, consider first forming a time-averaged estimate of the PSCM:

$$\hat{\mathbf{R}}_y(\theta) = \frac{1}{K} \sum_{k'=1}^{K} \mathbf{y}(k',\theta)\mathbf{y}^T(k',\theta), \tag{12.65}$$

where $\mathbf{y}(k', \theta)$ is the k'th parameterized observation vector and there are K total observations. Next, the two terms lacking closed forms are estimated according to

$$E\left[\ln(\eta/2)\right] \approx \frac{1}{K} \sum_{k'=1}^{K} \ln \left[\frac{1}{2}\mathbf{y}^T(k', \theta)\,\hat{\mathbf{R}}(\theta)\mathbf{y}(k', \theta)\right], \quad (12.66)$$

$$E\left[\ln K_P\left(\sqrt{2\eta}\right)\right] \approx \frac{1}{K} \sum_{k'=1}^{K} \ln K_P \sqrt{2\mathbf{y}^T(k', \theta)\,\hat{\mathbf{R}}(\theta)\mathbf{y}(k', \theta)}. \quad (12.67)$$

The empirical joint entropy is then found by substituting the time-averaged quantities of (12.65)–(12.67) into the theoretical expression (12.64). As before, the parameter θ which minimizes the resulting joint entropy is chosen as the location estimate [32].

Notice that the minimum entropy estimators consider more than just second-order features of the observation vector $\mathbf{y}(k, \theta)$. The performance of this and all previously described algorithms depends ultimately on the sensitivity of the statistical criterion (i.e., joint Laplacian entropy) to the parameter θ, particularly to parameters which lead to a large noise presence in $\mathbf{y}(k, \theta)$.

12.6 Sparse Representation of the PSCM

In real applications, the computational complexity of a particular algorithm needs to be taken into account. The advantage of the algorithms presented in this chapter is the utilization of additional microphones to increase robustness to noise and reverberation. On the other hand, all algorithms inherently require a search of the parameter space to determine the optimal θ. In this section, we propose a sparse representation of the PSCM in terms of the observed cross-correlation functions across the array.

In practice, the cross-correlation functions across all microphone pairs are computed for a frame of incoming data. This is typically performed in the frequency-domain by taking the inverse Fourier transform of the cross-spectral density (CSD):

$$R_{y_n y_m}(\tau) = \frac{1}{L} \sum_{l=0}^{L-1} Y_n^*(l)Y_m(l)e^{j2\pi \frac{l}{L}\tau}, \quad (12.68)$$

where $Y_n^*(l)Y_m(l)$ is the instantaneous estimate of the CSD between channels n and m at discrete frequency $\frac{l}{L}$, superscript * denotes complex conjugate, and

$$Y_n(l) = \sum_{k=0}^{L-1} y_n(k) e^{-j2\pi \frac{l}{L} k} \tag{12.69}$$

is the L-point fast Fourier transform of the signal at microphone n evaluated at discrete frequency $\frac{l}{L}$. From the N^2 cross-correlation functions, the various PSCMs must be constructed and then evaluated to determine the location estimate $\hat{\theta}$:

$$[\mathbf{R}_y(\theta)]_{nm} = R_{y_n y_m} [\mathcal{F}_{nm}(\theta)] . \tag{12.70}$$

Thus, the task is to construct $\mathbf{R}_y(\theta)$ from the cross-correlation functions $R_{y_n y_m}(\tau)$.

Notice that the cross-correlation functions are computed prior to forming the various PSCMs. Moreover, for a given microphone pair, the cross-correlation function usually exhibits one or more distinct peaks across the relative delay space. Instead of taking into account the entire range of τ, it is proposed in [33] to only take into account the highly-correlated lags when forming the PSCM.

The conventional search technique relies on the forward mapping between the parameter θ and the resulting relative delay τ:

$$\tau_{nm} = \mathcal{F}_{nm}(\theta) \tag{12.71}$$

is the relative delay experience between microphones n and m if the source is located at θ. The problem with forming the PSCMs using the forward mapping is that the entire parameter space must be traversed before the optimal parameter is selected. Moreover, there is no *a priori* information about the parameter that can be utilized in reducing the search. Consider instead the inverse mapping from the relative delay τ to the set of locations which experience that relative delay at a given microphone pair:

$$\mathcal{F}_{nm}^{-1}(\tau) = \{\theta | \mathcal{F}_{nm}(\theta) = \tau\} . \tag{12.72}$$

For the microphone pair (n, m), define the set $\mathcal{C}_{nm}(p)$ which is composed of the $2p$ lags directly adjacent to the peak value of $R_{y_n y_m}(\tau)$:

$$\mathcal{C}_{nm}(p) = \{\hat{\tau}_{nm} - p, \ldots, \hat{\tau}_{nm} - 1, \hat{\tau}_{nm}, \hat{\tau}_{nm} + 1 \ldots, \hat{\tau}_{nm} + p\} , \tag{12.73}$$

where

$$\hat{\tau}_{nm} = \arg\max_{\tau} R_{y_n y_m}(\tau). \tag{12.74}$$

The set $\mathcal{C}_{nm}(p)$ hopefully contains the most correlated lags of of the cross-correlation function between microphones n and m. Consider nonlinearly processing the cross-correlation functions such that

Table 12.1 Localization using the sparse PSCM.

Compute:

for all microphone pairs (n, m)

$$R_{y_n y_m}(\tau) = \frac{1}{L} \sum_{l=0}^{L-1} Y_n^*(l) Y_m(l) e^{j2\pi \frac{l}{L}\tau}$$

$$\hat{\tau}_{nm} = \arg\max_\tau R_{y_n y_m}(\tau)$$

$$\mathcal{C}_{nm}(p) = \{\hat{\tau}_{nm} - p, \ldots, \hat{\tau}_{nm} - 1, \hat{\tau}_{nm}, \hat{\tau}_{nm} + 1 \ldots, \hat{\tau}_{nm} + p\}$$

Initialization:

for all θ, $\mathbf{R}_y(\theta) = \mathbf{0}_{N \times N}$

Search:

for all microphone pairs (n, m)

    for all $\tau \in \mathcal{C}_{nm}(p)$

        look up $\mathcal{F}_{nm}^{-1}(\tau)$

        for all $\theta \in \mathcal{F}_{nm}^{-1}(\tau)$

            update : $[\mathbf{R}_y(\theta)]_{nm} = [\mathbf{R}_y(\theta)]_{nm} + R_{y_n y_m}(\tau)$

$\hat{\theta} = \arg\max_\theta f[\mathbf{R}_y(\theta)]$

$$R'_{y_n y_m}(\tau) = \begin{cases} R_{y_n y_m}(\tau), & \tau \in \mathcal{C}_{nm}(p) \\ 0, & \text{otherwise} \end{cases}. \tag{12.75}$$

The resulting elements of the PSCM are given by

$$\left[\mathbf{R}'_y(\theta)\right]_{nm} = R'_{y_n y_m}\left[\mathcal{F}_{nm}(\theta)\right]$$

$$= \begin{cases} R_{y_n y_m}\left[\mathcal{F}_{nm}(\theta)\right], & \mathcal{F}_{nm}(\theta) \in \mathcal{C}_{nm}(p) \\ 0, & \text{otherwise} \end{cases}. \tag{12.76}$$

The modified PSCM $\mathbf{R}'_y(\theta)$ is now sparse provided that the sets $\mathcal{C}_{nm}(p)$ represent a small subset of the feasible relative delay space for each microphone pair.

Table 12.1 describes the general procedure for implementing a localization algorithm based on the sparse representation of the PSCM. As a comparison, Table 12.2 describes the corresponding algorithm but this time employing the forward mapping from location to relative delay. The conventional search involves iterating across the typically large location space. On the other hand, the sparse approach introduces a need to identify the peak lag of each cross-correlation function, albeit avoiding the undesirable location search.

Table 12.2 Localization using the PSCM.

Compute:
    for all microphone pairs (n, m)

$$R_{y_n y_m}(\tau) = \frac{1}{L} \sum_{l=0}^{L-1} Y_n^*(l) Y_m(l) e^{j2\pi \frac{l}{L}\tau}$$

Initialization:
    for all θ, $\mathbf{R}_y(\theta) = \mathbf{0}_{N \times N}$

Search:
    for all locations θ
        for all microphone pairs (n, m)
            look up $\tau = \mathcal{F}_{nm}(\theta)$
                update : $[\mathbf{R}_y(\theta)]_{nm} = [\mathbf{R}_y(\theta)]_{nm} + R_{y_n y_m}(\tau)$

$$\hat{\theta} = \arg\max_\theta f[\mathbf{R}_y(\theta)]$$

12.7 Linearly Constrained Minimum Variance

All approaches described thus far have focused on the relationship between the location of the acoustic source and the resulting relative delays observed across multiple microphones. Such techniques are purely spatial in nature. Notice that the resulting algorithms consider a temporally instantaneous aperture, in that previous samples are not appended to the vector of received spatial samples.

A truly spatiotemporal approach to acoustic source localization encompasses both spatial and temporal discrimination: that is, the aperture consists of a block of temporal samples for each microphone pair. The advantage of including previous temporal samples in the processing of each microphone is that the resulting algorithm may distinguish the desired signal from the additive noise by exploiting any temporal differences between them. This is the essence of the linearly constrained minimum variance (LCMV) adaptive beamforming method proposed by Frost in 1972 [34], which is equivalent to the generalized sidelobe canceller of [35], both of which are nicely summarized in [36]. The application of the LCMV scheme to the source localization algorithm is presented in [37].

The parameterized spatiotemporal aperture at the array is written as

$$\bar{\mathbf{y}}(k, \theta) = \left[\mathbf{y}(k, \theta)\ \mathbf{y}(k-1, \theta) \cdots \mathbf{y}(k-L+1, \theta) \right]^T, \qquad (12.77)$$

where we have appended the previous $L - 1$ time-aligned blocks of length N to the aperture. With the signal model of (12.1) and assuming uniform attenuation coefficients, the parameterized spatiotemporal aperture is given by

$$\bar{\mathbf{y}}(k,\theta) = \bar{\mathbf{s}}(k-\tau,\theta) + \bar{\mathbf{v}}(k,\theta), \tag{12.78}$$

where

$$\bar{\mathbf{s}}(k,\theta) = \left[\, \mathbf{s}(k,\theta)\ \mathbf{s}(k-1,\theta)\ \cdots\ \mathbf{s}(k-L+1,\theta)\,\right]^T,$$
$$\bar{\mathbf{v}}(k,\theta) = \left[\, \mathbf{v}(k,\theta)\ \mathbf{v}(k-1,\theta)\ \cdots\ \mathbf{v}(k-L+1,\theta)\,\right]^T.$$

A location-parameterized multichannel finite impulse response (FIR) filter is formed according to

$$\mathbf{h}(\theta) = \left[\, \mathbf{h}_0^T(\theta)\ \mathbf{h}_1^T(\theta)\ \cdots\ \mathbf{h}_{L-1}^T(\theta)\,\right]^T, \tag{12.79}$$

where

$$\mathbf{h}_l(\theta) = \left[\, h_{l1}(\theta)\ h_{l2}(\theta)\ \cdots\ h_{lN}(\theta)\,\right]^T \tag{12.80}$$

is the spatial filter applied to the block of microphone signals at temporal sample $k-l$.

The question remains as to how to choose the multichannel filter coefficients such that the resulting steered spatiotemporal filter output allows one to better localize the source. In [37], it is proposed to select the weights such that the output of the spatiotemporal filter to a plane wave propagating from location θ is a filtered version of the desired signal:

$$\mathbf{h}^T(\theta)\,\mathbf{s}(k-\tau,\theta) = \sum_{l=0}^{L-1} f_l s(k-\tau-l). \tag{12.81}$$

In order to satisfy the desired criterion of (12.81), the multichannel filter coefficients should satisfy

$$\mathbf{c}_l^T(\theta)\mathbf{h}(\theta) = f_l, \quad l = 0, 1, \ldots, L-1, \tag{12.82}$$

where

$$\mathbf{c}_l(\theta) = \left[\, \mathbf{0}_N^T\ \cdots\ \mathbf{0}_N^T\ \underbrace{\mathbf{1}_N^T}_{l\text{th group}}\ \mathbf{0}_N^T\ \cdots\ \mathbf{0}_N^T\,\right]^T$$

is a vector of length NL corresponding to the lth constraint, and $\mathbf{0}_N$ is a vector of N zeros. The L constraints of (12.82) may be neatly expressed in matrix notation as

$$\mathbf{C}^T(\theta)\mathbf{h}(\theta) = \mathbf{f}, \tag{12.83}$$

where

$$\mathbf{C}(\theta) = \left[\, \mathbf{c}_0(\theta)\ \mathbf{c}_1(\theta)\ \cdots\ \mathbf{c}_{L-1}(\theta)\,\right] \tag{12.84}$$

is the constraint matrix and

$$\mathbf{f} = \begin{bmatrix} f_0 & f_1 & \cdots & f_{L-1} \end{bmatrix}^T \tag{12.85}$$

is the constraint vector.

The spatiotemporal filter output is given by

$$z(k, \theta) = \mathbf{h}^T(\theta)\bar{\mathbf{y}}(k, \theta). \tag{12.86}$$

For each candidate location θ, we seek to find the multichannel weights $\mathbf{h}(\theta)$ which minimize the total energy of the beamformer output subject to the N linear constraints of (12.84):

$$\hat{\mathbf{h}}(\theta) = \arg \min_{\mathbf{h}(\theta)} \mathbf{h}^T(\theta)\mathbf{R}_{\bar{y}}(\theta)\mathbf{h}(\theta) \quad \text{subject to} \quad \mathbf{C}^T(\theta)\mathbf{h}(\theta) = \mathbf{f}, \tag{12.87}$$

where

$$\mathbf{R}_{\bar{y}}(\theta) = E\left\{ \bar{\mathbf{y}}(k, \theta)\bar{\mathbf{y}}^T(k, \theta) \right\} \tag{12.88}$$

is the parameterized spatiotemporal correlation matrix (PSTCM), which is given by

$$\mathbf{R}_{\bar{y}}(\theta) = \begin{bmatrix} \mathbf{R}_y(\theta, 0) & \mathbf{R}_y(\theta, -1) & \cdots & \mathbf{R}_y(\theta, -L+1) \\ \mathbf{R}_y(\theta, 1) & \mathbf{R}_y(\theta, 0) & \cdots & \mathbf{R}_y(\theta, -L+2) \\ \vdots & \vdots & \ddots & \vdots \\ \mathbf{R}_y(\theta, L-1) & \mathbf{R}_y(\theta, L-2) & \cdots & \mathbf{R}_y(\theta, 0) \end{bmatrix},$$

where it should be pointed out that $\mathbf{R}_y(\theta, 0)$ is the PSCM.

The solution to the constrained optimization problem of (12.87) can be found using the method of Lagrange multipliers:

$$\hat{\mathbf{h}}(\theta) = \mathbf{R}_{\bar{y}}^{-1}(\theta)\mathbf{C}(\theta)\left[\mathbf{C}^T(\theta)\mathbf{R}_{\bar{y}}^{-1}(\theta)\mathbf{C}(\theta)\right]^{-1}\mathbf{f}. \tag{12.89}$$

Having computed the optimal multichannel filter for each potential source location θ, the estimate of the source location is given by

$$\hat{\theta}_s = \arg \max_{\theta} \hat{\mathbf{h}}^T(\theta)\mathbf{R}_{\bar{y}}(\theta)\hat{\mathbf{h}}(\theta),$$

meaning that the source estimate is given by the location which emits the most steered (and temporally filtered) energy.

12.7.1 Autoregressive Modeling

It is important to point out that with $L = 1$ (i.e., a purely spatial aperture), the LCMV method reduces to the MVDR method if we select $\mathbf{f} = f_0 = 1$. In this case, the PSTCM and PSCM are equivalent. Moreover, the constraint imposed on the multichannel filtering is

$$\mathbf{h}^T\left(\theta\right)\mathbf{s}\left(k-\tau,\theta\right) = s\left(k-\tau\right), \tag{12.90}$$

meaning that we are attempting to estimate the sample $s(k-\tau)$ from a spatial linear combination of the elements of $\mathbf{s}(k-\tau,\theta)$. Notice that such a procedure neglects any dependence of $s(k)$ on the previous values $s(k-1), s(k-2), \ldots$ of the signal. A signal whose present value is strongly correlated to its previous samples is well-modeled by an autoregressive (AR) process:

$$s(k) = \sum_{l=1}^{q} a_l s\left(k-l\right) + w(k), \tag{12.91}$$

where a_l are the AR coefficients, q is the order of the AR process, and $w(k)$ is the zero-mean prediction error.

Applying the AR model to the desired signal in the LCMV localization scheme, the constraint may be written as

$$\mathbf{h}^T\left(\theta\right)\mathbf{s}\left(k-\tau,\theta\right) = \sum_{l=1}^{L-1} a_l s\left(k-\tau-l\right), \tag{12.92}$$

where we have substituted

$$\begin{aligned} f_0 &= 0, \\ f_l &= a_l, \quad l = 1, 2, \ldots, q, \\ L - 1 &= q. \end{aligned}$$

With the inclusion of previous temporal samples in the aperture, the LCMV scheme is able to temporally focus its steered beam onto the signal with the AR characteristics embedded by the coefficients in the constraint vector $\mathbf{f}$. Thus, the discrimination between the desired signal and noise is now both spatial (i.e., the relative delays differ since the source and interference are located at disparate locations) and temporal (i.e, the AR coefficients of the source and interference or noise generally differ).

It is important to point out that in general, the AR coefficients of the desired signal are not known *a priori*. Thus, the algorithm must first estimate the coefficients from the observed microphone signals. This can either be accomplished using conventional single-channel methods [38] or methods which incorporate the data from multiple channels [39].

Figure 12.6 depicts the ensemble averaged LCMV spatial spectra for the simulated data described previously. A temporal aperture length of $L = 20$ is employed. The AR coefficients are estimated from a single microphone by solving the Yule-Walker equations [38]. The PSTCM is regularized before performing the matrix inversion necessary in the method. The resulting spa-

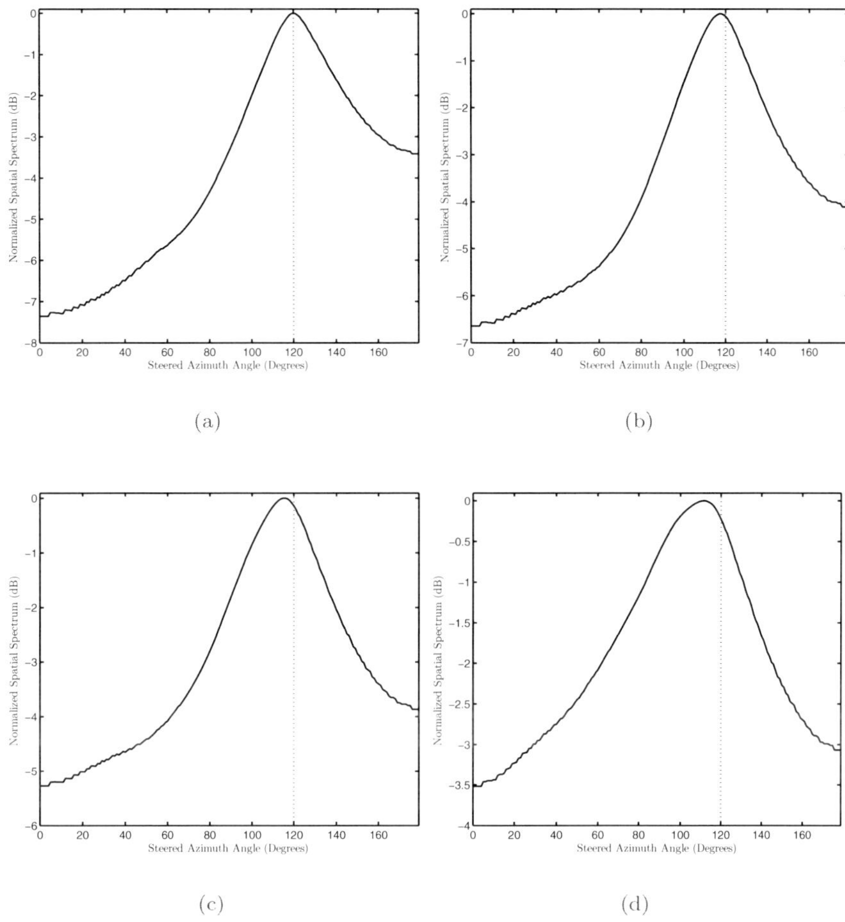

Fig. 12.6 Ensemble averaged LCMV spatial spectra: (a) 0 ms, (b) 100 ms, (c) 200 ms, and (d) 300 ms.

tial spectra are entirely free of any spurious peaks, albeit at the expense of a significant bias error.

12.8 Challenges

The techniques described in this chapter have been focused on integrating the information from multiple microphone in an optimal fashion to arrive at a robust estimate of the source location. While the algorithms represent some of the more sophisticated approaches in acoustic source localization, the problem remains challenging due to a number of factors.

- Reverberant signal components act as mirror images of the desired signal but originating from a disparate location. Thus, the additive noise component is strongly correlated to the desired signal in this case. Moreover, the off-diagonal elements of the PSCM at false parameters θ may be large due to the reverberant signal arriving from θ.
- The desired signal (i.e., speech) is non-stationary, meaning that the estimation of necessary statistics is not straightforward.
- The parameter space is large, and the current solutions to broadband source localization require an exhaustive search of the location space.

The reverberation issue is particularly problematic. In the worst-case scenario, a reflected component may arrive at the array with an energy greater than the direct-path. At this point, most localization algorithms will fail, as the key assumption of localization is clearly violated: the true location of the source does not emit more energy than all other locations. Unlike in beamforming, the reverberant signal must be viewed as interference, as the underlying location of the reverberant path is different from that of the source.

12.9 Conclusions

This chapter has provided a treatment of steered beamforming approaches to acoustic source localization. The PSCM and PSTCM were developed as the fundamental structures of algorithms which attempt to process the observations of multiple microphones in such a way that the effect of interference and noise sources is minimized and the estimated source location possesses minimal error.

Purely spatial methods based on the PSCM focus on the relationship between the relative delays across the array and the corresponding source location. By grouping the various cross-correlation functions into the PSCM, well-known minimum variance and subspace techniques may be applied to the source localization problem. Moreover, an information-theoretic approach rooted in minimizing the joint entropy of the time-aligned sensor signals was developed for both Gaussian and Laplacian signals incorporating higher-order statistics in the source localization estimate. While PSCM-based methods are amenable to real-time operation, additional shielding of the algorithm from interference and reverberation may be achieved by extending the aperture to include the previous temporal samples of each microphone. It was shown that the celebrated LCMV method may be applied to the source localization problem by modeling the desired signal as an AR process.

The inclusion of multiple microphones in modern communication devices is relatively inexpensive. The desire for cleaner and crisper speech quality necessitates multiple-channel beamforming methods, which in turn require the localization of the desired acoustic source. By combining the outputs of the microphones via the PSCM and PSTCM, cleaner estimates of the source location may be generated using one of the methods detailed in this chapter.

In addition to the algorithms presented in this chapter, the PSCM and PSTCM provide a neat framework for the development of future localization algorithms aimed at solving the challenging problem of acoustic source localization.

References

1. H. Krim and M. Viberg, "Two decades of array signal processing research: the parametric approach," *IEEE Signal Process. Mag.*, vol. 13, pp 67–94, July 1996.
2. D. H. Johnson, "The application of spectral estimation methods to bearing estimation problems," *Proc. IEEE*, vol. 70, pp. 1018–1028, Sept. 1982.
3. C. H. Knapp and G. C. Carter, "The generalized correlation method for estimation of time delay," *IEEE Trans. Acoust., Speech, Signal Process.*, vol. 24, pp. 320–327, Aug. 1976.
4. Y. Huang, J. Benesty, and G. W. Elko, "Microphone arrays for video camera steering," in *Acoustic Signal Processing for Telecommunication*, S. L. Gay and J. Benesty, eds., pp. 240–259. Kluwer Academic Publishers, Boston, MA, 2000.
5. Y. Huang, J. Benesty, and J. Chen, *Acoustic MIMO Signal Processing*. Springer-Verlag, Berlin, Germany, 2006.
6. Y. Huang, J. Benesty, and J. Chen, "Time delay estimation and source localization," in *Springer Handbook of Speech Processing*, J. Benesty, M. M. Sondhi, and Y. Huang, editors-in-chief, Springer-Verlag, Chapter 51, Part I, pp. 1043–1064, 2007.
7. D. B. Ward and R. C. Williamson, "Particle filter beamforming for acoustic source localization in a reverberant environment," in *Proc. IEEE ICASSP*, 2002, pp. 1777–1780.
8. D. B. Ward, E. A. Lehmann, and R. C. Williamson, "Particle filtering algorithms for tracking an acoustic source in a reverberant environment," *IEEE Trans. Speech, Audio Process.*, vol. 11, pp. 826–836, Nov. 2003.
9. E. A. Lehmann, D. B. Ward, and R. C. Williamson, "Experimental comparison of particle filtering algorithms for acoustic source localization in a reverberant room," in *Proc. IEEE ICASSP*, 2003, pp. 177–180.
10. A. M. Johansson, E. A. Lehmann, and S. Nordholm, "Real-time implementation of a particle filter with integrated voice activity detector for acoustic speaker tracking" in *IEEE Asia Pacific Conference on Circuits and Systems APPCCAS*, 2006, pp. 1004–1007.
11. C. E. Chen, H. Wang, A. Ali, F. Lorenzelli, R. E. Hudson, and K. Yao, "Particle filtering approach to localization and tracking of a moving acoustic source in a reverberant room," in *Proc. IEEE ICASSP*, 2006.
12. D. Li and Y. H. Hu, "Least square solutions of energy based acoustic source localization problems," in *Proc. International Conference on Parallel Processing (ICPP)*, 2004, pp. 443–446.
13. K. C. Ho and Ming Sun, "An accurate algebraic closed-form solution for energy-based source localization," *IEEE Trans. Audio, Speech, Language Process.*, vol. 15, pp. 2542–2550, Nov. 2007.

14. D. Ampeliotis and K. Berberidis, "Linear least squares based acoustic source localization utilizing energy measurements," in *Proc. IEEE SAM*, 2008, pp. 349–352.

15. T. Ajdler, I. Kozintsev, R. Lienhart, and M. Vetterli, "Acoustic source localization in distributed sensor networks," in *Proc. Asilomar Conference on Signals, Systems, and Computers*, 2004, pp. 1328–1332.

16. G. Valenzise, G. Prandi, M. Tagliasacchi, and A. Sarti, "Resource constrained efficient acoustic source localization and tracking using a distributed network of microphones," in *Proc. IEEE ICASSP*, 2008, pp. 2581–2584.

17. H. Buchner, R. Aichner, J. Stenglein, H. Teutsch, and W. Kellennann, "Simultaneous localization of multiple sound sources using blind adaptive MIMO filtering," in *Proc. IEEE ICASSP*, 2005, pp. III-97–III-100.

18. A. Lombard, H. Buchner, and W. Kellermann, "Multidimensional localization of multiple sound sources using blind adaptive MIMO system identification," in *Proc. IEEE International Conference on Multisensor Fusion and Integration for Intelligent Systems*, 2006, pp. 7–12.

19. D. H. Johnson and D. E. Dudgeon, *Array Signal Processing*. Englewood Cliffs, NJ: Prentice-Hall, 1993.

20. M. Omologo and P. G. Svaizer, "Use of the cross-power-spectrum phase in acoustic event localization," ITC-IRST Tech. Rep. 9303-13, Mar. 1993.

21. J. Dibiase, H.F. Silverman, and M.S. Brandstein, "Robust localization in reverberant rooms," in *Microphone Arrays: Signal Processing Techniques and Applications*, M. S. Brandstein and D. B. Ward, eds., pp. 157–180, Springer-Verlag, Berlin, 2001.

22. M. R. Schroeder, "New method for measuring reverberation time," *J. Acoust. Soc. Am.*, vol. 37, pp. 409–412, 1965.

23. J. B. Allen and D. A. Berkley, "Image method for efficiently simulating small-room acoustics," *J. Acoust. Soc. Am.*, vol. 65, pp. 943–950, Apr. 1979.

24. J. Krolik and D. Swingler, "Multiple broad-band source location using steered covariance matrices," *IEEE Trans. Acoust., Speech, Signal Process.*, vol. 37, pp. 1481–1494, Oct. 1989.

25. J. Capon, "High resolution frequency-wavenumber spectrum analysis," *Proc. IEEE*, vol. 57, pp. 1408–1418, Aug. 1969.

26. J. Dmochowski, J. Benesty, and S. Affes, "Direction-of-arrival estimation using the parameterized spatial correlation matrix," *IEEE Trans. Audio, Speech, Language Process.*, vol. 15, pp. 1327–1341, May 2007.

27. J. Dmochowski, J. Benesty, and S. Affes, "Direction-of-arrival estimation using eigenanalysis of the parameterized spatial correlation matrix," in *Proc. IEEE ICASSP*, 2007, pp. I-1–I-4.

28. R. O. Schmidt, *A Signal Subspace Approach to Multiple Emitter Location and Spectral Estimation*. Ph. D. dissertation, Stanford Univ., Stanford, CA, Nov. 1981.

29. R. O. Schmidt, "Multiple emitter location and signal parameter estimation," *IEEE Trans. Antennas Propag.*, vol. AP-34, pp. 276–280, Mar. 1986.

30. J. Dmochowski, J. Benesty, and S. Affes, "Broadband MUSIC: opportunities and challenges for multiple acoustic source localization," in *Proc. IEEE WASPAA*, 2007, pp. 18–21.

31. C. E. Shannon, "A mathematical theory of communication," *Bell Syst. Tech. J.*, vol. 27, pp. 379–423, 1948.

32. J. Benesty, Y. Huang, and J. Chen, "Time delay estimation via minimum entropy," *IEEE Signal Process. Lett.*, vol. 14, pp. 157–160, Mar. 2007.

33. J. Dmochowski, J. Benesty, and S. Affes, "A generalized steered response power method for computationally viable source localization," *IEEE Trans. Audio, Speech, Language Process.*, vol. 15, pp. 2510–2516, Nov. 2007.

34. O. L. Frost, III, "An algorithm for linearly constrained adaptive array processing," *Proc. IEEE*, vol. 60, pp. 926–935, Aug. 1972.

35. L. J. Griffiths and C. W. Jim, "An alternative approach to linearly constrained adaptive beamforming," *IEEE Trans. Antennas Propagat.*, vol. AP-30, pp. 27–34, Jan. 1982.
36. B. D. Van Veen and K. M. Buckley, "Beamforming: a versatile approach to spatial filtering," *IEEE ASSP Mag.*, vol. 5, pp. 4–24, Apr. 1988.
37. J. Dmochowski, J. Benesty, and S. Affes, "Linearly constrained minimum variance source localization and spectral estimation," *IEEE Trans. Audio, Speech, Language Process.*, vol. 16, pp. 1490–1502, Nov. 2008.
38. S. L. Marple Jr., *Digital Spectral Analysis with Applications*. Englewood Cliffs, NJ: Prentice Hall, 1987.
39. N. D. Gaubitch, P. A. Naylor, and D. B. Ward, "Statistical analysis of the autoregressive modeling of reverberant speech," *J. Acoust. Soc. Am.*, vol. 120, pp. 4031–4039, Dec. 2006.

Index

2D beamforming, 295

a posteriori error signal, 93
a priori error signal, 93
a priori SNR estimation, 142
acoustic echo, 90
acoustic echo cancellation (AEC), 25, 112
acoustic impulse response, 35, 90
acoustic source localization, 306, 313
   broadband MUSIC, 318
   linearly constrained minimum variance, 329
   maximum eigenvalue, 316
   minimum entropy, 320
   minimum variance distortionless response, 315
   steered response power, 313
acoustic transfer function (ATF), 226
adaptive estimation, 19
adaptive filter, 90
adaptive system identification, 72
affine projection algorithm (APA), 103, 104
analysis window, 5
anechoic plane wave model, 208
APLAWD speech database, 249
AR process, 112
array gain, 204
array manifold, 291
array processing, 229, 281
array signal processing, 199
associated Legendre function, 282
autoregressive modeling, 331

beamformer, 230
beamforming, 42, 199, 202, 229, 256, 281
beampattern, 207, 282
beamwidth, 288
blind source separation (BSS), 183, 256
blocking matrix, 42
broadband array, 199
broadband array gain, 204
broadband beamforming, 310
broadband beampattern, 207
broadband desired-signal-cancellation factor, 206
broadband desired-signal-distortion index, 220
broadband directivity factor, 210
broadband input SNR, 203
broadband MSE, 219
broadband noise-rejection factor, 206
broadband normalized MSE, 219
broadband output SNR, 203
broadband white noise gain, 211

Chebyshev polynomial, 288
clean speech model, 165, 171
codebook, 185
completeness condition, 56
confidence measure, 191
convergence of the misalignment, 95
convolutive transfer function (CTF), 40
cross-MTF approximation, 18
cross-spectrum matrix, 292
crossband filter, 6, 57

delay-and-sum beamformer, 153, 209, 256, 286
denoising, 151
dereverberation, 156
desired signal cancellation, 206
desired signal distortion, 219
desired-signal-cancellation factor, 206
desired-signal-reduction factor, 206

diffuse noise, 210
direction vector, 201
direction-of-arrival, 314
direction-of-arrival estimation, 282, 300
directivity, 210, 284
directivity index, 210, 284
distortion measure, 130
distortionless constraint, 287
Dolph-Chebyshev, 282
Dolph-Chebyshev beamformer, 288
double-talk, 91, 98, 116
double-talk detector (DTD), 91
dual source transfer function generalized
       sidelobe canceler (DTF-GSC), 258

echo cancellation, 88
EM algorithm, 174
error signal, 218
Euler angle, 296

far-field, 309
filter-and-sum beamformer, 256
Fisher information matrix, 192

Gaussian mixture model (GMM), 186
Gaussian scaled mixture model (GSMM),
       186
Gaussian signal, 323
Geigel DTD, 120
generalized eigenvalue problem, 258, 285
generalized sidelobe canceler (GSC), 226,
       257

hands-free communication, 33, 90
hypercardioid, 286

image method, 42
impulse response, 4
independent component analysis (ICA),
       184
input SNR, 235
inverse STFT, 5

Jacobi polynomial, 296
joint entropy, 320

Kalman filter, 106
Kalman gain vector, 109
Kullback-Leibler divergence, 192

Lagrange multiplier, 316
Laplacian signal, 324
least-mean-square (LMS), 19, 53, 72
Legendre polynomial, 284

linear convolution, 34
linear time-invariant system, 4, 35
linearly constrained minimum variance
       (LCMV), 227, 255
log-spectral distortion, 146
loudspeaker linearization, 50

magnitude squared coherence function
       (MSCF), 251
main lobe, 288
manifold matrix, 301
MAP estimator, 186
matrix inversion lemma, 231
maximum likelihood (ML), 165, 186
maximum SNR beamformer, 235
mean-squared error (MSE), 218, 230
median filter, 193
microphone, 286
microphone arrays, 199, 306
minimum mean-squared error (MMSE),
       154
minimum variance distortionless response
       (MVDR), 221, 226, 232, 290
misadjustment, 91, 93
misalignment vector, 95
MMSE estimator, 130, 154, 187
MMSE-LSA estimator, 131
model order selection, 13
monochromatic plane wave, 212
MTF approximation, 14
multichannel auto-regressive (MCAR), 155
multichannel Wiener filter, 226, 257
multiple-null beamformer, 293
multiplicative transfer function, 14, 63
multiplicative transfer function (MTF), 36,
       259
MUSIC, 300, 301, 320
mutual information, 321

narrowband array gain, 205
narrowband beampattern, 208
narrowband desired-signal-cancellation
       factor, 207
narrowband desired-signal-distortion
       index, 220
narrowband directivity factor, 210
narrowband input SNR, 203
narrowband MSE, 220
narrowband noise-rejection factor, 206
narrowband normalized MSE, 220
narrowband output SNR, 204
narrowband source localization, 307
narrowband white noise gain, 211
near-field, 310

near-field beamforming, 297
near-field radius, 299
noise reduction, 225
noise rejection, 206
noise-reduction factor, 206, 236
noise-rejection factor, 206
non-negative matrix factorization, 188, 194
non-parametric VSS-NLMS, 94, 97
nonlinear acoustic echo cancellation, 81
nonlinear system identification, 53
nonlinear undermodeling, 69
normal equations, 108
normalized LMS (NLMS), 20, 93
normalized misalignment, 112

OM-LSA estimator, 131
orthogonal triangular decomposition (QRD), 258
orthogonality theorem, 109
output SNR, 235

parameter estimation, 307
parameterized spatial correlation matrix, 311, 312
parameterized spatiotemporal correlation matrix, 331
particle filtering, 308
perceptual evaluation of speech quality (PESQ), 146
plane wave, 282
plane-wave decomposition, 285
power spectral density, 203
principal eigenvector, 317

quadratic distortion measure, 134
quadratic kernel, 57
quadratic log-spectral amplitude, 140
quadratic spectral amplitude, 137

Rayleigh quotient, 285, 317
recursive least-squares (RLS), 106, 109
regular beampattern, 284
regularization, 96, 106
relative transfer function (RTF), 33, 258
    identification, 34
reverberation, 35
reverberation model, 160
    multichannel autoregressive, 161
    multichannel moving average, 160
    simplified multichannel autoregressive, 163
reverberation time, 8
room acoustics model, 168
room impulse response, 293

segmental SNR, 144
set membership NLMS (SM-NLMS), 94
short-time Fourier transform (STFT), 5, 34–36
side lobe, 288
signal blocking factor (SBF), 42
signal-to-noise ratio (SNR), 203, 285
simultaneous detection and estimation, 131, 146
single sensor source separation, 185
single-talk, 97
sound field, 281
source separation
    AR-based, 187
    GSMM-based, 186
    multi-window, 190
sparse representation, 326
spatial aliasing, 212
    broadband signal, 217
    monochromatic signal, 215
spatial correlation matrix, 311
spatial filtering, 310
spatial smoothing, 303
spatiotemporal correlation matrix, 311
spatiotemporal filtering, 310
speech dereverberation, 151, 225
speech distortion weighted multichannel Wiener filter, 226, 230
speech enhancement, 129, 131, 256
speech presence probability, 131
spherical array processing, 282
spherical Bessel function, 283
spherical Fourier transform, 282
spherical Hankel functions, 283
spherical harmonics, 281, 282
spherical microphone array, 281
steered beamforming, 306
steering vector, 201, 292
subband, 2
superdirective beamforming, 210
synthesis window, 5
system identification, 2, 4, 33, 34, 51, 88, 90, 106

time delay estimation, 306
tracking, 112, 120
transfer function generalized sidelobe canceler (TF-GSC), 45, 257
transient noise, 140

under-modeling, 90, 98
under-modeling noise, 101

variable forgetting factor RLS, 106, 110, 111
variable step-size adaptive filter, 88
Volterra model, 51
Volterra system identification, 51
VSS-APA, 103, 105, 107
VSS-NLMS, 92, 96

weighted prediction error (WPE), 159
white noise gain, 211, 284
Wiener filter, 220
Wigner-d function, 296
Woodbury's identity, 231, 234